RATE COEFFICIENTS IN ASTROCHEMISTRY

ASTROPHYSICS AND SPACE SCIENCE LIBRARY

A SERIES OF BOOKS ON THE RECENT DEVELOPMENTS
OF SPACE SCIENCE AND OF GENERAL GEOPHYSICS AND ASTROPHYSICS
PUBLISHED IN CONNECTION WITH THE JOURNAL
SPACE SCIENCE REVIEWS

VOLUME 146
PROCEEDINGS

RATE COEFFICIENTS IN ASTROCHEMISTRY

PROCEEDINGS OF A CONFERENCE
HELD AT UMIST, MANCHESTER, U.K.
SEPTEMBER 21-24, 1987

Edited by

T. J. MILLAR

and

D. A. WILLIAMS

Department of Mathematics, UMIST, Manchester, U.K.

KLUWER ACADEMIC PUBLISHERS
DORDRECHT / BOSTON / LONDON

Library of Congress Cataloging in Publication Data

CIP

Rate coefficients in astrochemistry : proceedings of a conference held in UMIST, Manchester, U.K., September 21-24, 1987 / edited by T.J. Millar and D.A. Williams.
p. cm. -- (Astrophysics and space science library)
Includes index.
ISBN-13: 978-94-010-7851-1
1. Cosmochemistry--Congresses. 2. Chemical reaction, Rate of--Congresses. I. Millar, T. J., 1952- . II. Williams, D. A. (David Arnold), 1937- . III. Series.
QB450.R38 1988
523.02--dc19 88-12045
CIP

ISBN-13: 978-94-010-7851-1 e-ISBN-13: 978-94-009-3007-0
DOI: 10.1007/978-94-009-3007-0

Published by Kluwer Academic Publishers,
P.O. Box 17, 3300 AA Dordrecht, The Netherlands.

Kluwer Academic Publishers incorporates
the publishing programmes of
D. Reidel, Martinus Nijhoff, Dr W. Junk and MTP Press.

Sold and distributed in the U.S.A. and Canada
by Kluwer Academic Publishers,
101 Philip Drive, Norwell, MA 02061, U.S.A.

In all other countries, sold and distributed
by Kluwer Academic Publishers Group,
P.O. Box 322, 3300 AH Dordrecht, The Netherlands.

TABLE OF CONTENTS

PREFACE

"An attempt has been made to compute the numbers of certain molecules in interstellar space ..... A search for the bands of CH, OH, NH, CN and C_2 would appear to be promising"

P Swings and L Rosenfeld
Astrophysical Journal **86**,483(1937)

This may have been the first attempt at modelling interstellar chemistry. As with models today, the methods used lacked reliability, but the speculation was impressive! Mark Twain might well have said of this infant subject "One gets such wholesale returns of conjecture out of such a trifling investment of fact". The detection of unidentified lines around the period that Swings and Rosenfeld were writing provoked much interest, but even the most optimistic speculator could hardly have imagined developments which would occur during the next 50 years. By 1987 about 70 varieties of molecule had been identified in the interstellar and circumstellar regions. They range in complexity from simple diatomics such as H_2 and CO to such species as ethanol C_2H_5OH, acetone $(CH_3)_2CO$, and the largest interstellar molecule detected so far, cyano-penta-acetylene $HC_{11}N$. The study of these molecules in astronomy has developed enormously, especially over the last 20 years, and is now codified in the new subject of astrochemistry.

That such a variety of chemical species should exist in tenuous regions of the Galaxy is fascinating. However, their major importance is that they allow astronomers to probe regions of the gas that are otherwise inaccessible, and to study dark, dense regions where, for example, star formation may be occurring. To exploit the information contained in the observations one needs to develop models of the chemistry describing the great variety of chemical processes that can occur in interstellar and circumstellar regions. Many of these processes were correctly identified soon after the first interstellar detections, for example, in papers by Swings in 1942, by Kramers and ter Haar in 1946, and especially by Bates and Spitzer in 1951. This last paper includes almost all types of process we currently consider, including radiative and dissociative recombination, photoprocesses, ion-molecule reactions, surface reactions, and so forth.

With the rapid increase in the numbers of identified types of interstellar molecules in the early 1970s, comprehensive models of interstellar chemistry were developed by a number of authors, including Klemperer, Solomon and Herbst. These models showed that much important chemistry occurs in reactions between ions and molecules, with the ionization being maintained by cosmic rays or by photoionization. Such reactions can account for the observed interstellar ions (now numbering

10 species) and for the enhanced abundance of deuterium observed in many molecules in cold interstellar clouds. As the models developed in complexity the demands on knowledge of chemical rate coefficients became severe. Even so, sophisticated models, such as that developed by Black and Dalgarno for the molecular cloud toward the star Zeta Oph, had some remarkable successes in description and prediction.

These and later models emphasized the desperate need for accurate reaction rate data, and this perceived demand stimulated much experimental and theoretical work on appropriate atomic and molecular processes. Laboratory workers have risen magnificently to the challenge and many important rate coefficients have now been measured, and the temperature dependence of some has been determined. These remarkable advances have occurred largely through the development, especially by Smith and Adams in Birmingham, of the Selected Ion Flow Tube apparatus. New theoretical techniques have also been developed and exploited with the greater availability of computing power. The interaction between theoretician and experimentalist is often most fruitful, as the following pages show.

The models of interstellar chemistry not only provide a stimulus to the chemist; in several instances the failure of theoretical models when compared to interstellar observations has led to suggestions which have resulted in new laboratory measurements and new phenomena being identified. As an example we mention the large HCS^+/CS abundance ratios observed in cold clouds. These were interpreted using a theoretical approach to ion-molecule collisions which showed that, at low temperatures, ion-dipolar molecule collisions would have rate coefficients much greater than that predicted by the ADO theory or measured at room temperature, a result confirmed by subsequent experiments and new theoretical developments.

Because of the central importance of reaction rate data to models developed at UMIST and elsewhere we thought it appropriate to hold a conference to review recent progress in theoretical and experimental work in this area, and to set it in context with the astrochemical modelling. The aim was specifically to identify those reactions and processes, crucial in the modelling, that are uncertain and in need of study. UMIST seemed to us an appropriate venue for such a meeting, not only because it is a centre for astrochemical modelling, but also because it is an institution long devoted to scientific research, an institution with which Dalton - the founder of atomic theory - and his pupil Joule - a founder of thermodynamics - were connected in byegone years. The interdisciplinary nature of our subject would, we feel, have appealed to them.

Some 50 years after the origins of astrochemistry, therefore, the conference was held at UMIST (September 1987) with the title "Reactive rate coefficients in Astrophysics". Particular prominence was given to the 17 major invited reviews which were intended to cover the main areas of laboratory and theoretical chemistry and the astrochemical modelling.

Papers based on these reviews form the bulk of this book. A few contributed papers were also accepted and are included. We believe that taken together these papers form a useful and coherent account of today's experimental and theoretical techniques, give some indication of the present activity in astrochemical modelling, and contain many suggestions for a lifetime of future work.

We are especially grateful to the Science and Engineering Research Council for financial support through the Collaborative Computational Projects No. 6 (Heavy Particle Dynamics) and No. 7 (Analysis of Astronomical Spectra), to Drs N G Adams, D C Clary, D R Flower and E Herbst for advice on the scientific content of the meeting and to Dr A E Lynas-Gray, then the CCP7 Secretary, for administrative support.

T J Millar and D A Williams
Mathematics Department
UMIST
March 1988

LIST OF PARTICIPANTS

N.G. ADAMS, Dept. of Space Research, University of Birmingham, PO Box 363, Birmingham B15 2TT, UK

R.J. ALLAN, Daresbury Laboratory, Warrington WA4 4AD, UK

G.G. BALINT-KURTI, School of Chemistry, Univ. of Bristol, Bristol BS8 1TS UK

D.R. BATES, Dept. of Applied Mathematics and Theoretical Physics, The Queen's University, Belfast BT7 1NN, UK

D.L. BAULCH, Dept. of Physical Chemistry, Univ. of Leeds, Leeds LS2 9JT UK

A. BENNETT, Dept. of Mathematics, UMIST, PO Box 88, Manchester M60 1QD UK

J.H. BLACK, Steward Observatory, Univ. of Arizona, Tuscon, AZ 85721, USA

D.K. BOHME, Dept. of Chemistry, York University, 4700 Keele Street Downsview, Ontario, CANADA M3J 1P3

P.D. BROWN, Dept. of Mathematics, UMIST, PO Box 88, Manchester M60 1QD UK

S.B. CHARNLEY, Inst. fur extraterrestriche Physik, MPI fur Physik und Astrophysik, D-8046 Garching bei Munchen, WEST GERMANY

D.C. CLARY, University Chemical Laboratory, Lensfield Road, Cambridge CB2 1EW, UK

J.N.L. CONNOR, Dept. of Chemistry, Univ. of Manchester, Manchester M13 9PL, UK

C. COURBIN, Daresbury Laboratory, Warrington WA4 4AD, UK

A. DALGARNO, Harvard-Smithsonian Center for Astrophysics, 60 Garden Street, Cambridge, MA 02138, USA

C.E. DATEO, University Chemical Laboratory, Lensfield Road, Cambridge CB2 1EW, UK

J. DAVIDSSON, Dept. of Physical Chemistry GU, Chalmers University of Technology, S-412 96 Goteborg, SWEDEN

D.J. DEFREES, Molecular Research Institute, 701 Welch Rd, Suite 213, Palo Alto, CA 94305, USA

A.S. DICKINSON, Dept. of Physics, Univ. of Newcastle, Newcastle-upon-Tyne NE1 7RU, UK

C.H. DEPUY, Dept. of Chemistry, Univ. of Colorado, Campus Box 215, Boulder, CO 80309, USA

K. ERIKSSON, Astronomiska Observatoriet, Box 515, S-751 20 Uppsala SWEDEN

D.R. FLOWER, Dept. of Physics, The University, Durham DH1 3LE, UK

K. GILES, Dept. of Space Research, University of Birmingham, PO Box 363 Birmingham B15 2TT, UK

T. HASEGAWA, Canadian Institute for Theoretical Astrophysics, Univ. of Toronto, CANADA M5S 1A1

M.J. HENCHMAN, Dept. of Chemistry, Brandeis Univ., Waltham, MA 02254 USA

E. HERBST, Dept. of Physics, Duke Univ., Durham, NC 27706, USA

C.R. HERD, Dept. of Space Research, University of Birmingham, PO Box 363 Birmingham B15 2TT, UK

B. JANSEN OP DE HAAR, University Chemical Laboratory, Lensfield Road Cambridge CB2 1EW, UK

A.P. JONES, Dept. of Mathematics, UMIST, PO Box 88, Manchester M60 1QD UK

P.G. JUDGE, Dept. of Theoretical Physics, Univ. of Oxford, 1 Keble Road, Oxford OX1 3NP, UK

J. LE BOURLOT, DAMAp, Observatoire de Meudon, Place Jules Janssen F-92195 Meudon Principal Cedex, FRANCE

A.E. LYNAS-GRAY, Dept. of Physics and Astronomy, University College, Gower Street, London WC1 6BT, UK

R. McCARROLL, Univ. P. et M. Curie, Lab. de Physique et Optique Corpusculaires, Tour 12-E5, F-75252 Paris Cedex, FRANCE

B.J. McINTOSH, Dept. of Space Research, University of Birmingham, PO Box 363, Birmingham B15 2TT, UK

J.B. MARQUETTE, Laboratoire d'Aerothermique, 4ter Routes des Gardes F-92190 Meudon, FRANCE

G. MAUCLAIRE, LPCR - Bat. 350, Univ. de Paris-Sud, F-91405 Orsay Cedex, FRANCE

T.J. MILLAR, Dept. of Mathematics, UMIST, PO Box 88, Manchester M60 1QD UK

S. MILLER, Dept. of Physics and Astronomy, University College, Gower Street, London WC1 6BT, UK

T.S. MONTEIRO, Dept. of Physics, The University, Durham DH1 3LE, UK

L.A.M. NEJAD, Insitute of Astronomy, Madingley Road, Cambridge CB3 OHA UK

G. PINEAU DES FORETS, DAPHE, Observatoire de Paris, F-92195 Meudon Principal Cedex, FRANCE

J.M.C. RAWLINGS, Dept. of Mathematics, UMIST, PO Box 88, Manchester M60 1QD, UK

E. ROUEFF, DAMAp, Observatoire de Meudon, Place Jules Janssen, F-92195 Meudon Principal Cedex, FRANCE

B.R. ROWE, Laboratoire d'Aerothrmique, 4ter Routes des Gardes, F-92190 Meudon, FRANCE

C.M. SHARP, Institut fur Astrophysik, Karl-Schwarzschild-Str. 1, D-8046 Garching bei Munchen, WEST GERMANY

D. SMITH, Dept. of Space Research, University of Birmingham, PO Box 363, Birmingham B15 2TT, UK

I.W.M. SMITH, Dept. of Chemistry, University of Birmingham, PO Box 363, Birmingham B15 2TT, UK

N.D. TWIDDY, Dept. of Physics, University College, Penglais, Aberystwyth SY23 3BZ, UK

P. VALIRON, Groupe d'Astrophysique, CERMO, B.P. 68, F-38402 St. Martin d'Heres Cedex, FRANCE

E.F. VAN DISHOECK, Insitute for Advanced Study, School of Natural Sciences, Princeton, NJ 08450, USA

Y.-P. VIALA, DEMIRM, Observatoire de Meudon, F-92195 Meudon Principal Cedex, FRANCE

J.C. WHITEHEAD, Dept. of Chemistry, Univ. of Manchester, Manchester M13 9PL, UK

D.C.B. WHITTET, School of Physics and Astronomy, Lancashire Polytechnic, Preston PR1 2TQ, UK

U. WILHELMSSON, Dept. of Physical Chemistry GU, Chalmers University of Technology, S-412 96 Goteborg, SWEDEN

D.A. WILLIAMS, Dept.of Mathematics, UMIST, PO Box 88, Manchester M60 1QD UK

THEORY OF REACTIVE COLLISIONS AT LOW TEMPERATURES

D. C. Clary†
University Chemical Laboratory
Lensfield Rd., Cambridge CB2 1EW, UK
and Joint Institute for Laboratory Astrophysics
University of Colorado, Boulder, CO 80309-0440, USA

ABSTRACT. A description is given of calculations of chemical reaction rate coefficients at low temperatures (<100 K). Fast reactions relevant to interstellar chemistry are emphasized. Several comparisons with experimental data are made. The review concentrates on ion-molecule reactions, although neutral reactions are also briefly discussed.

1. INTRODUCTION

An understanding of the chemical reactions occurring in interstellar clouds is fundamental for theories of star formation and evolution, and for the interpretation of many astronomical observations [1]. With so many molecules now being observed in interstellar clouds, chemical reaction models which can explain how these molecules are produced and destroyed are becoming increasingly more valuable. The most modern chemical reaction networks that have been proposed involve following the concentration of several hundred atomic and molecular species as a function of time, and reliable temperature-dependent rate coefficients for several thousand reactions are a vital requirement in such simulations. The role of ion-molecule reactions has been shown to be of particular importance in these networks as these reactions can have very large rate coefficients at the low temperatures of interstellar clouds [2]. Furthermore, a more limited number of neutral species, particularly radicals and open-shell atoms, can have large rate coefficients at low temperatures [3]. Since only a relatively small number of reactions have been studied in the laboratory at the temperatures relevant to interstellar chemistry, theory plays an important role in producing many of the required rate coefficients.

Two questions often asked by interstellar chemists are: How reliable are the rate coefficients produced by theory? And, can one

---

†1987-88 Visiting Fellow, Joint Institute for Laboratory Astrophysics, National Bureau of Standards and University of Colorado

T. J. Millar and D. A. Williams (eds.), Rate Coefficients in Astrochemistry, 1–16.

use a rate coefficient measured or calculated at 300 K in interstellar simulations at temperatures less than 20 K? The answers to these questions can only be found by extensive comparisons of rate coefficients produced by both experiment and theory for a wide range of temperatures. This is largely the aim of this review. We emphasize the reactions of ions with dipolar molecules as these reactions show the largest temperature dependencies and provide stringent tests on theory. There are also many ion-dipole reactions of vital importance in interstellar chemistry [4].

In Section 2 we describe various theories that have been proposed for reactions that are fast at low temperatures. Valuable contributions have been made by many research workers and, for the important case of ion-dipolar molecule reactions, we do make comparisons between several different theories that have been developed.

Sections 3 and 4 give several comparisons between theory and experiment for rate coefficients, particularly involving ion-dipolar molecule reactions. For convenience and continuity, we take the liberty of presenting results obtained largely in our own calculations but we emphasize that several other researchers have made a variety of useful calculations in this field.

In Section 5 we present our concluding remarks and, in particular, attempt to summarize the situations for which the theoretical results will be particularly reliable and useful. We also describe the theoretical and experimental work that will be needed in the future.

2. THEORY

2.1. Interaction Potentials

The rates of chemical reactions are determined by the potential energy surfaces for the systems in question. The most rigorous way of obtaining potential energy surfaces for reactions is to perform ab initio electronic structure calculations. Although there have been significant developments in electronic structure theory in recent years [5], very few reliable potential energy surfaces have been produced that cover the whole reaction space. However, in the case of very fast reactions, the reaction rate is often determined by the purely attractive long-range part of the potential where the distance between the reacting particles is large [6]. These long-range potentials can often be constructed from knowledge of just the multipole moments (especially dipole and quadrupole moments), polarizabilities and charges on the interacting particles [7]. For example, the leading terms in the long-range interaction potential between a particle with charge q and molecule with isotropic polarizability α and dipole moment μ_D are

$$V(R,\theta) = -\frac{\alpha q^2}{2R^4} - \frac{q\ \mu_D \cos\theta}{R^2} \qquad (1)$$

where R is the distance between the charge and (point) dipole and θ is the angle that the charge-dipole vector makes with the direction of the dipole. As another example, the leading term in the long-range interaction potential for two molecules with dipole moments μ_1 and μ_2 is

$$V(R,\theta_1,\theta_2,\phi_1,\phi_2) = -\frac{\mu_1\mu_2}{R^3}\,[2\cos\theta_1\cos\theta_2 - \sin\theta_1\sin\theta_2\cos(\phi_1-\phi_2)] \tag{2}$$

where (θ_i,ϕ_i) describe the orientation of dipole i. Even when the required molecular parameters in equations (1) and (2) are not given in the literature it is now normally quite straightforward to calculate them using ab initio electronic structure packages.

Therefore, if the chemical reaction rate coefficient is determined just by these long range forces it is only necessary to solve the equations of motion for the particles moving on these long-range potentials (Hamilton's equations in classical mechanics and, ideally, the Schrödinger equation in quantum mechanics). Then collisional cross sections are obtained and, by Maxwell-Boltzmann averaging over collision energy and, if necessary, the initial rotational states of the reacting molecules, we obtain the required rate coefficients. However, solving these dynamical equations of motion is not trivial, especially using quantum mechanics. The reason for this difficulty is that the potentials of equations (1) and (2) depend on angles which couple the rotational states of the reacting molecule or molecules together. Thus a multi-channel dynamics theory is required for a proper treatment of the problem. Indeed, as we shall illustrate in Sections 3 and 4 this strong angular coupling introduces dramatic rotational effects on the reaction rate coefficients which, in turn, have a strong influence on the temperature dependence of the rate coefficients. Thus it is important to calculate initially rotationally selected rate coefficients $\{k_j\}$ before the final rate coefficient k is obtained.

2.2. The Capture Approximation

It is useful to illustrate how the dynamics problem is solved by referring to Figure 1. This illustrates the typical potential energy profile for a fast reaction dominated by long-range forces. It is essential that the reaction is exothermic. Superimposed on the electronic potential is the centrifugal term

$$V_c(R) = \frac{J(J+1)}{2\mu R^2} \tag{3}$$

where J is the total angular momentum for the collision and μ is the collisional reduced mass. In Figure 1, it can be seen that there is an effective centrifugal barrier at the point marked B. Providing that there are no secondary barriers in the shorter range region of

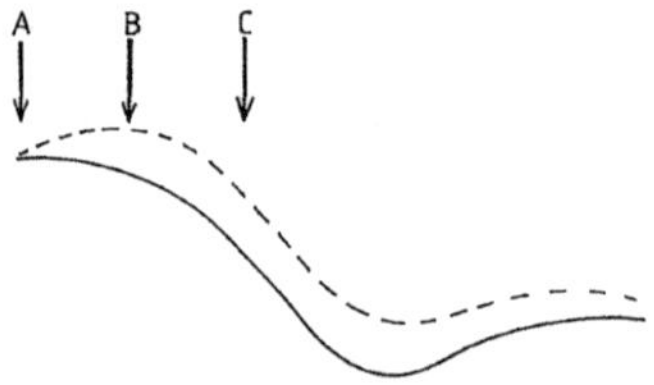

Figure 1. Potential energy profile for a fast reaction (solid line). Dotted line shows effective potential obtained by adding the centrifugal potential.

the potential it is only necessary to solve the equations of motion for the dynamics over the region between the arrows A and C. It is assumed that once the incoming particle has passed through or over the centrifugal barrier and has reached point C, it always goes on to react. Thus a capture approximation is assumed. The larger the exothermicity of the reaction, the less likely will be the possibility of secondary barriers in the potential reflecting the reaction particle back to reactants. Thus the capture approximation is particularly appropriate for many proton transfer reactions and for reactions of ions such as He^+, C^+ and N^+ for which the parent atoms have large ionization potentials. However, it is only by comparing calculated rate coefficients or cross sections with experiment that it is possible to be absolutely certain that a reaction will go at the capture rate.

2.3. Quantum Mechanical Calculations

The above considerations concerning the capture approximation have enabled us to develop general quantum mechanical theories for these types of reactions dominated by long-range intermolecular forces [8-10]. Because it is only necessary to consider the solution of the Schrödinger equation in the entrance channel of the reaction, all the accurate and approximate quantum theories that have been developed for non-reactive scattering [11] can be applied to the reactive scattering problem considered here by applying boundary conditions simply at the points A and C of Figure 1. By expanding the wavefunction for the problem in terms of a coupled translational and rotational basis set it is possible, for this restricted region, to solve the so-called close-coupling equations numerically on the computer. For polyatomic systems, molecules with small rotor constants and molecule-molecule reactions the size of the rotational basis set can be prohibitively large. However it is then possible to use the various sudden approximations (for a review see [11]) to reduce the number of rotational basis functions required (but still retaining some rotational coupling). This enables quantum calculations to be performed even on polyatomic molecule problems. However, these types of accurate quantum calculations are not cheap in computer time and, fortunately, a simple approximation arises from the exact theory that has been found to be accurate in many cases.

2.4 Rotationally Adiabatic Theory

The quantum mechanical Hamiltonian for the reaction contains the term

$$H_r = Bj^2 + \ell^2/2\mu R^2 + V(R,\theta) \qquad (4)$$

where, for clarity, we have assumed a particle-rigid rotor interaction. Here B is the molecular rotor constant, j^2 is the rotational angular momentum operator and ℓ^2 is the operator representing the orbital motion of the particle about the molecule. Unless the dipole moments or polarizabilities are very sensitive to the vibrational motion of the molecule it is accurate to neglect the effect of vibrations on the reaction rate for these types of long-range reactions. For a given J and fixed values of R, H_r can be diagonalized using a spaced-fixed basis set of combined rotor and orbital functions [12]. This yields a set of rotationally adiabatic potential curves $\varepsilon^J_{jl}(R)$ which are labeled by the (j,l) state that they tend to as R gets large.

An example of the rotationally adiabatic potential curves for the $He^+ + HC\ell$ (ion-dipole) reaction is given in Figure 2. It can be seen that the curves for each (j,l) state are widely separated in the region of the centrifugal maxima. This is due to the very strong anisotropy of the ion-dipole reaction as indicated in equation (1). The diagram at once suggests a "rotationally adiabatic" approximation in which it is assumed that the reaction probability for each initial (j,l) state is determined simply by calculating the reaction probability for the individual $\varepsilon^J_{jl}(R)$ curve. Furthermore, since many J values contribute to the reaction cross section, tunneling will not be important (except at temperatures much less than 1 K) and a classical capture approximation can be applied in which it is assumed that the reaction probability is 1 if the collision energy is above the centrifugal barrier in $\varepsilon^J_{jl}(R)$ and is zero otherwise.

This is a very simple procedure that enables the reaction cross sections and rate coefficients to be determined simply by a small number of matrix diagonalizations. Furthermore, by applying an approximation such as the centrifugal sudden approximation (CSA) in which the coupling in the projection quantum numbers of the rotational basis sets is neglected [11], the rotationally adiabatic potential curves for J > 0 are obtained from those calculated for J = 0 simply by adding the centrifugal term of equation (3).

We have given this combined rotationally adiabatic capture, centrifugal sudden approximation the (rather unfortunate) acronym ACCSA (or AC for short). It is important to stress that the ACCSA computations can be performed with a computer of minor power whereas the accurate quantum mechanical calculations described in Section 2.3 normally need to be done on a supercomputer such as a CRAY. Furthermore, the rate coefficients obtained using a capture theory such as the ACCSA should be upper bounds to the exact results. The methods described above are general for any type of fast ion or neutral reaction dominated by long-range intermolecular forces. Several other related theories have been specially developed for ion-dipolar

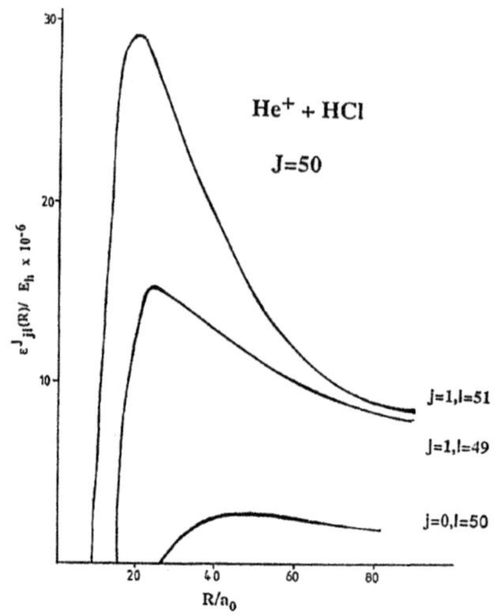

Figure 2. Rotationally adiabatic potentials $\varepsilon^J_{jl}(R)$ for the He^+ + HCℓ reaction with J = 50 [10].

molecule reactions. These include the perturbed rotational states approximation of Sakimoto and Takayanagi [13], the statistical adiabatic channel model of Troe [14] and an adiabatic invariance method of Bates and Morgan [15]. These theories also provide the $\{k_j\}$ and, by making extra approximations, they can yield analytical formulae for the rate coefficients. In Section 3 we will make a comparison between the results of some of these theories. We should also mention that classical trajectory computations can also be carried out on these long-range reactions [16] and statistical theories [17] have been successfully applied also.

2.5 Simple Theories

Even simpler theories have been proposed for reactions dominated by long-range forces, and, in particular, ion-molecule reactions. Bowers [6] and co-workers have developed theories that average over the angular dependence of the potential and ignore the explicit dependence of the rate coefficients on the initial rotational states. This enables simple formulae to be obtained. Thus, in the "Average Dipole Orientation Approximation" (ADO) the ion-dipole rate coefficient has the form

$$k = \frac{2\pi|q|}{\mu^{1/2}} \{\alpha^{1/2} + C(\alpha,\mu_D,T)\ \mu_D[2/(\pi k_B T)]^{1/2}\} \tag{5}$$

where C is a tabulated constant. The theory is improved slightly [6] by including approximately the conservation of angular momentum (the AADO theory). Furthermore, for ion-quadrupole collisions the same approach can be applied to yield the "Average Quadrupole Orientation Approximation" (AQO) [6].

An even simpler approach that gives insight into the temperature dependences of the rate coefficients for these reactions is to ignore the angular dependence of the potential completely. Then simple classical theory can be used [6] to find the maximum impact parameter, and hence cross section, for potentials of the form $-R^{-n}$. This gives a temperature dependence in the rate coefficient of the form $T^{-(2/n)+1/2}$. Thus n must be less than 4 for the rate coefficient to have a negative temperature dependence. This implies at once that it is the ion-dipole reactions, for which the potential depends on $-R^{-2}$, that will give the strongest negative temperature dependence.

In Table I we list the various types of long-range interaction potential, their $-R^{-n}$ dependence and the rate coefficient temperature dependences associated with these $-R^{-n}$ dependences. We also give the ratio between the rate coefficients at 10 K and 300 K predicted from this very simple theory. Since this very simple approach neglects the rotational motion of the molecule these results should only be considered as an approximate guideline. However, they do suggest that it is only the ion-dipole reaction rate coefficients that will deviate strongly (larger than a factor of two) from the room temperature rate coefficient and thus, from the point of view of interstellar chemistry, it is important for theoreticians to concentrate on the ion-dipole reactions. Therefore, most of the results of the next section refer to ion-dipole reactions. We note that the charge-quadrupole interaction does suggest a slight negative temperature dependence [18] but experiments have so-far failed to confirm this result [19].

Table I. Temperature dependencies of rate coefficients from $-R^{-n}$ potentials

Interaction	$-R^{-n}$	T dependence of k	k(10 K)/k(300 K)
Charge-Induced Dipole	$-R^{-4}$	T^{0}	1
Charge-Dipole	$-R^{-2}$	$T^{-1/2}$	5.5
Charge-Quadrupole	$-R^{-3}$	$T^{-1/6}$	1.8
Dipole-Dipole	$-R^{-3}$	$T^{-1/6}$	1.8
Dipole-Quadrupole	$-R^{-4}$	T^{0}	1

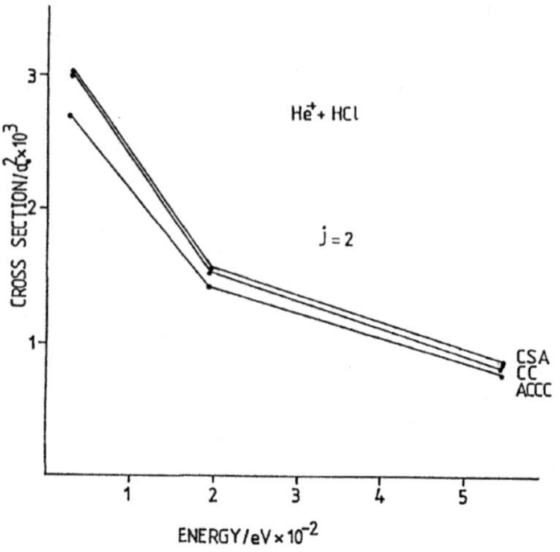

Figure 3. Cross sections for the He^+ + HCℓ (j=2) reaction as a function of collision energy [10].

3. CALCULATIONS ON ION-DIPOLE REACTIONS

In this section we, first of all, compare various calculations for ion-dipole reaction cross sections and rate coefficients, and then go on to make several comparisons of theory with experimental data.

In Figure 3 we show cross sections for the He^+ + HCℓ(j) charge transfer reaction calculated [10] using the accurate quantum theory (denoted CC) described in Section 2.3, a CSA approximation to this theory and the rotationally adiabatic capture theory (denoted AC) described in Section 2.4. It can be seen that the agreement between all the results is very good and this provides an excellent test of the AC theory for this type of reaction.

In Table II we compare rate coefficients calculated [20] for the He^+ + HCℓ reaction using three different theories - the ACCSA, the statistical adiabatic channel model (SACM) of Troe [14] and classical trajectory calculations [16]. The trajectory calculations have been parameterized to give the simple formula

$$\begin{aligned} k &= k_L \ \{0.3371\ y\ [B/k_BT]^{1/2} + 0.62\} \quad \text{for } k_BT/B \le y^2/8 \\ &= k_L \ \{0.0475\ (y[B/k_BT]^{1/2} + 0.7198)^2 + 0.9754\} \quad \text{for } k_BT/B > y^2/8 \end{aligned} \tag{6}$$

where $k_L = 2\pi(\alpha/\mu)^{1/2}$ is the Langevin rate coefficient and $y = \mu_D/(\alpha B)^{1/2}$.

Table II. Calculations of rate coefficients for the He^+ + HCℓ reaction

Temp/K	k/cm^3 s^{-1} molec^{-1} × 10^{-9} ACCSA	SACM	Trajectory
27	8.4	8.6	8.6
68	5.5	5.6	5.9
300	3.2	3.0	3.5

It can be seen that the agreement between all three methods is excellent. Furthermore, the adiabatic invariance method of Bates and Morgan [15] also gives good agreement with these theories for this temperature range as does the perturbed rotational state approximation [13]. Our quantum-mechanical results thus verify the trajectory formula of equation (6) and we recommend that it be used in interstellar simulations for reactions that go at the capture rate. However, to be sure that a reaction does go at the capture rate it is important to have experimental data, at least at room temperature.

In Figure 4 we compare the ACCSA rate coefficients for the H_3^+ + HCN → H_2 + H_2CN^+ proton transfer reaction with those obtained in a selected ion flow tube (SIFT) apparatus [21]. Also shown are the ADO rate coefficients for this reaction. It can be seen that the agreement between the ACCSA and experimental rate coefficients is very good and the strong negative temperature dependence predicted even by simple theory (see Section 2.5) is obtained. Although the ADO theory does give good agreement with experiment at room temperature, the agreement is not so good at lower temperatures. This is because the ADO theory does not account explicitly for the rotational state

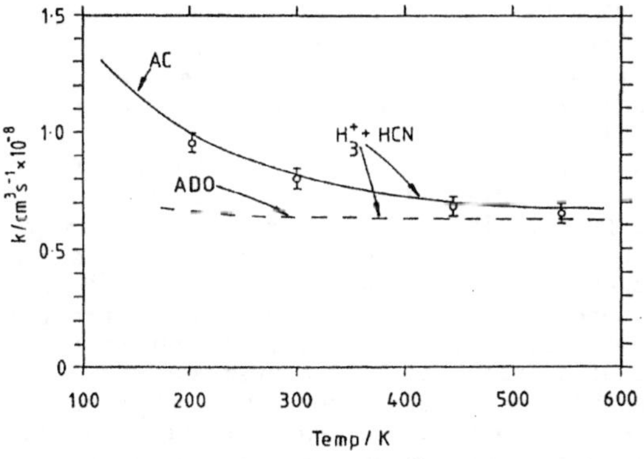

Figure 4. Rate coefficients for the H_3^+ + HCN reaction. Straight line shows AC calculations, open circles SIFT experiments and dotted line the ADO rate coefficients [21].

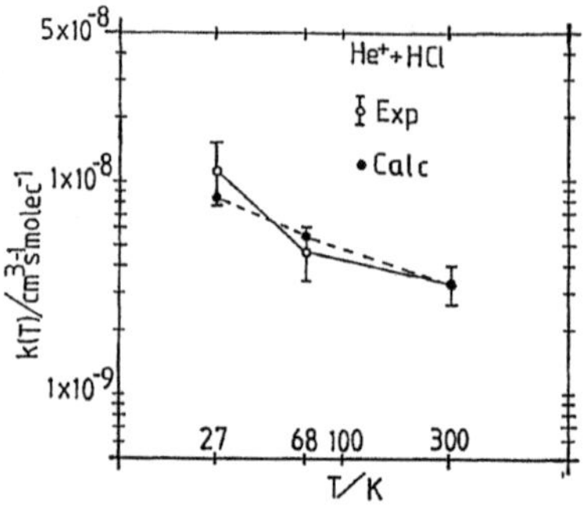

Figure 5. Rate coefficients for the He^+ + HCℓ reaction. AC (closed circles) and CRESU (open circles) rate coefficients are shown [20].

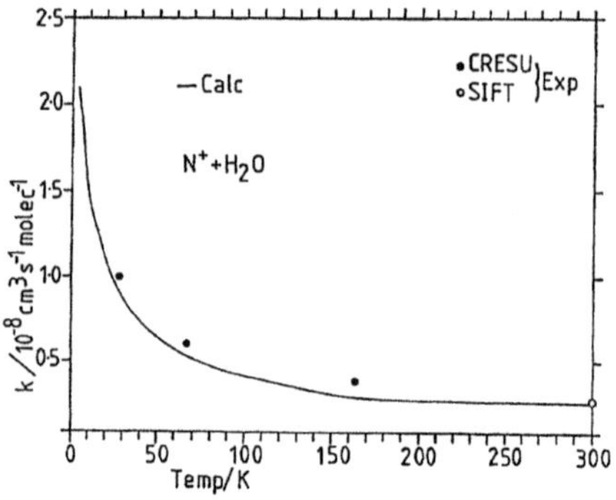

Figure 6. Rate coefficient for the N^+ + H_2O reaction. Straight line shows AC calculations [22], circles the experimental measurements.

dependence on the rate coefficients, unlike the ACCSA and several of the other theories. Thus the ADO rate coefficients should be used with caution in interstellar simulations.

The CRESU experiments at Meudon have enabled rate coefficients to be obtained for ion-dipole reactions at temperatures as low as 27 K (see the chapter by Rowe et al. in this book). This provides an excellent data base for testing theory. Figure 5 shows a comparison of ACCSA and CRESU rate coefficients for the He^+ + HCℓ reaction. It can be seen that the agreement between theory and experiment is excellent. Similar good agreement is found for several other charge transfer reactions [20] including He^+ + SO_2,H_2S, C^+ + SO_2,H_2S and N^+,C^+ + H_2O (shown in Figure 6). These theoretical/experimental comparisons thus confirm the strong negative temperature dependence predicted by theory for these strongly exothermic charge transfer reactions involving ions for which the parent atoms have high ionization potentials. They also confirm that these kinds of reactions can go at the capture rate.

Another important feature arising from the theory is the sensitivity of the rotationally selected rate coefficients k_j to the initial rotational state j. This is illustrated in Figure 7 for the N^+ + NH_3 reaction [22]. It can be seen that the k_j decrease strongly as j is increased. This effect is due to the fact that the minimum path of the potential energy surface involves the ion aligning collinearly with the dipole. If the molecule is rotating it becomes harder for this line of approach to be achieved and the reaction probability is accordingly reduced. Also shown in Figure 7 as a dotted line is the Maxwell-Boltzmann average of the rate coefficients $\{k_j\}$. This has a much steeper temperature dependence than any of the individual k_j and thus the temperature dependence of the k is directly related to the Maxwell-Boltzmann average over j. In interstellar simulations it is sometimes necessary to use the k_j with j = 0 as radiative decay at low temperatures can leave many molecules in their ground state. In this case it will be important to use rotationally selected rate coefficients obtained from a theory such as the ACCSA.

We should note that Troe has shown [14] that there should be a leveling off in the ion-dipole rate coefficients at very low temperatures (<5 K). However, for most systems of interest, this effect occurs at temperatures too low to be of importance to interstellar chemistry (for more details see reference [14]).

After considering reactions (mainly strongly exothermic charge transfers and proton transfers) which do go at the capture rate, it is important to discuss those reactions that do not. Table III gives a comparison between ACCSA and experimental SIFT [21] rate coefficients for the exchange reactions $C^+ + HCN \rightarrow C_2N^+ + H$ and $O^+ + HCN \rightarrow HCO^+ + N$. It can be seen that the ACCSA rate coefficients slightly overestimate the experimental result in this case. The same is true for the $C^+ + HC\ell \rightarrow C\ell C^+ + H$ exchange reaction studied in the CRESU experiments [20]. The reason for the overestimation of the rate coefficients by theory is probably related to the fact that these exchange reactions have a secondary barrier in the potential energy surface in the shorter-range region which can reflect the ion back to reactants and reduce the overall reaction rate. Accurate ab initio computations are required to probe the nature of these effects.

A very clear and systematic study of these secondary barrier effects in ion-dipole exchange reactions is provided by an experimental and theoretical study of the solvated reaction [23] $OH^-(H_2O)_n + CH_3C\ell \rightarrow CH_3OH + C\ell^-(H_2O)_n$. Here it was found that the ratio of the experimental to ACCSA rate coefficients decreased sharply as n was increased. Ab initio computations confirm that the secondary barrier in the potential does rise dramatically as the number of water molecules attached to the ion is increased [24].

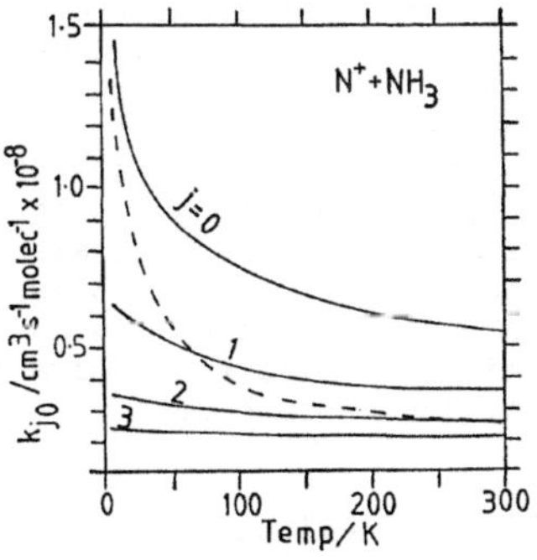

Figure 7. Rate coefficients k_j for the $N^+ + NH_3$ reaction obtained in AC calculations.

Table III. Rate coefficients for the C^+ + HCN and O^+ + HCN exchange reactions in units of 10^{-9} cm^3 s^{-1} $molec^{-1}$ [21].

Reaction	Temp/K 205	300	440	540
C^+ + HCN → C_2N^+ + H	3.4[a] 5.5[b]	3.1 4.6	3.0 4.0	2.9 3.7
O^+ + HCN → HCO^+ + N	3.7[a] 5.0[b]	3.5 4.2	3.3 3.6	3.1 3.4

[a]SIFT experiments
[b]ACCSA calculations

Most of the experimental/theoretical comparisons for ion-dipole reactions have been done on small molecules, but it is of great importance to interstellar chemistry to determine whether the theories are applicable to reactions involving more complicated polyatomic molecules. With this in mind a combined experimental and theoretical study has been carried out [25] on the charge-transfer reaction of N^+ with the three isomers of dichloro-ethylene ($C_2H_2C\ell_2$). One of these isomers (trans-1,2) has no dipole moment while the cis-1,2 and 1,1 isomers have appreciable dipoles. The CRESU experiments gave the rate coefficients at 27, 68 and 163 K while the SIFT experiments yielded the coefficient at 300 K. Figure 8 compares the ACCSA rate coefficients with the experiments. It can be seen that the reaction of the non-polar trans isomer does go at the Langevin rate. However, the polar isomers have rates overestimated by the ACCSA and there is only a slight negative temperature dependence in the experimental rate coefficient. It is clear that the dipole moment of the polar isomers (all of which have very similar polarizabilities) has little effect on the reaction. One explanation of this effect is that the most favorable angle of approach of the ion to the molecule for charge transfer is out of the molecular plane, into the π orbitals of the $C_2H_2C\ell_2$ and at this angle of approach the dipole moment has a zero component. Similar experimental/theoretical comparisons are also obtained for the H_3^+ proton transfer reactions with the $C_2H_2C\ell_2$ isomers. It is clear that more theoretical/experimental comparisons on the reactions of ions with dipolar polyatomic molecules are needed to determine which types of reactions go at the capture rate.

4. CALCULATIONS ON FAST NEUTRAL REACTIONS

Advances in experimental techniques have shown [3] that some reactions of free radicals and open shell atoms have quite large rate coefficients (>10^{-11} cm^3 s^{-1} $molec^{-1}$). These rates are not as large as those seen for ion-dipole reaction (10^{-8} - 10^{-9} cm^3 s^{-1} $molec^{-1}$)

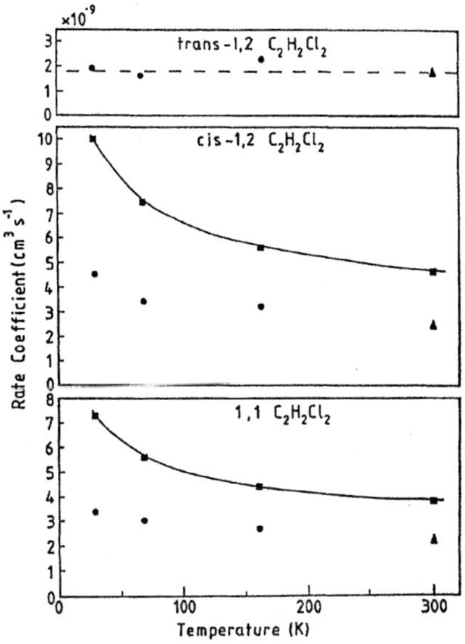

Figure 8. Rate coefficients for the N^+ + $C_2H_2C\ell_2$ reactions. Squares show AC calculations, circles CRESU experiments and triangles SIFT experiments [25].

but they are still appreciable enough in certain cases to be important for interstellar chemistry. Reactions involving dipole-dipole and dipole-quadrupole interactions, in particular, can have quite large rates. The accurate quantum theory described in Section 2.3 and the rotationally adiabatic capture theory outlined in Section 2.4 have been applied to these types of reactions [8]. An example is shown in Figure 9 where ACCSA calculations [26] of rate coefficients for the

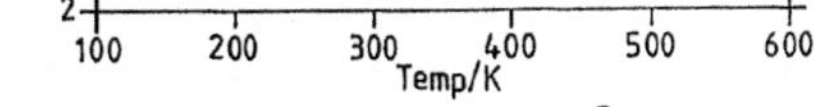

Figure 9. Rate coefficients for the $O(^3P)$ + OH reaction. Straight line shows AC calculations [26], circles show experiment [27].

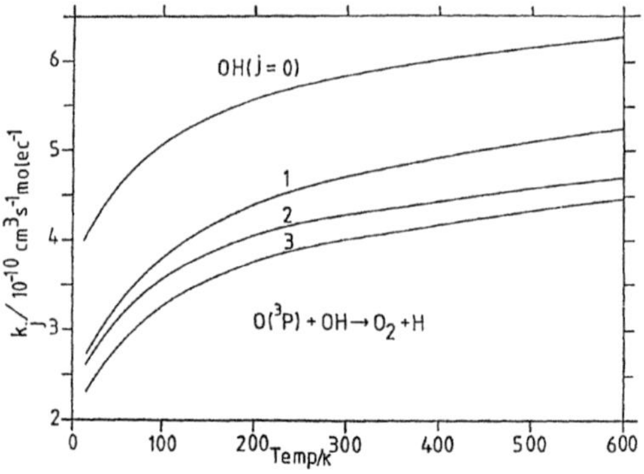

Figure 10. AC calculations of rate coefficients k_j for the $O(^3P)$ + OH reaction.

$O(^3P) + OH \rightarrow O_2 + H$ reaction are compared with experiment [27] for temperatures greater than 250 K. The agreement is quite good and there is a weak negative temperature dependence in both experimental and calculated rate coefficients. This negative temperature dependence is related to two effects. First, as shown in Figure 10, the k_j decrease with increasing j for a given temperature and, consequently, the Boltzmann average over j does produce a slight negative temperature dependence in k. Second, the species involved in the reaction are open shell and it is necessary to divide by the electronic partition function to obtain the final rate coefficient and this also contributes to the negative temperature dependence. For a fully rigorous theoretical treatment of a reaction such as this it would be necessary to state select the initial electronic states of the reactants in the calculations and follow the different electronic potential energy surfaces involved.

Unfortunately, the experimental methods for fast neutral reactions have not yet developed to enable rate coefficients to be obtained at interstellar temperatures (less than 100 K). Thus it has not yet been possible to test theory in the rigorous way that has been done for ion-dipole reactions. For reviews of experiment and theory in this area see references [3] and [8].

5. CONCLUSIONS

It should be clear to the reader that it is now possible to apply quantum mechanical-based capture theories to a wide variety of reactions dominated by long-range intermolecular forces. For reactions controlled by shorter range forces the calculations are much harder to do and have only been carried out on a small number of very specialized systems [28]. Given that the capture calculations can be performed, can we make any positive statements as to which reactions go at the capture rate and which do not? We have seen that many strongly exothermic charge transfer and proton transfer reactions involving ions and dipolar molecules do go at the capture rate but, even in this class of reactions, there are some that do not. Thus, to be absolutely sure that a reaction does go at the capture rate, experiments are necessary in each case. The question that then can be asked is, where can theory be useful? The answer to this question is that if a rate coefficient is measured accurately at room temperature and the rate coefficients agree well with a rotationally selected theory then the results obtained using that theory should be reliable for the lower temperature range which is of most importance to interstellar chemistry and where the experiments are difficult to perform. Since reactions involving dipolar molecules have the strongest negative temperature dependence, it is these reactions that should be considered most carefully when extrapolating room temperature rate coefficients down to lower temperatures.

Looking to the future, it is clear that more experimental results will be very valuable for reactions at lower temperatures. The CRESU experiments are a great advance in this direction and it would be of considerable interest if this type of technique could be extended to neutral reactions. We have shown how knowledge of the rotationally selected rate coefficients k_j is important for understanding the temperature dependencies of the overall reaction rate coefficients k. Measurement of the k_j or rotationally selected cross sections for these fast reactions presents a very exciting challenge to experiments in the future. It is likely that electronic state dependence on rate coefficients will also be important for the temperature dependencies of some fast reactions and this is an almost untouched area in both experiment and theory which might have significant consequences for interstellar chemistry. Finally, the recent advances in ab initio quantum chemistry, reactive scattering dynamics and the revolution in supercomputers will mean that substantially more highly rigorous theoretical calculations and predictions will come through in the near future that could be of great significance for interstellar chemistry.

Acknowledgments

This work was supported by the SERC and the EEC. I would like to acknowledge very stimulating experimental/theoretical collaborations with D. Smith, N. G. Adams and B. R. Rowe and his co-workers. This review was written while the author was a Visiting Fellow at the Joint Institute for Laboratory Astrophysics. The author would like to give his thanks to everyone at JILA for their kind and stimulating hospitality.

REFERENCES

[1] W. W. Duley and D. A. Williams, Interstellar Chemistry, 1984, Academic Press, London.
[2] A. Dalgarno and J. H. Black, Rep. Prog. Phys. 39 (1976) 573
[3] M. J. Howard and I. W. M. Smith, Prog. React. Kinet. 12 (1983) 55.
[4] N. G. Adams, D. Smith, and D. C. Clary, Astrophys. J. 296 (1985) L31.
[5] C. E. Dykstra Ed., Advanced Theories and Computational Approaches to the Electronic Structure of Molecules, Reidel, Dordrecht, 1984.
[6] T. Su and M. T. Bowers, in Gas Phase Ion Chemistry, Vol. 1, 1979, ed. M. T. Bowers, Chap. 3.
[7] A. D. Buckingham, Adv. Chem. Phys. 12 (1967) 107.
[8] D. C. Clary, Molec. Phys. 53 (1984) 3.
[9] D. C. Clary, Molec. Phys. 54 (1985) 605.
[10] D. C. Clary and J. P. Henshaw, Faraday Disc. Chem. Soc. 84, in press.
[11] D. C. Clary, J. Phys. Chem. 91 (1987) 1718.
[12] A. M. Arthurs and A. Dalgarno, Proc. R. Soc. London Ser. A 256 (1960) 540.
[13] K. Sakimoto and K. Takayanagi, J. Phys. Soc. Japan 48 (1980) 2076.
[14] J. Troe, J. Chem. Phys. 87 (1987) 2773.
[15] W. L. Morgan and D. R. Bates, Astrophys. J. 314 (1987) 817.
[16] T. Su and W. J. Chesnavich, J. Chem. Phys. 76 (1982) 5183.
[17] R. A. Barker and D. P. Ridge, J. Chem. Phys. 64 (1976) 4411.
[18] D. R. Bates and I. Mendas, Proc. R. Soc. London A 402 (1985) 245.
[19] C. Rebrion, J. B. Marquette, B. R. Rowe, N. G. Adams, and D. Smith, Chem. Phys. Lett. 136 (1987) 495.
[20] C. Rebrion, J. B. Marquette, B. R. Rowe, and D. C. Clary, Chem. Phys. Lett., in press.
[21] D. C. Clary, D. Smith, and N. G. Adams, Chem. Phys. Lett. 119, (1985) 320.
[22] D. C. Clary, J. Chem. Soc., Faraday Trans. 2 83 (1987) 139.
[23] P. M. Hierl, A. F. Ahrens, M. Henchman, A. A. Viggiano, J. F. Paulson, and D. C. Clary, J. Am. Chem. Soc. 108 (1986) 3142.
[24] K. Ohta and K. Morokuma, J. Phys. Chem. 89 (1985) 5845.
[25] C. Rebrion, J. B. Marquette, B. R. Rowe, C. Chakravarty, D. C. Clary, N. G. Adams, and D. Smith, J. Phys. Chem., in press.
[26] D. C. Clary and H.-J. Werner, Chem. Phys. Lett. 112 (1984) 346.
[27] M. J. Howard and I. W. M. Smith, J. Chem. Soc. Faraday Trans.2 77 (1981) 997.
[28] D. C. Clary, Ed., The Theory of Chemical Reaction Dynamics, Reidel, Dordrecht, 1986.

RADIATIVE ASSOCIATION

David R. Bates
Department of Applied Mathematics and Theoretical Physics
Queen's University of Belfast
Belfast BT7 1NN
United Kingdom

and

Eric Herbst
Department of Physics
Duke University
Durham, NC 27706
USA

ABSTRACT. Radiative association reactions are reactive processes in which two smaller gas phase species collide to form a larger molecule while emitting a photon. These reactions are thought to be important in the synthesis of molecules in both diffuse and dense interstellar clouds. Models of interstellar clouds require the rate coefficients of a variety of radiative association reactions as input yet few experimental studies of these processes have been undertaken. Therefore, the role of theory in the determination of radiative association rate coefficients is paramount. Most experimental studies of association reactions are at sufficiently high pressure that the mechanism for association is collisional rather than radiative. Yet even collisional (ternary) association studies yield valuable information about radiative association processes. In this review, we consider the nature of association reactions - both radiative and ternary - and discuss experimental and theoretical approaches to the determination of rate coefficients of radiative association reactions.

1. INTRODUCTION

The role of radiative association reactions in the chemistry of diffuse and dense interstellar clouds has been recognized by a sizable number of scientists including Bates (1951) and Solomon and Klemperer (1972), for the synthesis of diatomic molecules; and Williams (1972), Black and Dalgarno (1973), Herbst and Klemperer (1973), Smith and Adams (1978), and Huntress and Mitchell (1979), for the synthesis of polyatomic species. In radiative association, two species collide to form an unstable molecule normally called a "collision complex" which becomes stabilized through radiating sufficient energy. If rapid enough, this process is a good mechanism for the synthesis of polyatomic molecules in the low density interstellar medium. Two major types of radiative association reactions in

T. J. Millar and D. A. Williams (eds.), Rate Coefficients in Astrochemistry, 17–40.

interstellar clouds are thought to exist. Both of these involve collisions between positive ions and neutral molecules to form intermediate complexes. Such collisions are normally without activation energy and so may be quite fast at the low temperatures characteristic of the ambient interstellar medium. On the other hand, most collisions between neutral species that form complexes involve considerable activation energy and are slow under normal interstellar conditions; reactions involving atoms and radicals are an exception. But, no radiative association reactions in this latter class appear to be important in interstellar clouds.

In the first class of interstellar radiative association reactions, the neutral reactant is molecular hydrogen and the process of radiative association acts as a detour when normal ion-molecule reactions cannot occur. As one example, the reaction

$$C^+ + H_2 \text{ -----> } CH^+ + H \qquad (1)$$

is known to be endothermic by 0.39 eV and therefore does not occur under normal conditions in the low temperature (10 K - 70 K) interstellar medium. However, the radiative association reaction

$$C^+ + H_2 \text{ -----> } CH_2^+ + h\nu \qquad (2)$$

can occur and is thought to be quite important as an initial step in the "fixation" of carbon into hydrocarbons (Black and Dalgarno 1973). The current best estimate for the rate coefficient of process (2) under interstellar conditions is $1(-16) \leq k_2 \leq 1(-15)$ $cm^3\ s^{-1}$; more than five orders of magnitude below the rate coefficient for impact or close collisions (Herbst 1982a). Still, the large abundance of molecular hydrogen renders this an important process. Another example of this type of radiative association reaction is

$$CH_3^+ + H_2 \text{ -----> } CH_5^+ + h\nu. \qquad (3)$$

This reaction is important, despite its small rate coefficient, because of the endothermicity of all channels leading to two products (e.g., $CH_4^+ + H$).

In the second class of radiative association reactions, the neutral species involved is a heavy molecule and the process of radiative association leads to the formation of complex molecular ions (Smith and Adams 1978; Huntress and Mitchell 1979). An example is the reaction

$$CH_3^+ + H_2O \text{ -----> } CH_3OH_2^+ + h\nu, \qquad (4)$$

which forms protonated methyl alcohol in dense interstellar clouds. Since all neutral species other than hydrogen are minor constituents of interstellar clouds, processes such as (4) must be rapid to be of importance. Indeed, the current best estimate for the rate coefficient of this process at 10 K is 2(-9) $cm^3\ s^{-1}$, which is close to the collisional limiting value (Herbst 1985a). Reactions in this second class are important in models of dense interstellar clouds in which the gas phase chemistry of complex molecules is included (Leung, Herbst, and Huebner 1984).

Despite its importance in interstellar cloud chemistry, the process of radiative association has not received much attention in the laboratory. The reason for this lack of attention is that it is difficult to study because, unless the gas pressure is quite low, radiative association either competes with or is totally overwhelmed by ternary association, a process in which the collision complex is stabilized by inelastic collisions with neutral molecules. A few radiative association reactions have been studied at low pressure in the laboratory and some information concerning these processes can be extracted from studies of analogous ternary association reactions. It is necessary to discuss the detailed mechanism by which both radiative and ternary association reactions proceed in order to understand the experimental difficulties involved in studying radiative association and to understand the relationship between the two types of association reactions.

2. MECHANISM FOR ASSOCIATION REACTIONS

Consider a collision of two species - labelled A^+ and B - to form a collision complex AB^{+*} which can stabilize itself by emission of radiation or by an inelasic collision with a gas molecule C. The process can be divided into the following steps:

$$
\begin{aligned}
A^+ + B \;&\underset{k_d}{\overset{k_f}{\rightleftharpoons}}\; AB^{+*} \\
AB^{+*} \;&\xrightarrow{k_r}\; AB^+ + h\nu \\
AB^{+*} + C \;&\xrightarrow{k_c}\; AB^+ + C
\end{aligned}
\qquad (5)
$$

where each step has an associated rate coefficient k. The rate coefficient subscripts stand for complex formation (f), complex redissociation (d), complex radiative stabilization (r), and complex collisional stabilization (c). In general, these rate coefficients do not have single values for a given association process but depend on quantum numbers and energies. For example, k_f depends on the internal quantum states of the reactants and their collision energy.

An overall rate law for the formation of stable AB^+ can be obtained if the steady-state approximation is applied to the concentration of the complex; in other words, the approximation is made that the formation rate and overall destruction rate of the complex are equal. This condition is quickly reached in most experimental situations. Then, the formation rate of AB^+ is given by the equation

$$d[AB^+]/dt = k_{eff}[A^+][B] \qquad (6)$$

where the symbols [] refer to number density and the effective rate coefficient k_{eff} is defined by the relation

$$k_{eff} = \{k_f(k_r + k_c[C])\}/\{k_d + k_r + k_c[C]\}. \qquad (7)$$

In the numerator of expression (7) are the formation rate of the complex and the two rates of complex stabilization. In the denominator are the three rates of complex destruction. Note that k_{eff} is not truly a constant since it depends on the number density of species C. Note also, as mentioned above, that k_{eff} cannot really be regarded as the overall rate coefficient for association until it is averaged over all parameters on which the individual rate coefficients depend.

It is convenient at this stage to ignore this last word of caution and to regard the individual partial rate coefficients in expression (7) as entities with specific values. An advantage is that pressure regimes in which one association mechanism dominates may then be easily found. The idea of pressure regimes, amplified below, survives the inclusion of a more realistic range of values for each partial rate coefficient; such inclusion however tends to increase the transition ranges between the various pressure regimes.

2.1. Pressure Regimes and Experimental Approaches

Pressure regimes exist if the rate for complex redissociation k_d exceeds the rate for radiative stabilization k_r which, as will be discussed below, is thought to be probably at least around 1(3) s^{-1}. In this circumstance every complex formed will not be stabilized by the radiative mechanism alone. The first of three pressure regimes to be considered is the low pressure or radiative regime, which is defined by the relation $k_r >> k_c[C]$. Here radiative stabilization is more rapid than collisional stabilization. If one assumes collisional stabilization to occur at the collisional limiting value of ≈ 1(-9) $cm^3\ s^{-1}$ and k_r to be 1(3) s^{-1}, the inequality becomes $[C] << 1(12)\ cm^{-3}$. Under these conditions, equation (7) reduces to the much simpler relation

$$k_{eff} \approx \{k_f/k_d\}k_r \tag{8}$$

where the effective rate coefficient for association is independent of the density of the gas [C] and is radiative in nature. The right-hand-side of equation (8) can then be labeled k_{ra} where "ra" stands for radiative association.

As the density [C] increases, a transition region occurs in which both ternary and radiative association are important. Eventually, [C] becomes sufficiently large that the criterion $k_c[C] >> k_r$ is reached. Then ternary or collisional association dominates and (as long as $k_d >> k_c[C]$) equation (7) reduces to

$$k_{eff} \approx \{k_f/k_d\}k_c[C] \tag{9}$$

where the effective rate coefficient for association is now linearly dependent on density. The rate coefficient expression multiplying [C] is normally referred to as k_{3b}, the rate coefficient for ternary (three-body) association, and is expressed in units of $cm^6\ s^{-1}$.

Finally, as the density increases still further, a limit is reached in which all complexes are collisionally stabilized. This limit - achieved when $k_c[C] >> k_d$ - results in the saturated regime where $k_{eff} = k_f$; that is, any complex formed is stabilized. In Figure 1 the pressure regimes for association are depicted in a log-log plot of k_{eff} vs. [C]. To compute k_{eff}, the rate coefficients k_f, k_d, k_r, and k_c have been set at the values 1(-9) $cm^3\ s^{-1}$, 1(7) s^{-1}, 1(3) s^{-1}, and 1(-9) $cm^3\ s^{-1}$, respectively. Only the rate coefficient for complex redissociation is totally arbitrary; the rate coefficients for complex formation and stabilization have been set at standard collisional values. The curve in figure 1 shows the pressure regimes. At low gas densities the rate coefficient is constant at its radiative value, at somewhat higher densities it begins to increase with increasing density, eventually achieving a linear dependence, and at still higher densities, the dependence on density begins to lessen until complete saturation is achieved.

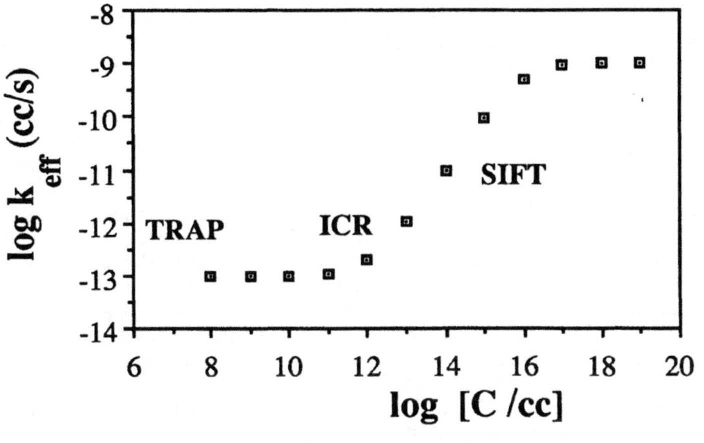

Figure 1. A log-log plot of k_{eff} vs. [C].

Superimposed on the plot of k_{eff} vs. [C] in Fig. 1 are some of the experimental techniques used to study ion-molecule association reactions in the laboratory. The low pressure ion trap technique (designated TRAP) of Dunn and co-workers (see, e.g. Barlow, Dunn, and Schauer 1984a,b) operates at sufficiently low densities that there is no doubt that only radiative association is being observed. Unfortunately, these experiments are difficult and only a few have been performed. In only one - the radiative association reaction between CH_3^+ and H_2 (Barlow, Dunn, and Schauer 1984a,b) - was an actual value rather than an upper limit determined. A strength of this technique is that it can be utilized at temperatures as low as 10 K if H_2 is used as the neutral reactant.

At higher densities, a valuable technique involves the ion cyclotron resonance (ICR) apparatus which has been utilized for a large number of normal ion-molecule reactions of importance to interstellar chemistry by Huntress and co-workers (see the compendium of Anicich and Huntress 1986),and can also be utilized for association reactions. Unfortunately, the gas density in ICR experiments is sufficiently high that both radiative and ternary association must be considered. These can be separated out by a linear plot of k_{eff} vs [C] rather than a log-log plot. At pressures sufficiently low that saturation is not a problem (complex redissociation is the dominant complex destruction mechanism) equation (7) simplifies to

$$k_{eff} = k_{ra} + k_{3b}[C] \tag{10}$$

where k_{ra} and k_{3b} have been defined above. Averaging of k_{ra} and k_{3b} over quantum states and collision energies should not affect the validity of equation (10) unless the range of these rate coefficients is such that some values approach saturation, in which case a more complicated density dependence occurs. Extrapolation of a measured linear plot of k_{eff} vs. [C] to zero density should allow determination of k_{ra}. Using an ICR apparatus, Kemper, Bass, and Bowers (1985) have shown that the rate coefficient for radiative association between CH_3^+ and HCN at room temperature is less than 5(-12) cm^3 s^{-1} by such an extrapolation although previous ICR work had indicated a much larger value of ≈1(-10) cm^3 s^{-1}(Bass *et al.* 1981). Huntress and collaborators are currently analyzing data from several reactions that should lead to the determination of radiative association rate coefficients by this technique or some variant of it (McEwan 1987). In addition, Gerlich and Kaefer (1987, 1988) have utilized it to interpret experiments on several association reactions in a high pressure trap.

The dominant technique in association reaction studies has been the SIFT (selected ion flow tube) technique and its modifications, pioneered by Smith and Adams (see, e.g., Smith and Adams 1979) and also used by Bohme and co-workers (Raksit and Bohme 1983; Bohme and Raksit 1985), McEwan and co-workers (Knight *et al.* 1986), and Lindinger and co-workers (Saxer *et al.* 1987). This technique operates at gas densities of ≈ 1(16) cm^{-3}, typically with helium as the bath gas, and cannot be used to obtain radiative association rates directly. Indeed, the gas density in a SIFT is high enough that saturation can occur in systems with large association rate coefficients. Many ternary association reactions have been measured by this technique. The results can be used to infer radiative association rate coefficients indirectly in the following manner. Comparison of equations (8) and (9) shows that the rate coefficients for radiative and ternary association share a common factor of k_f/k_d and differ only in the stabilization rate coefficient. If one measures k_{3b} and estimates k_c, one obtains k_f/k_d from relation (9). Estimation of k_r then yields k_{ra} via equation (8). The analysis thus requires knowledge of k_c and k_r. Some evidence exists concerning the likely size of k_c (see Section 3.3), which may be taken to have the Langevin value given in eq. (27) below. In any event, the major uncertainty in the analysis is not k_c, but k_r, the radiative stabilization rate, which must be determined theoretically. As will be discussed below, the current estimate that $k_r \approx 1(3)$ s^{-1} for most radiative association reactions may well be incorrect.

If one requires k_{ra} at a different, usually lower, temperature from that of the SIFT measurements, it is best to proceed with the aid of theory which is reliable as far as the temperature variation is concerned. Theory provides the only means of allowing for the excitation conditions in the interstellar medium being different from those in the laboratory. The difference may be great: for example the molecular hydrogen in dense interstellar clouds is thought to have its rotational levels in true thermal equilibrium whereas the molecular hydrogen used in most laboratory experiments consists of the normal 3 to 1 ortho to para mixture.

3. THEORETICAL TREATMENTS

3.1. Thermal Model

A simple approach to the calculation of both ternary and radiative association reaction rate coefficients for polyatomic systems was given by Herbst (1979; 1980a). This approach can be called the "thermal" model because its premise is that the ratio of the formation rate to the redissociation rate of the collision complex is equal to the ratio which exists if the system is at thermal equilibrium. In this approach, one can only calculate the *ratio* of the rate coefficients k_f and k_d so that it can be used only in a well-defined pressure regime - either the radiative (eq. 8) or the ternary (eq. 9) - unless an additional asumption is made regarding one of the two rate coefficients.

In the thermal model, the ratio k_f/k_d is

$$k_f/k_d = q(AB^{+*}) / \{q(A^+)\, q(B)\} \tag{11}$$

where the q's are molecular partition functions per unit volume (Hill 1960) which are sums over molecular energy levels E_i weighted by appropriate Boltzmann factors:

$$q = (1/V)\Sigma_i \exp(-E_i/kT) \tag{12}$$

where V represents volume. The partition functions can be expressed as products of translational and internal factors, where the translational energy levels are those of a particle in a box and the internal levels are determined by electronic, vibrational, and rotational motions. Performing this separation leads to the relation

$$k_f/k_d = h^3\{2\pi\mu kT\}^{-3/2}\, q^{int}(AB^{+*})/\{\, q^{int}(A^+)\, q^{int}(B)\, \} \tag{13}$$

where the superscript "int" refers to internal motion, h is Planck's constant, and μ is the reduced mass of the reactants. For reactants at room temperature and below, normally the only internal motion with energy levels close enough together to lead to partition functions that possess a temperature dependence is rotation. The vibrational partition is unity at these temperatures and the electronic partition function is equal to the degeneracy of the ground electronic state g_e. Exceptions occur for electronic fine structure states and large floppy molecules with low frequency vibrations (Viggiano 1984). With the further simplifying assumption that the spacings between rotational levels are smaller than kT (this permits the sum in eq. (12) to be approximated by an integral), internal (rotational) partition functions for linear and non-linear reactant molecules are easily obtained (Hill 1960):

$$q^{int} = g_e kT/B \qquad \text{(linear molecules)} \tag{14a}$$

$$q^{int} = g_e \pi^{1/2} (kT)^{3/2} / \{ABC\}^{1/2} \quad \text{(non-linear molecules)} \tag{14b}$$

where A, B, and C are so-called rotational constants that depend on the inverse of the moments of inertia along principal axes (Townes and Schawlow 1955) and formula (14b) is given for the least symmetric (asymmetric top) case. Because of the Pauli Principle, the rotational partition functions must be modified for molecules with selected symmetry elements. In the simplest treatment, this modification takes the form (Hill 1960)

$$q'^{\,int} = q^{int}\, g_n/\sigma \tag{15}$$

where g_n is the nuclear spin degeneracy and σ is the so-called symmetry number, or number of rotational symmetry elements that leave the molecule unchanged. More detailed modifications may be needed for molecules at lower temperature. The case of H_2 is especially important. In the simplistic treatment above, $g_n = 4$ (each nucleus having two spin states), $\sigma = 2$, and $q'^{int} = 2\, q^{int}$. In reality, however, H_2 has two sets of rotational level stacks that do not communicate with one another - the odd (ortho) J levels (J = 1,3,...) which possess a composite nuclear spin I = 1, and the even (para) J levels (J = 0,2,4...) which possess a composite nuclear spin I = 0. If nuclear spin were not conserved in association reactions, all one need do to compute q'^{int} for H_2 exactly would be to obtain the sum

$$q'^{int}(H_2) = \Sigma_{J\ odd}\, 3(2J+1)\exp(-E_J/kT) + \Sigma_{J\ even}\,(2J+1)\exp(-E_J/kT) \qquad (16)$$

where 2J+1 is the rotational degeneracy, $g_e=1$, and the factor of three in the ortho sum is due to nuclear spin degeneracy. This expression reduces to the value obtained from (15) at high temperatures but becomes larger at low temperatures. However, use of eq. (16) is not a total solution to the problem since ortho and para hydrogen are really different species. It is therefore best to consider them separately and obtain separate values for k_f/k_d, which can then be appropriately averaged.

How does one compute the partition function of the collision complex? Even though the complex is an unstable molecule, its partition function can be calculated in the same manner as that of a stable molecule with the proviso that only internal energy levels above the dissociation limit of the molecule be included. Those energy levels below the dissociation limit belong to the stable molecule and not to the complex. Unlike the case of the reactants, however, the vibrational level spacing in the complex is much smaller than kT. Indeed, these levels are so close together that they can be treated continuously and one can define a vibrational energy density of states. A simple analytical formula for the vibrational energy density of states of a polyatomic molecule has been given by Whitten and Rabinovitch (1963) using the harmonic oscillator approximation. This approximation is used with at most slight modifications in all of the theories discussed here. It relates the vibrational energy density of states ρ_v to the harmonic frequencies ν_i, the vibrational energy E_{vib}, and the zero-point energy E_z of the complex:

$$\rho_v = (E_{vib} + aE_z)^{s-1} / \{\Gamma(s)\, \Pi_i\, (h\nu_i)\} \qquad (17)$$

where a is an empirically derived factor ($0<a<1$), Γ is the gamma function, and s, the number of vibrational degrees of freedom of the complex, is

$$s = 3N - 6 \qquad (18)$$

for non-linear molecules, where N is the number of atoms in the complex.

In the thermal model, the partition function of the complex is

$$q^{int}(AB^{+*}) = \int_{E=0}^{\infty} \rho_{vr}\,(E + D_o)\exp(-E/kT)dE \qquad (19)$$

where both vibrational and rotational levels of the complex are included in the energy density of states ρ_{vr} (Whitten and Rabinovitch 1964; Forst 1973; Troe 1977b) and the integral is over all internal energy states of the complex above the dissociation limit D_o. The expression for ρ_{vr} is similar to but somewhat more complex than that given for ρ_v above. To be complete, it should also contain both an electronic-nuclear spin degeneracy factor and a symmetry number. The integration of eq. (19) can be accomplished simply for the normal case in which $D_o >> kT$. In this limit, the exponential dependence of the integrand dominates and

$$q^{int}(AB^{+*}) \approx \rho_{vr}(D_o)kT. \tag{20}$$

Using equations (13) and (20), the standard thermal formula for k_f/k_d can be derived:

$$k_f/k_d = h^3\{2\pi\mu kT\}^{-3/2}\,\rho_{vr}(D_o)kT\,/\,\{\,q'^{\,int}(A^+)\,q'^{\,int}(B)\,\}. \tag{21}$$

Assuming that the internal partition functions of the reactants do not include significant electronic and vibrational temperature dependence and that the rotational energy spacings are smaller than kT, the derived temperature dependence of k_f/k_d (see eq.'s 14 and 21) is of the simple form

$$k_f/k_d \propto T^{-n}\ ;\ n = (r_{A^+} + r_B + 1)/2 \tag{22}$$

where the r's are the number of rotational degrees of freedom for the respective reactants (r=2 or 3 for a linear or non-linear species, respectively.) Thus, if both reactants are non-linear, the predicted inverse temperature dependence is $T^{-3.5}$, a very severe dependence indeed. The predicted severe inverse temperature dependence lessens dramatically as kT becomes smaller than the rotational level spacing and one must use a summation rather than an integration for the rotational levels in eq. (12).

The value of k_f/k_d depends on the energy density of vibrational-rotational states of the complex which is a very rapidly increasing function both of D_o (the dissociation energy) and of N (the number of atoms in the complex). Consequently, k_f/k_d for, say, a $D_o = 3$ eV, N = 7 complex is likely to be orders of magnitude greater than k_f/k_d for, say, a $D_o = 0.5$ eV, N = 4 complex.

Before a discussion of some quantitative results, it is useful to compare the thermal model with a more refined model of Bates (1979a,b, 1980), here termed the "modified thermal" model.

3.2. Modified Thermal Model

In this approach, it is recognized that not all complex states can be reached in binary collisions and moreover that the long range attraction increases the collision rate of the reactants. The reactants A^+ and B are treated as structureless particles that come together with orbital angular momentum quantum number J. For potentials containing an r^{-n} dependence where n>2, the orbital momentum leads to an effective potential which contains a centrifugal barrier at long range (cf. Levine and Bernstein 1974). Bates (1979a) originally used a model that had been introduced by Troe (1977a,b) for neutral reactants. In a modification, Bates (1979b, 1980) and also Herbst (1980b,c) took the long range attraction between the ion and the polarizable neutral to be the so-called Langevin interaction:

$$V(r) = -e^2\alpha/2r^4 \tag{23}$$

where r is the A^+-B separation, e is the electronic charge, and α is the polarizability.

For given J, close collisions only take place if the energy of relative motion E is sufficient to pass over the centrifugal barrier. An equivalent alternative statement is that for given E close collisions only take place if J is less than the value that would cause an unpassable barrier: for interaction (23) this requires

$$J \leq 2\pi\,(8\mu^2\alpha e^2 E)^{1/4} / h \ . \tag{24}$$

On multiplying $\rho_v\,(E + D_o)$ of eq. (17) by the statistical weight 2J+1 and Boltzmann factor exp(-E/kT) and first integrating over the J that satisfy condition (24) and then over all E it is found that the thermal model eq. (20) is replaced by

$$q^{int}\,(AB^{+*}) \approx 4\pi^2\,(kT)^{3/2}\,\rho_v(D_o)\,\mu\,(2\pi\alpha e^2)^{1/2} / h^2. \tag{25}$$

Comparison of eq. (20) and (25) shows that there is a $T^{1/2}$ difference in the temperature variation. There is also a difference, that depends on the polarizability α, on the magnitude. This arises because the modified thermal model allows for the collecting action of the long range attraction. If the neutral reactant has a large permanent dipole moment, there is less difference from the thermal model as regards the temperature variation but the effect of the collecting action of the long range attraction is much more marked.

The modified thermal model is accurate enough to justify the trouble of taking the dependence of ρ_{vr} on energy and on rotational quantum numbers (cf. Whitten and Rabinovitch 1964; Forst 1973, and Troe 1977b) into account. If this is done the integrations over J and E can no longer be carried out analytically. It should be noted that the modfied thermal model is designed for treating only radiative association of thermal systems or ternary association in the low density region where third order kinetics prevail.

3.3. Comparison With Ternary Data

The thermal and modified thermal approaches so far discussed are incomplete since the expressions for the radiative and ternary rate coefficients also contain a stabilization rate coefficient - k_r for the radiative case and k_c for the ternary case. In order to compare theory with experiment, one must calculate these additional quantities. In this section, we consider k_c and compare theoretical and laboratory data for ternary association, the process normally studied in the laboratory. Theoretical approaches to k_r are discussed in section 4.

Numerous experimental studies on ternary association (see, for example, Cates and Bowers 1980; Johnsen, Chen, and Biondi 1980; Kemper, Bass, and Bowers 1985) show that the value of the three-body rate coefficient depends to some extent on the identity of the third body with He, the species used most frequently as the third body, being relatively inefficient. This dependence must arise from k_c, the rate coefficient for collisional stabilization of the complex. Although the reality is doubtless less simple (see Troe 1977a or Bass *et al.* 1981) it is common practice to write

$$k_c = \beta_c k_{coll} \tag{26}$$

where k_{coll} is the rate coefficient for close collisions and β_c is the effective probability that a single collision casues stabilization. For a non-polar gas k_{coll} may be taken to be the Langevin rate coefficient

$$k_L = 2\pi(\alpha e^2/\mu)^{1/2} \ . \tag{27}$$

Troe (1977b) has related β_c to the mean energy change $<\Delta E>$ that the complex experiences in a collision through

$$\beta_c / (1 - \beta_c^{1/2}) = -<\Delta E>/kT \ . \tag{28}$$

Table I gives some derived values of β_c for assumed values of $<\Delta E>$ in the likely range.

TABLE I PROBABILITY OF STABILIZATION IN A COLLISION

	$-<\Delta E>$ (kcal mol^{-1})						
	0.1	0.2	0.4	0.7	1.0	1.5	3.0
T (K)				β_c			
50	0.38	0.54	0.69	0.79	0.84	0.89	0.94
75	0.30	0.45	0.60	0.72	0.78	0.84	0.91
100	0.25	0.38	0.54	0.66	0.73	0.80	0.89
150	0.19	0.30	0.45	0.57	0.65	0.73	0.84
200	0.15	0.25	0.38	0.51	0.59	0.67	0.80
300	0.11	0.19	0.30	0.42	0.50	0.59	0.73
400	0.09	0.15	0.25	0.36	0.43	0.52	0.67

It is seen that β_c is predicted to decrease as T is increased and that the effect is most pronounced if $|<\Delta E>|$ is small. Herbst (1982b) has developed a rather complex theory that also gives β_c to decrease as T is increased.

Table II makes some comparisons between theory and experiment. The experimental results in the first row require comment. Johnsen, Chen, and Biondi (1980) did not observe CH_2^+ since it reacts rapidly

$$CH_2^+ + H_2 \text{ -----> } CH_3^+ + H \ . \tag{29}$$

Supposing (as is probable) that process (29) proceeds at the Langevin rate 1.6(-9) cm^3 s^{-1} Johnsen, Chen, and Biondi (1980) argued that β_c is unity in (CH_2^{+*}) - H_2 collisions. With this value of β_c, the experimental and modified thermal values of n should agree, and they do. It is now possible to derive the values of β_c for the reaction in the second row of Table II (for which the Langevin rate coefficient is 6(-10) cm^3 s^{-1}). In comparing the measured ternary rate coefficients (remembering that those in the first row are half what they would be if hydrogen molecules were distinguishable) we find that β_c for He is 0.55

TABLE II TERNARY ASSOCIATION: EXPERIMENT AND THEORY

Temperature Dependence of k_{3b} (80 K - 300 K) expressed as T^{-n} and Values at 300 K ($cm^6 s^{-1}$)

	n			k_{3b}		
System	Expt.	Thermal	Mod.	Expt.	Thermal[a]	Mod.[a]
$C^+ + H_2$ in H_2	0.5[b]	1.1[c]	0.6[c]	2.8(-29)[b]	8.7(-30)[c]	3.1(-30)[c]
$C^+ + H_2$ in He	1.2[be]	1.1[c]	0.6[c]	5(-30)[b]	6.6(-30)[f]	2.3(-30)[f]
$CH_3^+ + H_2$ in He	2.3[d]	3.0[g]	2.5[h]	1.1(-28)[d]	8.4(-28)[g]	3.4(-28)[h]; 3.7(-28)[i]
$CH_3^+ + CO$ in He	2.4[d]	3.0[g]	2.5[h]	2.4(-27)[d]	1.8(-26)[g]	6.4(-27)[h]; 1.8(-26)[i]

a in the calculations β_c was formally set to unity.
b Johnsen, Chen, and Biondi (1980)
c Herbst (1981); theoretical results in first row take H_2 indistinguishability into account.
d Adams and Smith (1981)
e Fehsenfeld, Dunkin, and Ferguson (1974)
f based on Herbst (1981) and $k_L = 6(-10)$ $cm^3 s^{-1}$
g Herbst (1979)
h Herbst (1980b)
i Bates (1983a)

and 0.22 at 80 K and 300 K, respectively. The agreement with the variation in the appropriate column of Table I is satisfactory. Other measurements by Cates and Bowers (1980) suggest that β_c for He is 0.31 at 300 K. The accord is good.

In view of the variation of β_c with T the agreement between the experimental and modified thermal values of n in the last two rows of Table II must be regarded as fortuitous. It is probable that the experimental ternary rate coefficients concerned are smaller than they should be at 80 K, and to a less extent at 300 K, because of the ambient gas not being at a low enough density for the kinetics to be truly third order (Bates 1986b). This could explain the discrepancy.

The absolute values of theoretical ternary rate coefficients depend critically on the dissociation energies which are usually known with precision. They also depend on the vibration frequencies. *Ab initio* quantal calculations are the best source of these when they are available. They should in general be fairly accurate but are subject to systematic error which can be significant since the expression for ρ_{vr} contains the product of what may be a large number of frequencies. Again, one (or more) of the frequencies may be so low that it is best replaced by a free rotation when the complex carries the full association energy. The replacement would be expected to change the other frequencies (cf. Bates 1986c) but the effect has not been investigated quantitatively. In the absence of results from *ab initio* quantal calculations it is necessary to estimate the frequencies from those for similar complexes. The order-of-magnitude agreement shown in Table II is typical.

There is little doubt that such discrepancies as exist between good experimental results and the modified thermal theory of Bates are due to errors in the chosen values of the complex's parameters. While the modifed thermal theory is successful the restrictions on its use mentioned at the end of Section 3.2 must be remembered. Phase space theory is more troublesome but is of general applicability. The accuracy that can be achieved is again limited by the uncertainties in the parameters of the complex involved.

3.4. Phase Space Approach To Association Reactions

This approach, pioneered by Bowers and co-workers for association reactions (see, e.g. Bass, Chesnavich, and Bowers 1979; Bass *et al.* 1981; Bass and Jennings 1984; Bass and Bowers 1987) has also been used by Herbst (1981; 1985b,c,d; 1987) and, in somewhat simplified form, by Bates (1983a;1985a,b;1986a,b,c). It is a state-to-state microcanonical formulation in which reactants in individual quantum states collide with a fixed collision energy to form a complex defined by its total energy, angular momentum, and vibrational energy density of states. The complex can then be stabilized or redissociate into a variety of "product" (normally the product species are the same as the reactant species) quantum states consistent with conservation of energy and angular momentum. Once the cross section for formation of the complex from any set of reactant quantum states is specified, usually via the Langevin model for non-polar neutrals or one of several models for polar neutrals (e.g. Morgan and Bates 1987; Bates and Morgan 1987) the dissociation of the complex back into those states is also specified via microscopic reversibility. The "state-to-state" value for the effective association rate coefficient at any gas density is given by the expression

$$k_{eff}(J_A,J_B,E_{coll} \Rightarrow J,E) = k_f(J_A,J_B,E_{coll} \Rightarrow J,E)\{k_r + k_c[C]\} / \{k_d(J,E) + k_r + k_c[C]\} \quad (30)$$

where k_{eff}, k_r, and k_c have been defined previously; J_A, J_B, and J are the angular momentum quantum numbers for quantum states of A^+, B, and the complex, respectively; E_{coll} is the reactant collision energy; and E is the total complex energy. The dissociation rate coefficient for the complex k_d is the sum of the dissociation rates into all accessible product states. Specification of electronic and vibration quantum numbers has been deleted for simplicity but in calculations one most sometimes consider dissociation of the complex into excited vibrational states. A detailed expression for $k_d(J,E)$ is given by Herbst (1985b). Once k_{eff} has been calculated, it must be summed over complex angular momentum states J and averaged over collision energy and internal reactant quantum states. As shown by Herbst (1981), this rather tedious process reduces to the modified thermal result if the system is thermal and in a well-defined pressure regime.

Use of the phase space theory for non-thermal, low pressure (radiative) systems is important in both interstellar shocks (Herbst 1985b) and the ambient interstellar medium where rotational energy is often subthermal (Bates 1983a). Use of this theory for ternary systems in which there may be departure from third order kinetics has been explored by Bass and Jennings (1984) and Bates (1986c) who conclude that such departure is more of a problem in SIFT experiments than is customarily realized. The reason is that k_d is a strong function of the quantum number J (Bass, Chesnavich and Bowers 1979); the large variety of k_d values obtained leads to a less sharply defined density dependence for k_{eff} than shown in Figure 1. Bates (1986b) has given a useful application: he developed parametric formulae whereby the low ambient gas density limit to the ternary association rate (which is what is needed in connection with radiative association) may easily be found from the measured rate coefficient at a known ambient gas density even if the kinetics are not third order.

Use of the phase space theory in yet another context has been undertaken by Herbst (1985c,d; 1987) and Bass *et al.* (1983) who have investigated why ternary association reactions are sometimes observed to compete in systems with normal exothermic channels, a result which is contrary to expectation (Bates 1983b). To treat these systems, eq. (30) must include an additional complex dissociation rate channel and can no longer reduce to the modified thermal treatment upon averaging. It is found that association can compete with normal exothermic reaction channels if the potential energy surface has barriers in the exit channel that although not large enough to prohibit normal products are large enough to slow the rate of complex dissociation into products. This rate is found to be a strong function of complex angular momentum with large amounts of angular momentum slowing the dissociation rate considerably. Thus, association or redissociation into reactants is favored for high angular momentum collisions and dissociation into products for low angular momentum collisions. An important system is the reaction between CH_3^+ and NH_3 in which association to form $CH_3NH_3^+$ competes with two exothermic channels (Herbst 1985c,d). The latest experimental measurements of k_{eff} at pressures high enough to be near saturation in the association channel are in excellent agreement with the phase space calculations (Herbst 1985d; Saxer *et al.* 1987). An important conclusion from this work is that radiative association reactions can occur at an appreciable rate under interstellar conditions despite the existence of a competitive exothermic channel *if the potential energy surface leading to that channel has a sufficiently high barrier*. On the other hand, the existence of an exothermic channel does depress the value of the association rate coefficient from what it would be in the absence of an exothermic channel unless the barrier is very high. An important example of this depression has been discussed by Herbst (1987) and involves the formation of oxygen-containing organic molecules in giant interstellar clouds. Herbst (1987) now feels that the calculated radiative association rate coefficients are not large enough to produce the observed abundances of species such as dimethyl ether.

4. THE RATE OF RADIATIVE STABILIZATION

In order to calculate radiative association rate coefficients to the accuracy obtained in the ternary case or to use ternary data to estimate radiative association rate coefficients, it is necessary to determine the radiative stabilization rate of the complex k_r. Before attempting to calculate the radiative stabilization rate the specific mechanism involved must be known. Until quite recently, it was thought (Herbst 1982c, 1985a) that the only *general* mechanism for radiative stabilization was vibrational relaxation. According to this line of reasoning, the complex would normally be formed in its ground electronic state since the ground electronic states of most important reactants are non-degenerate and correlate with only one potential surface of the complex. If the complex is formed in its ground state, the only apparent mechanism for radiative stabilization is emission from the quasi-continuum of closely-spaced vibrational-rotational levels above the dissociation limit to discrete, stable levels below the dissociation limit. Given that the reactants possess thermal energy, the complex is formed with an energy on the order of several kT above the dissociation limit. Loss of this amount of energy requires emission of a photon of at least this much energy; i.e., an infra-red photon. Infra-red emission in molecules is normally associated with transitions between vibrational states and recent theoretical treatments of vibrational emission from highly excited states have been undertaken neglecting rotational contributions (Herbst 1982c, 1985a; Bates 1986d). Experimental work on emission from highly vibrationally excited polyatomic molecules, especially unstable ones, is needed.

The rate coefficient for radiative stabilization k_r of a polyatomic complex can be expressed by the equation

$$k_r = \Sigma_{\{n\}} P_{\{n\}} \Sigma_{\{m\}} A_{\{n\}\text{->}\{m\}} \qquad (31)$$

in which $P_{\{n\}}$ is the probability that the complex is in state {n} of a large number of accessible vibrational states, and $A_{\{n\}\text{->}\{m\}}$ is the Einstein A coefficient for spontaneous emission from initial state {n} to final state {m}, which must be below the dissociation limit. The double sum is over all possible final and initial states. In the usual level of approximation, states {n} and {m} can be regarded as sets of weakly coupled harmonic oscillators involving collective vibrational motions (normal modes). Then each of these states can be represented by a set of occupation numbers of the normal modes; viz., n_1, n_2,...for state {n}. If the probabilities $P_{\{n\}}$ can be regarded as equal and the transition moment can be regarded as purely dipolar, it can be shown (Herbst 1982c; Bates (1986d) that k_r reduces to the following equation:

$$k_r = (E/s) \sum_{i=1}^{s} A^{(i)}_{1\text{->}0} /(h\, v_i) \qquad (32)$$

in which s is the number of vibrational degrees of freedom, E is the total complex energy ($E \approx D_o$), v_i is the vibrational frequency of the ith normal mode, and $A_{1\text{->}0}{}^{(i)}$ is the spontaneous emission rate of the fundamental transition of the ith normal mode. These fundamental spontaneous emission rates can be obtained from absolute intensity data on infra-red transitions compiled over the years by infra-red spectroscopists. Unfortunately, most if not all of these data were taken using neutral molecules, not the ionic species involved in interstellar radiative association reactions. However, quantal calculations have provided us with some information on dipole moment derivatives (on which the transition probabilities in eq. (32) depend (Herbst 1985a; Botschwina 1987). It would appear that protonated ions have at least some normal modes with large intensities so that their overall emission rates in the infra-red are quite high, typically an order of magnitude higher than neutral molecules. Indeed, use of limited quantal data in equation (32) suggests that for vibrational energies of a few eV, values of k_r in the vicinity of 1(3) s^{-1} may not be atypical. Herbst (1985a) has used this value of k_r in a compilation of calculated radiative association rate coefficients needed for models of interstellar clouds. We will refer to it as the "standard" value.

It has been understood for some time that vibrational emission is not the only mechanism by which radiative stabilization of the complex can occur. If the complex is formed in an excited electronic state, then emission to stable vibrational levels of the ground electronic state is a possibility. Since electronic transitions typically occur at frequencies larger than vibrational transitions and since the spontaneous transition coefficient contains a factor in which the frequency of emission is cubed, electronic stabilization can be much more rapid than vibrational stabilization if there is a large transition dipole moment between the two electronic states. Calculations on the radiative association between C^+ and H_2 have shown that there appears to be an efficient pathway for production of the CH_2^{+*} complex in its excited $^2B_1$ state, which can then emit to stable levels of the ground state with a rate coefficient $k_r \approx 1(5)\ s^{-1}$, two orders of magnitude faster than the vibrational rate (Herbst, Schubert, and Certain 1977; Herbst 1982a). The resulting radiative association rate coefficient is sufficiently large to account for the observed rate of CH production in diffuse interstellar clouds (Black and Dalgarno 1977).

Despite the detailed calculations on the radiative association of CH_2^+, it had not been appreciated that electronic emission could be a general process for complex stabilization until quite recently. This idea awaited one of the few experiments undertaken in the field of radiative association - the low temperature ion trap experiment by Dunn and co-workers on the radiative association of CH_3^+ and H_2 (Barlow, Dunn, and Schauer 1984 a,b). Although these scientists measured a rate coefficient at 13 K that was at first thought to be less than an order of magnitude higher than theoretical values, the discrepancy worsened considerably when Bates (1986d) deduced that because ortho hydrogen is an energy source the CH_3^+ ions were heated to above 13 K: for example he calculated that their translational temperature was about 50 K and that they have much internal energy. Using SIFT data of Smith, Adams, and Alge (1982) to normalize his theory so that $\beta_c = 0.3$ at 300 K, Bates (1986d) calculated that to reproduce the trap data requires a value for k_r of $3.5(4)\ s^{-1}$, much larger than the standard vibrational rate. He then made the reasonable inference that electronic emission must be involved and gave reasons why a low-lying excited state of CH_5^+ may exist. The ground state of CH_5^+ correlates with the reactants $CH_3^+ + H_2$. Bates (1986d) argued that the electronic surface that correlates with CH_4 + H^+ and $CH_3 + H_2^+$ may have a deep well, that this well may be reached by reactants in the ground state well by crossing; and that an electronic transition between the two states concerned may occur. This naturally led him to consider the possibility of such stabilization in other radiative association processes (Bates 1987a,b). Bates (1987c) also examined data on the association energies of polyatomic ions and concluded that species having isomers that are quite close in energy are not uncommon.

Despite lack of information on relevant excitation energies and transition dipole moments, Herbst and Bates (1987) attempted to quantify Bates' ideas with the objective of estimating how much enhancement relative to vibrational relaxation an electronic transition might give. They considered two cases (Case 1, Case 2) in their modelling. In Case 1 reactants A^+ and B come together along the ground state potential surface and can cross over onto an excited surface if it is energetically accessible. An electronic transition from the excited to the ground state that stabilizes the system may occur. Denoting the rate of this transition by $k_e(1)$ and the rate of vibrational relaxation by k_v (previously denoted by k_r), the mechanism causes an enhancement of the stabilization rate by the factor

$$\Phi = 1 + (k_e(1)/k_v)K \tag{33}$$

where K is an equilibrium coefficient relating the numbers of excited and ground state complexes. Eq. (33) yields a maximum enhancement - it pertains only in the absence of saturation at the collisional limit. The energy density of vibrational states increases rapidly with the total energy that is available for vibration. It is hence less in an electronically excited state than in the ground state. Consequently, K is less than unity and is a rapidly decreasing function of the electronic excitation energy. The stabilization rate due to the electronic transition alone is $k_e(1)K$ in Case 1.

Before proceeding it is worth recalling how $k_e(1)$ is calculated. One starts with eq. (31) as for vibrational emission. However the spontaneous emission rates $A_{\{n\}->\{m\}}$ now represent transitions from vibrational levels {n} of the excited electronic state to stable levels {m} of the ground electronic state. They can be expressed via the standard relation

$$A_{\{n\}->\{m\}} = Ch^3 \nu^3_{\{n\}->\{m\}} |\langle\{m\}|\{n\}\rangle|^2 \quad (34)$$

where C includes the electronic transition dipole moment squared and some fundamental constants, and $\nu_{\{n\}->\{m\}}$ represents the frequency of the emitted photon in a transition from state {n} to state {m}. The size of the spontaneous emission rate is critically dependent on the vibrational overlap squared (the Franck-Condon factor). For the diagonal Franck-Condon array ($\langle\{m\}|\{n\}\rangle = \delta_{\{m\}\{n\}}$) that pertains when the two electronic states have the same structural parameters, the only non-zero spontaneous emission coefficients in eq. (34) are those for which the vibrational quantum numbers in {n} and {m} are identical. In this case, $k_e(1)$ reduces to the simple form

$$k_e(1) = C\varepsilon_0{}^3 \quad (35)$$

where ε_0 is the zero-point excitation energy of the excited electronic state. Assuming that the transition dipole moment is constant, the dependence of the enhancement factor Φ on ε_0 is then determined by a trade-off: $k_e(1)$ increases as ε_0 cubed whereas K decreases as ε_0 increases. The result is that an enhancement maximum occurs at excitation energies greater than zero (Herbst and Bates 1987). For more normal situations in which non-zero off-diagonal Franck-Condon exist, the situation is more complex but detailed calculations show that typically the excitation energy at which enhancement is a maximum is reduced, often to zero.

In Case 2 the radiative association rate involves no curve crossing process. Rather the reactants A^+ and B react along the ground state surface only. However, if an excited state is sufficiently low-lying that *it comes below the energy of the reactants for certain configurations of the molecular geometry*, then emission from the ground state to lower vibrational levels of the excited electronic state can occur. These levels are stable to dissociation of the complex but will subsequently emit to still lower levels of the ground electronic state. Now it is clear that in the limit of zero off-diagonal Franck-Condon factors, Case 2 enhancement cannot occur since if states {n} and {m} have the same energy with respect to their respective potential minima and the same normal mode frequencies, the only possible electronic transition will be *absorption* from the ground state to the excited state with the absorbed photon having an energy equal to ε_0. However, if the structures of the two electronic states are quite different, the possibility of large off-diagonal Franck-Condon factors exists and sizable emission rates can occur, due principally to transitions to very low-lying vibrational states of the excited electronic state. For Case (2) enhancement there is no equilibrium coefficient in the expression for Φ since emission occurs directly from the ground electronic state and the enhancement factor is

$$\Phi = 1 + k_e(2)/k_v \tag{36}$$

where $k_e(2)$ is the rate of electronic emission from the ground state to lower, stable levels of a low-lying excited state. Careful examination indicates that $k_e(2)$ will tend to be at its largest when the excited state has zero excitation energy.

Herbst and Bates (1987) have performed a series of model calculations. In these they let k_v have the standard value 1(3) s^{-1}. There is much uncertainty regarding k_e which is proportional to $|D(tr)|^2$ where $|D(tr)|$ is the transition dipole moment involved. Charge transfer transitions in ionized molecules may be quite strong (Bates 1987a). Herbst and Bates (1987) tentatively took $|D(tr)|$ to be 1 Debye. They regarded this as a reasonable rather than the maximum likely value. For comparison it may be noted that $|D(tr)|$ is typically about 5 Debye for strong atomic transitions. The calculations show that if $|D(tr)|$ is 1 Debye enhancements Φ in the range 10-100 can occur either through the Case 1 or Case 2 mechanisms or a combination of both depending on the excitation energy and the magnitudes of the Franck-Condon factors involved.

5. HOW ACCURATE ARE CALCULATED RADIATIVE ASSOCIATION RATES?

How accurate are the tabulations of calculated rate coefficients for radiative association reactions in the current literature (e.g. Herbst 1985a) as updated by Herbst (1985c, 1987) for systems with competing exothermic channels and by Bates (1987b) for the $CH_3^+ + H_2$ system? Many of these rate coefficients are quite large at interstellar cloud temperatures implying that the processes are important in cloud models. A perhaps overly optimistic and outdated view is shown in Fig. 2 for systems in which no experimental radiative association data are available. A distinction must be made between systems for which ternary association data are available and systems for which such data are lacking. If ternary association data exist the theory can be normalized to the experimental data and the major problem in calculating radiative association rate coefficients is determining the radiative stabilization rate. For protonated ionic complexes of principal interest in interstellar chemistry, current *ab initio* quantal calculations suggest that k_r for the vibrational stabilization mechanism is on the order of 1(3) s^{-1} for complexes with bond energies of a few eV. This number should be good to an order of magnitude and leads to an uncertainty in k_{ra} of this amount. In addition, calculated radiative association rate coefficients at low temperature are often saturated at the collisional limiting value and are thus insensitive to k_r. If low-lying electronic states are prevalent, however, radiative stabilization rates in excess of 1(3) s^{-1} are possible. Unless these states are quite low-lying (less than 1 eV excitation energy), enhancements in the stabilization rate by more than a factor of 10 may not be common. However, the additional uncertainty in k_r engendered by this mechanism increases the uncertainty in calculated values of k_{ra} by perhaps a factor of 10 for systems that are not at or near saturation.

If ternary association has not been studied for a particular system then there is always the possibility of activation energy. If activation energy does exist, the radiative association rate coefficient is so small as to be unimportant. If activation energy does not exist, there is still the problem of an accurate determination of the density of vibrational states of the complex. Experience on our part in the calculation of ternary association rate

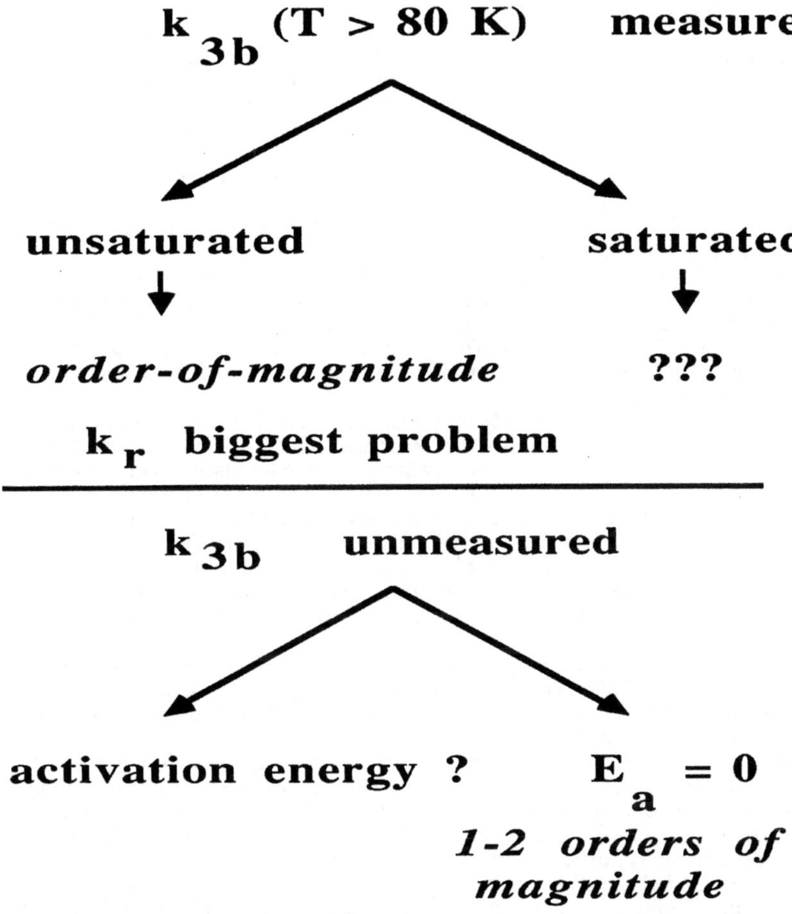

Figure 2. An assessment of uncertainties in calculated radiative association rate coefficients.

coefficients and their comparison with experimental values suggests that this density can be estimated accurately to an order of magnitude even if quantal calculations on complex vibrational frequencies are not available as long as the dissociation energy is known accurately. Since this parameter is often known to within a few tenths of an eV, the uncertainty exclusive of k_r is perhaps one order of magnitude. The overall uncertainy in

k_{ra} is doubtless larger but is in the range 1-2 orders of magnitude if k_r is known to an order of magnitude and worse if k_r is known with less accuracy. It is to be hoped that this situation in theory will be improved by *ab initio* quantal calculations of excited electronic states which will help in determining the dominant mechanism for radiative stabilization.

What about comparison between theory and experiment for those few systems in which laboratory values for or upper limits to radiative association rate coefficients are available? Two systems - $C^+ + H_2$ - and $CH_3^+ + H_2$ - have been studied both by a low pressure ion trap method, in which k_{ra} is determined directly, and a higher pressure trap method, in which k_{ra} is determined via extrapolation of a linear plot of k_{eff} vs density to zero density. The results are shown in Table III below along with theoretical phase space results derived from papers of both Herbst and Bates. The experiments utilize "normal" hydrogen, in which the ortho/para ratio is 3/1, so that the results do not pertain to the interstellar medium where H_2 is thought to be in true thermodynamic equilibrium. In addition, care must be taken not to attempt to direct comparison with the published theoretical results on $C^+ + H_2$ (Herbst 1982a) which utilizes sub-thermal excitation of C^+. The third system in Table III is the CH_3^+ + HCN reaction, for which ICR data and a phase space calculation exist.

The low temperature data on $C^+ + H_2$ by Luine and Dunn (1985) is in the form of an upper limit which is only somewhat above the theoretical result. At higher temperatures, the result of Gerlich and Kaefer (1987) is somewhat below the best theoretical value by a factor of ≈ 3. The experimental result, however, is preliminary. Since the $C^+ + H_2$ system was calculated to proceed via an excited electronic state of CH_2^{+*} (Case 1 enhancement), ternary data cannot easily be utilized to "normalize" the theoretical approaches since the ternary rate coefficient is more sensitive to the vibrational state density in the ground electronic state of CH_2^{+*}. It should also be mentioned that Case 2 enhancement of the radiative association rate of this system is also possible and should be investigated.

The situation regarding the CH_3^+ + HCN radiative association reaction is that the experimental upper limit of 5(-12) $cm^3\ s^{-1}$ obtained by Kemper, Bass, and Bowers (1985) is not much below the phase space value of Bass *et al.* (1981) if the standard vibrational relaxation rate of 1(3) s^{-1} is used.

The $CH_3^+ + H_2$ data from two different experiments are enigmatic. The low pressure ion trap experiment of Barlow, Dunn, and Schauer (1984a,b) has been claimed by Bates (1986d) to have the ions at an effective temperature much higher than the 13 K of the ambient gas. He deduced a stabilization rate k_r of 3.5(4) s^{-1}. This value of k_r leads to a radiative association rate k_{ra} of 3(-13) $cm^3\ s^{-1}$ at 80 K, the temperature in the high pressure trap experiment of Gerlich and Kaefer (1988). However, the value of k_{ra} determined by Gerlich and Kaefer (1988), again a preliminary value, is less than 3(-15) $cm^3\ s^{-1}$ which is almost two orders of magnitude lower.

One possibility to be examined is that Bates (1986d) erred in his interpretation of the experiment of Barlow *et al.* (1984a,b). In this event, the standard vibrational relaxation rate of 1(3) s^{-1} gives a k_{ra} at 13 K just a factor of four below the value measured by Barlow *et al.* (1984a,b) and also gives a k_{ra} at 80 K between two and three times the upper limit measured by Gerlich and Kaefer (1988). However the factors are in *opposite* directions. Theory gives the temperature variation of k_{ra} reliably. It predicts that in going from 13 K to 80 K k_{ra} should fall by a factor of only 3 whereas the measured values fall by a factor of more than about 40 which would mean that the experimental error in one or

TABLE III RADIATIVE ASSOCIATION: EXPERIMENT AND THEORY

System	Experimental k_{ra}	T(K)	Theoretical k_{ra}	
	Rate Coefficient k_{ra} ($cm^3\ s^{-1}$)			
$C^+ + H_2$	<1.5(-15)[a]	10	4(-16)[b]	
	3(-17)[c]	230	1(-16)[b]	
$CH_3^+ + HCN$	<5(-12)[d]	300	1(-11)[e]	
		k_r =	1(3) s^{-1}	3(4) s^{-1}
$CH_3^+ + H_2$	1.1(-13)[f]	13 (?)	2.6(-14)[g]	9.3(-13)[g]
		hot ions (?)	3.1(-15)[g]	1.1(-13)[g]
	<3(-15)[h]	80 (?)	8.2(-15)[g]	2.9(-13)[g]
		hot ions (?)	- - -	- - -

[a] Luine and Dunn (1985) [b] Herbst (1982a) [c] Gerlich and Kaefer (1987)
[d] Kemper *et al.* (1985) [e] Bass *et al.* (1981) [f] Barlow *et al.* (1984a,b)
[g] Bates (1987b) [h] Gerlich and Kaefer (1988)

both investigations is much greater than represented. Gerlich and Kaefer (1988) also measured the 80 K rate coefficient for

$$CH_3^+ + H_2 + He \text{ -----> } CH_5^+ + He \qquad (37)$$

and

$$CH_3^+ + H_2 + H_2 \text{ -----> } CH_5^+ + H_2\ . \qquad (38)$$

They found k_{37} to be 1.7(-27) $cm^6\ s^{-1}$ which is in quite good agreement with the value of 3.5(-27) $cm^3\ s^{-1}$ that Bates (1986b) inferred from the SIFT data of Smith *et al.* (1982). Their measurements require that at 80 K β_c for H_2 is about five times β_c for He so that even if $\beta_c(H_2)$ were unity, $\beta_c(He)$ would be only 0.2 which is a low value (see Table I).

Consideration must be given to the possibility that the results of Gerlich and Kaefer

(1988) are affected by ortho hydrogen heating (despite the accord with the SIFT data). The issue should be decided by their proposed measurements using para hydrogen.

6. SUMMARY

Radiative association reactions at low temperatures are thought to be of importance in the chemistry of both diffuse and dense interstellar clouds. Yet very few of these reactions have been studied in the laboratory, chiefly because unless the pressure is very low, radiative association must compete with ternary association. Indeed, in most laboratory experiments, ternary association is dominant and any information on the corresponding radiative process must be inferred indirectly. Because of the paucity of experimental work, theoretical treatments are essential. Given some knowledge of the reactants and molecule formed in the association reaction, theoretical treatments achieve order-of-magnitude accuracy in the determination of ternary association rate coefficients, and should give the temperature dependence reliably. This latter point is crucial because it allows confidence to be placed on a theoretical rate coefficient versus temperature curve that has been normalized with the aid of a good measurement at one temperature. The most accurate theories are the modified thermal and phase space approaches. The former is easier to utilize and is applicable when the system is truly thermal. For systems that are non-thermal the more computationally tedious phase space approach is necessary as it is in determining the association rate coefficient for systems with competing exothermic channels.

Although theoretical treatments of ternary association rates are quite successful, their success in determining radiative association rates is limited. The key problem is the calculation of the radiative stabilization rate of the collision complex. The mechanism by which photons are emitted is uncertain. The few measurements so far undertaken do not give a clear guide. The situation will doubtless be resolved with both *ab initio* quantal calculations and a new generation of experiments including some that measure the intensity of the radiation emitted in the stabilization process.

ACKNOWLEDGMENTS

D. R. B. thanks the U. S. Air Force for support under grant AFOSR-85-0202. E. H. thanks the National Science Foundation (U.S.) for support of his work via grant AST-8513151.

REFERENCES

Adams, N. G. and Smith, D. 1981, *Chem. Phys. Letters*, **79**, 563.
Anicich, V. G. and Huntress, W. T., Jr. 1986, *Ap. J. Suppl. Ser.*, **62**, 553.
Barlow, S. E., Dunn, G. H., and Schauer, K. 1984a, *Phys. Rev. Letters*, **52**, 902.
Barlow, S. E., Dunn, G. H., and Schauer, K. 1984b, *Phys. Rev. Letters,* **53**, 1610.
Bass, L. M. and Bowers, M. T. 1987, *J. Chem. Phys.*, **86**, 2611.
Bass, L. M., Cates, R. D., Jarrold, M. F., Kirchner, N. J., and Bowers, M. T. 1983, *J. Am. Chem. Soc.*, **105**, 7024.

Bass, L. M., Chesnavich, W. J., and Bowers, M. T. 1979, *J. Am. Chem. Soc.*, **101**, 5493.
Bass, L. M. and Jennings, K. R. 1984, *Int. J. Mass Spectrom. Ion Proc.*, **58**, 307.
Bass, L. M., Kemper, P. R., Anicich, V. G., and Bowers, M. T. 1981, *J. Am. Chem. Soc.*, **103**, 5283.
Bates, D. R. 1951, *M. N. R. A. S.*, **111**, 303.
Bates, D. R. 1979a, *J. Chem. Phys.*, **71**, 2318.
Bates, D. R. 1979b, *J. Phys. B*, **12**, 4135.
Bates, D. R. 1980, *J. Chem. Phys.*, **73**, 1000.
Bates, D. R. 1983a, *Ap. J.*, **270**, 564.
Bates, D. R. 1983b, *Ap. J. Letters*, **267**, L121.
Bates, D. R. 1985a, *Ap. J.*, **298**, 382.
Bates, D. R. 1985b, *J. Chem. Phys.*, **83**, 4448.
Bates, D. R. 1986a, *Chem. Phys. Letters*, **123**, 187.
Bates, D. R. 1986b, *J. Chem. Phys.*, **84**, 6233.
Bates, D. R. 1986c, *J. Chem. Phys.*, **85**, 2624.
Bates, D. R. 1986d, *Phys. Rev. A*, **34**, 1878.
Bates, D. R. 1987a, *Interstellar Cloud Chemistry Revisited* in *Modern Applications of Atomic and Molecular Processes*, ed. A. E. Kingston (London: Plenum).
Bates, D. R. 1987b, *Ap. J.*, **312**, 363.
Bates, D. R. 1987c, *Int. J. Mass Spectrom. Ion Proces.* (in press, Eldon Ferguson Issue).
Bates, D. R. and Morgan, W. L. 1987, *J. Chem. Phys.*, **87**, 2611.
Black, J. H. and Dalgarno, A. 1973, *Ap. Letters*, **15**, 79.
Black, J. H. and Dalgarno, A. 1977, *Ap. J. Suppl. Ser.*, **34**, 405.
Bohme, D. K. and Raksit, A. B. 1985, *M. N. R. A. S.*, **213**, 717.
Botschwina, P. 1987, *Tenth Colloquium On High Resolution Molecular Spectroscopy*, Dijon, France, Talk E2.
Cates, R. D. and Bowers, M. T. 1980, *J. Am. Chem. Soc.*, **102**, 3994.
Fehsenfeld, F. C., Dunkin, D. B., and Ferguson, E. E. 1974, *Ap. J.*, **188**, 43.
Forst, W. 1973, *Theory of Unimolecular Reactions* (New York: Academic).
Gerlich, D. and Kaefer, G. 1987, 5th International Swarm Seminar, Birmingham, U. K.
Gerlich, D. and Kaefer, G. 1988, Symposium on Atomic and Surface Physics, La Plagne, France.
Herbst, E. 1979, *J. Chem. Phys.*, **70**, 2201.
Herbst, E. 1980a, *Ap. J.*, **237**, 462.
Herbst, E. 1980b, *J. Chem. Phys.*, **72**, 5284.
Herbst, E. 1980c, *Ap. J.*, **241**, 197.
Herbst, E. 1981, *J. Chem. Phys.*, **75**, 4413.
Herbst, E. 1982a, *Ap. J.*, **252**, 810.
Herbst, E. 1982b, *Chem. Phys.*, **68**, 323.
Herbst, E. 1982c, *Chem. Phys.*, **65**, 182.
Herbst, E. 1985a, *Ap. J.*, **291**, 226.
Herbst, E. 1985b, *Astr. Ap.*, **153**, 151.
Herbst, E. 1985c, *Ap. J.*, **292**, 484.
Herbst, E. 1985d, *J. Chem. Phys.*, **82**, 4017.
Herbst, E. 1987, *Ap. J.*, **313**, 867.
Herbst, E. and Bates, D. R. 1987, submitted to *Ap. J.*
Herbst, E. and Klemperer, W. 1973, *Ap. J.*, **185**, 505.
Herbst, E., Schubert, J. G., and Certain, P. R. 1977, *Ap. J.*, **213**, 696.

Hill, T. L. 1960, *An Introduction to Statistical Thermodynamics* (Reading, Mass.: Addison-Wesley).
Huntress, W. T., Jr. and Mitchell, G. F. 1979, *Ap. J.*, **231**, 456.
Johnsen, R., Chen, A. K., and Biondi, M. A. 1980, *J. Chem. Phys.*, **73**, 3166.
Kemper, P. R., Bass, L. M., and Bowers, M. T. 1985, *J. Phys. Chem.*, **89**, 1105.
Knight, J. S., Freeman, C. G., McEwan, M. J., Smith, S. C., Adams, N. G., and Smith, D. 1986, *M. N. R. A. S.*, **219**, 89.
Leung, C. M., Herbst, E., and Huebner, W. F. 1984, *Ap. J. Suppl.*, **56**, 231.
Levine, R. D. and Bernstein, R. B. 1974, *Molecular Reaction Dynamics* (London: Oxford).
Luine, J. A. and Dunn, G. H. 1985, *Ap. J. Letters*, **299**, L67.
McEwan, M. J. 1987, private communication.
Morgan, W. L. and Bates, D. R. 1987, *Ap. J.*, **318**, 817.
Raksit, A. B. and Bohme, D. K. 1983, *Int. J. Mass Spectrom. Ion Proces.*, **55**, 69.
Saxer, A., Richter, R., Villinger, H., Futrell, J. H., and Lindinger, W. 1987, *J. Chem. Phys.*, **87**, 2105.
Smith, D. and Adams, N. G. 1978, *Ap. J. Letters*, **220**, L87.
Smith, D. and Adams, N. G. 1979, in *Gas Phase Ion Chemistry*, ed. M. T. Bowers (New York: Academic), Vol. 1, p.1.
Smith, D., Adams, N. G., and Alge, E. 1982, *J. Chem. Phys.*, **77**, 1261.
Solomon, P. M. and Klemperer, W. 1972, *Ap. J.*, **178**, 389.
Townes, C. H. and Schawlow, A. L. 1955, *Microwave Spectroscopy* (New York: McGraw-Hill)
Troe, J. 1977a, *J. Chem. Phys.*, **66**, 4745.
Troe, J. 1977b, *J. Chem. Phys.*, **66**, 4758.
Viggiano, A. A. 1984, *J. Chem. Phys.*, **81**, 2639.
Whitten, G. Z. and Rabinovitch, B. S. 1963, *J. Chem. Phys.*, **38**, 2466.
Whitten, G. Z. and Rabinovitch, B. S. 1964, *J. Chem. Phys.*, **41**, 1883.
Williams, D. A. 1972, *Ap. Letters*, **10**, 17.

DISSOCIATIVE RECOMBINATION: POLYATOMIC POSITIVE ION REACTIONS WITH ELECTRONS AND NEGATIVE IONS

David R. Bates
Department of Applied Mathematics and Theoretical Physics
Queen's University of Belfast
Belfast BT7 1NN
United Kingdom

and

Eric Herbst
Department of Physics
Duke University
Durham, NC 27706
USA

ABSTRACT. The neutral products arising from dissociative recombination reactions between polyatomic positive ions and electrons are discussed from a theoretical point of view. It is argued that polyatomic ions are composed of valence bonds and that the dissociative recombination process normally involves the rupture of one of these bonds. This viewpoint leads to markedly different products from the predictions of previous theories. Recombination between a polyatomic positive ion and a negative PAH ion is also considered and it is noted that such reactions may lead to dissociation of the positive ion.

1. RECOMBINATION WITH ELECTRONS

The theory of dissociative recombination between diatomic positive ions and electrons was formulated many years ago (Bates 1950). The process takes place through a radiationless transition in which the free electron enters a bound orbital of a state having a repulsive potential energy curve (Figure 1) so that the two atoms move apart and thereby prevent the inverse radiationless transition from occurring:

$$XY^+ + e \text{ -----> } X + Y . \tag{1}$$

It was shown that if the initial vibrational wave function is $\chi(r)$ and the final vibrational wave function may be represented by the delta function of Winans and Stueckelberg (1928), then the recombination coefficient at electron temperature T is given by

$$\alpha_e = h^3 A g(n) F / \{2\, [2\pi m k T]^{3/2}\, g(i)\} \tag{2}$$

T. J. Millar and D. A. Williams (eds.), Rate Coefficients in Astrochemistry, 41–48.

with

$$F = \int |\chi(r)|^2 dr/d\varepsilon \exp(-\varepsilon/kT) d\varepsilon \qquad (3)$$

where A is the rate of the radiationless transition (which is here assumed to be the rate limiting step as it commonly is), g(i) and g(n) are the statistical weights of the states of the ionized and neutral molecule that are involved, and ε is the energy supplied by the electron in a vertical transition between the two potential energy surfaces. The dimensionless entity F is, in effect, a Franck-Condon factor. If the potential energy surfaces interact near the minimum of the ionic surface, F is typically about kT when expressed in eV. Smaller values of F, and therefore relatively smaller values of α_e, occur in cases like He_2^+ where the interaction is well away from the minimum.

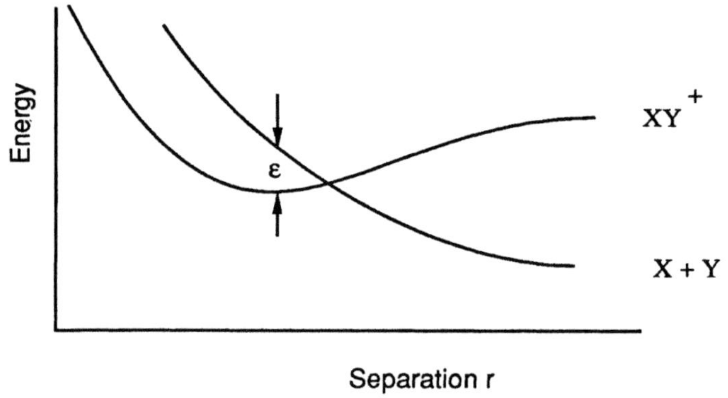

Figure 1. Potential energy surfaces involved in process (1).

The radiationless transition may break even a strong multiple bond. This is evident from experimental data (see for example Massey 1982) on α_e for diatomic ions. Instructive calculations on

$$NO^+ + e \longrightarrow N + O \qquad (4)$$

have been done by Bardsley (1968). In the ground state of NO^+ the bonding orbitals outnumber the anti-bonding orbitals by six and the dissociation energy has the high value of 10.9 eV. Process (4) occurs through a radiationless transition in which the free electron enters an anti-bonding orbital and a bound electron is simultaneously switched from a bonding to an anti-bonding orbital. This still leaves three more bonding orbitals than anti-bonding orbitals but nevertheless the potential energy surface is repulsive in the relevant region and only weakly attractive at greater separations.

1.1. Polyatomic Ions

Herbst (1978) has formulated a statistical theory of dissociative recombination reactions involving polyatomic ions. The physical basis of this theory is that the trajectory of the separating neutral fragments undergoes a complex motion before complete bond rupture is attained. Despite its simplicity, the theory requires extensive computer calculations to predict neutral product branching ratios. In general, breakage of single bonds is preferred to breakage of multiple bonds. Green and Herbst (1979) looked at the process of dissociative recombination somewhat differently and inferred that the greater relative motion between those sets of neutral products in which one fragment is a hydrogen atom would lead to their being favored over neutral products in which both fragments are heavy. This second approach has been used in several model calculations of interstellar clouds. A model advocated by Bates (1986, 1987a) and discussed below is different. According to him a polyatomic ion may be regarded as containing a single ionized atom having the valency of the isoelectronic neutral atom and a number of neutral atoms each having its customary valency, the whole being held together by localized valence bonds. In dissociative recombination, one of the valence bonds of the ionized atom is broken by a radiationless transition that changes the attraction associated with the bond into a repulsion in consequence of which two fragments of the system separate, thereby stabilizing the neutralization as in the diatomic case.

Ab initio quantal computations like those of Ave *et al.* (1976) or Millar *et al.* (1987) show that charge dispersal may be very pronounced in polyatomic ions. Any doubt that this might cast on the validity of the model of Bates has been dismissed by a recent investigation (Bates 1987b) that focused attention on the remarkable consistency of the derived values of the effective bond energy $B[X^+G_i]$ between an atomic ion X^+ and a group G_i, for example the methyl group. Letting A[ ] denote the association energy of the molecular species indicated in the brackets, the effective bond energies were defined to be such that

$$A[X^+G_1\text{---}G_n] \equiv \Sigma_i B[X^+G_i] + \Sigma_i A[G_i] \ . \qquad (5)$$

If Y_i is the atom of the group G_i to which X^+ is valence bound, $B[X^+G_i]$ is the sum of the valence bond energy $B[X^+Y_i]$ and the energy due to the long range interactions between X^+ and the other atoms of the group. Association energies for use on the left of identity (5) were obtained from the table of measured heats of formation compiled by Lias *et al.* (1984); those for use on the right were obtained taking the bond contributions to the heats of formation of neutral systems to be as recommended by Benson and Buss (1958).

Table I shows some results obtained by Bates (1987b). The procedure he followed is straightforward. Having obtained $B[N^+H]$ from $A[H_4N^+]$, he assumed that the value is independent of the species of ion. He then obtained $B[N^+Me]$ (Me = methyl) from each of $A[H_3N^+Me]$, $A[H_2N^+Me_2]$, and $A[HN^+Me_3]$. The three values differ only slightly. Assuming that their mean, 110.6 kcal mol^{-1}, is the value in all ions he proceeded in like manner. The five values of $B[N^+Et]$ (Et = ethyl) are in excellent agrement. Moreover, the increase in going from $B[N^+Me]$ through $B[N^+Et]$ and $B[N^+nPr]$ to $B[N^+nBu]$ (nPr = n-propyl; nBu = n-butyl) can reasonably be attributed to an increase in the polarization energy. These and other results support the simple valence bond model of polyatomic ions, at least for bonds between atomic ions and functional groups in saturated species.

TABLE I VALENCE BOND MODEL OF POLYATOMIC IONS

Ion	Bond N^+G	Derived Bond Energy $B[N^+G]$ kcal mol^{-1}
H_4N^+	N^+H	126.4
H_3N^+Me	N^+Me	110.4
$H_2N^+Me_2$		110.6
HN^+Me_3		110.7
H_3N^+Et	N^+Et	114.4
$H_2N^+Et_2$		113.4
H_2N^+MeEt		114.2
HN^+Me_2Et		114.1
HN^+MeEt_2		113.8
H_3N^+nPr	N^+nPr	115.4
H_3N^+nBu	N^+nBu	116.4

Note: Me = CH_3, Et = C_2H_5, nPr = n-C_3H_7, nBu = n-C_4H_9

1.2. Branching Ratios

Information on what are the products of dissociation of vibrationally cold H_3O^+ is of some importance in interstellar cloud chemistry (Lepp, Dalgarno, and Sternberg 1987). According to the approach outlined above, the radiationless transition that occurs would be expected to affect only one of the valence bonds so that the course of the reaction should be

$$H_3O^+ + e \longrightarrow H_2O + H \quad (6)$$

rather than

$$H_3O^+ + e \longrightarrow OH + 2H \text{ (or } H_2\text{)} , \quad (7)$$

a result in conflict with the statistical approach. There is enough energy for the OH to be in one of its low excited states but the other products must be in their ground electronic states. The free electron can only enter the 2p orbital of O which reduces the number of its unpaired electrons by one and therefore correspondingly reduces its valency so that a single H atom is shed. Any suggestion that the H_2O formed in channel (6) might have 4.8 eV or more internal energy and thus be able to dissociate into H_2 + O may be dismissed (Bates 1987a) if the process by which H_3O dissociates is a direct one.

Unlike the previous case, several channels are commonly open in the approach of

Bates. For example, the dissociative recombination of $CH_3OH_2^+$ may proceed via either

$$CH_3OH_2^+ + e \text{ -----> } CH_3OH + H \qquad (8)$$

or

$$CH_3OH_2^+ + e \text{ -----> } CH_3 + H_2O \ , \qquad (9)$$

depending upon which bond of the O^+ ion is broken. The branching ratios depend on the radiationless transition probability A and on the Franck-Condon factors of eq. (3) and hence cannot easily be predicted. In so far as CH_5^+ can be regarded as having the structure $CH_3^+ \cdot H_2$, the dissociative recombination channels are predicted to be

$$CH_3^+ \cdot H_2 + e \text{ -----> } CH_2 + H + H_2 \qquad (10)$$

and

$$CH_3^+ \cdot H_2 + e \text{ -----> } CH_3 + H_2 \qquad (11)$$

but *not*

$$CH_3^+ \cdot H_2 + e \text{ -----> } CH_4 + H . \qquad (12)$$

Here the weak polarization bond is assumed to break on neutralization whether or not the valence bond also breaks.

If a polyatomic ion can be represented by two valence bond structures of comparable energy, it is necessary to take into account that a good representation of the wave function includes a contribution from both structures. This increases the number of possible dissociative recombination channels. Thus, for $HCNH^+$ the possible channels are

$$HC^+{=}NH + e \text{ -----> } H + CNH \qquad (13)$$

$$\text{ -----> } HC + NH \qquad (14)$$

and

$$HC{\equiv}N^+H + e \text{ -----> } HCN + H \qquad (15)$$

$$\text{ -----> } CH + NH . \qquad (16)$$

Channel (14) or (16) is not a major channel in previous theories. The breaking of multiple bonds involving heavy atoms is unique to the theory of Bates and has serious consequences in interstellar chemistry (see the discussion by Herbst in this volume).

Bates (1987a) has considered some other dissociative recombination processes but for the reasons mentioned immediately after eq. (9) could not provide reliable estimates of the branching ratios. Millar *et al.* (1987) have attempted to estimate these ratios for a wide variety of polyatomic ions in order to calculate the sensitivity of gas phase models of dense interstellar clouds to them. These authors have also preferred to utilize *ab initio* quantal computations of the charge distributions on positive ions rather than adopt the more simple valence bond approach.

2. RECOMBINATION WITH NEGATIVE IONS

Omont (1986) has outlined the evidence for the presence of polycyclic aromatic hydrocarbons in the interstellar medium and given a full acccount of their properties. He pointed out that PAH^- ions may be more abundant than free electrons in dense clouds. Lepp and Dalgarno (1987) recognized that in this circumstance the recombination of positive ions occurs through mutual neutralization with PAH^- ions and have discussed the consequential changes in the chemistry. Omont (1986) gives the rate coefficient for the process to be approximately

$$\alpha_i = 4 \times 10^{-7} (N_C/50)^{1/2} (10/M_i)^{1/2} (10/T)^{1/2} \text{ cm}^3 \text{ s}^{-1} \tag{17}$$

where N_C is the number of C atoms in PAH and M_i is the mass of the positive ion on the amu scale. This expression is really the rate coefficient for collisions that involve an actual impact.

The transfer of an electron takes place on impact if it has not already taken place before impact at a crossing between the ionic potential surface along which the approach begins and a potential surface belonging to neutral products, PAH and M, in

$$PAH^- + M^+ \text{ -----> } PAH + M \quad . \tag{18}$$

The probability P of transfer at a crossing where PAH^- and M^+ are a distance R_x apart is given by the Landau-Zener (Landau 1932; Zener 1932) formula

$$P = \exp\{-M_i^{1/2} f(R_x) |H_{if}(R_x)|^2\} \tag{19}$$

where $f(R_x)$ is a slowly varying function which need not be displayed and $H_{if}(R_x)$ is the interaction matrix element which increases rapidly as R_x decreases. The distance R_x is measured from M^+ to the closest atom on PAH, this being where the interaction keeps the excess electron. To a sufficient approximation

$$R_x = 14Å / \Delta E \tag{20}$$

where

$$\Delta E = (I - A) \text{ eV} \quad . \tag{21}$$

I is the ionization potential of M (which may be in an excited state) and A is the electron affinity of PAH (about 1 eV; Omont 1986). Calculations (Bates and Boyd 1956) show that in the case of atomic positive and negative ions, P is small if ΔE is less than about 1 eV but is close to unity if ΔE is greater than about 2 eV. It therefore seems likely that the transfer of an electron in process (18) occurs mainly at crossings rather than on impact. Transfer of an electron at a crossing is usually presumed not to change the structure of either collision partner. Changes are possible.

If the positive ion in process (18) is a polarization-bound complex like $R^{+}\cdot H_2$, mutual neutralization leads to dissociation:

$$PAH^- + R^{+\cdot}H_2 \text{ -----> } PAH + R + H_2 \quad (22)$$

because there is no binding between R and H_2.

Bates and Boyd (1956) observed that transfer of an electron to an ionic species AB^+ may yield AB in an unstable state. Their formulation gives the rate coefficient for

$$PAH^- + AB^+ \text{ -----> } PAH + A + B \quad (23)$$

to be

$$\alpha_i = 1.1 \times 10^{-6} (10/M_i)^{1/2}(10/T)^{1/2} \Delta E^{-1} \quad cm^3 \, s^{-1} \quad (24)$$

if it be supposed that there is only a single crossing, that ΔE is greater than about 2 eV, and that this makes P unity as for atomic ions. Note that I of eq. (21) is here the vertical ionization potential of AB when in an unstable state at the AB^+equilibrium internuclear distance. It is convenient to denote it by I(D) and to denote the ionization potential of a stable state of AB by I(S).

Results (17) and (24) happen to agree when N_C is approximately 50 because the radius of the PAH palette

$$a = 0.9 \, (N_C)^{1/2} \text{ Å} \quad (25)$$

and R_x of eq. (20) then happen to be about equal.

It is difficult to predict whether or not an encounter between PAH^- and a particular positive ion leads to dissociation. Lack of information on I(D) is one reason. Another is lack of information on excited stable states of the neutral molecule. A state for which I(S) is less than I(D) but yet is large enough to make P almost unity would prevent the crossing at which dissociation could occur from being reached. In the important case of DCO^+ it is not possible to predict whether the course of the more important crossing is

$$PAH^- + DCO^+ \text{ -----> } PAH + DCO \quad (26)$$

or

$$PAH^- + DCO^+ \text{ -----> } PAH + D + CO \, . \quad (27)$$

If course (26) is followed DCO would strike PAH with only around 2 eV kinetic energy of relative motion but would carry excitation energy which would be released on impact. Hence D would have only around 0.2 eV kinetic energy of relative motion which is favorable to the possibility of its becoming bound onto a PAH radical site instead of remaining bound to CO.

ACKNOWLEDGMENTS

D. R. B. thanks the U. S. Air Force for support under grant AFOSR-85-0202. E. H. thanks the National Science Foundation (U. S.) for support of his work via grant AST-8513151.

REFERENCES

Ave, D. H., Webb, H. M., and Bowers, M. T. 1976, *J. Am. Chem. Soc.* , **98**, 311.
Bardsley, J. N. 1968, *J. Phys. B* , **1**, 365.
Bates, D. R. 1950, *Phys. Rev.* , **78**, 492.
Bates, D. R. 1986, *Ap. J. Letters* , **306**, L45.
Bates, D. R. 1987a, in *Modern Applications of Atomic and Molecular Physics* , ed. A. E. Kingston (London: Plenum).
Bates, D. R. 1987b, *Int. J. Mass Spectrom. Ion Proc.* (in press).
Bates, D. R. and Boyd, T. J. M. 1956, *Proc. Phys. Soc. A* , **69**, 1956.
Benson, S. W. and Buss, J. H. 1958, *J. Chem. Phys.* , **29**, 546.
Green, S. and Herbst, E. 1979, *Ap. J.* , **229**, 121.
Herbst, E. 1978, *Ap. J.* , **222**, 508.
Landau, L. 1932, *Z. Phys. Sowjet* , **2**, 46.
Lepp, S. and Dalgarno, A. 1987, *Ap. J.* (in press).
Lepp, S., Dalgarno, A., and Sternberg, A. 1987, *Ap. J.* , **321**, 383.
Lias, S., Liebman, J. F., and Levin, R. D. 1984, *J. Phys. Chem. Ref. Data* , **13**, 695.
Massey, H. S. W., 1982, in *Applied Atomic Collision Physics* , eds. H. S. W. Massey, E. W. McDaniel, and B. Bederson (New York: Academic), Vol. 1, p. 40.
Millar, T. J., DeFrees, D. J., McLean, A. D., and Herbst, E. 1987, *Astr. Ap.* (in press).
Omont, A. 1986, *Astr. Ap.* , **164**, 159.
Winans, J. G. and Stueckelberg, E. C. G. 1928, *Proc. Nat. Acad. Amer.* , **14**, 867.
Zener, C. 1932, *Proc. Roy. Soc. London A* , **137**, 696.

PHOTODISSOCIATION AND PHOTOIONIZATION PROCESSES

Ewine F. van Dishoeck
Princeton University Observatory
Princeton, NJ 08544

ABSTRACT. The photodissociation and photoionization processes of species of astrophysical interest are reviewed, with emphasis on recent developments. Depth–dependent photodestruction rates by the interstellar radiation field are presented for a number of atoms and small molecules. The photodissociation processes of CO are discussed with reference to new laboratory data on the absorption cross sections.

1. INTRODUCTION

In any astrophysical region where ultraviolet photons can penetrate, photodissociation and photoionization are major destruction processes of the neutral species. Even for molecular ions, photodissociation can be an important removal mechanism if the fraction of electrons or other negatively charged species in the region is sufficiently low. Photodissociation plays a dominant role in the chemistry of diffuse interstellar clouds and in the outer parts of dense clouds (Eddington 1926; Bates and Spitzer 1951). More recently, it has been realized that cosmic–ray induced photons can significantly affect the chemistry inside dense clouds (Prasad and Tarafdar 1983; Sternberg, Dalgarno and Lepp 1987; Gredel, Lepp and Dalgarno 1987). The outer envelopes of mass–losing stars are exposed to the general interstellar radiation field, so that photodissociation plays an important role in the circumstellar chemistry (Goldreich and Scoville 1976; Omont 1987). Finally, the solar radiation field is responsible for the production of many of the observed radicals in cometary atmospheres (Wurm 1934; Huebner and Carpenter 1979). In order to interpret or predict the observed abundances of molecules in these regions, an accurate knowledge of the photodestruction rates is a prerequisite. This review will be limited mostly to photoprocesses induced by the interstellar radiation field.

2. PHOTODISSOCIATION MECHANISMS

2.1 *Small molecules*

Although photodissociation is usually referred to as a relatively simple process, it can proceed in various ways. They have been discussed extensively recently by van Dishoeck (1987*a*) and Kirby and van Dishoeck (1988), and will be only briefly reiterated here. The various processes are summarized in Figure 1 for a diatomic molecule. For small

T. J. Millar and D. A. Williams (eds.), Rate Coefficients in Astrochemistry, 49–72.

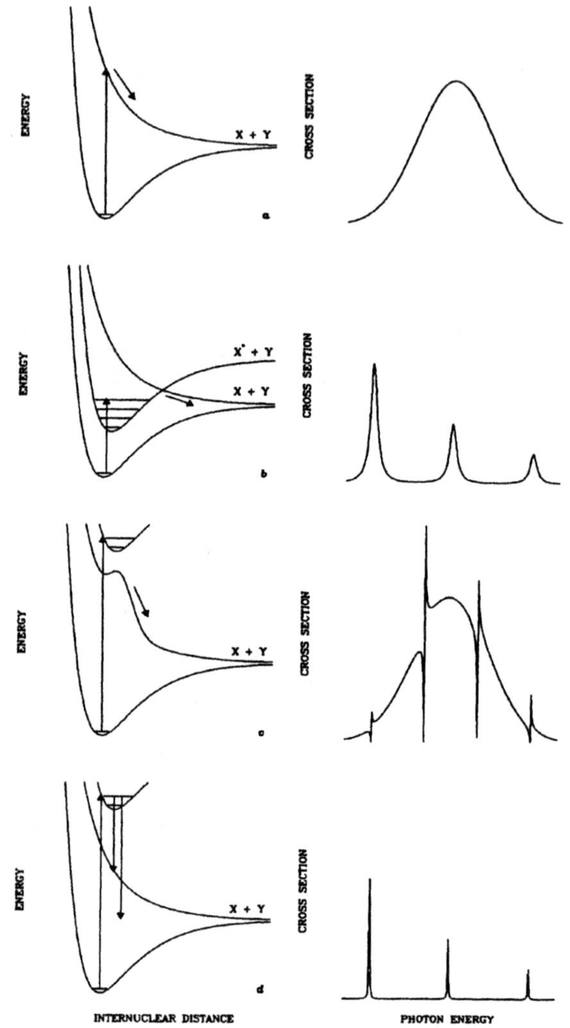

Figure 1. *Potential energy curves and characteristic cross sections for the processes of (a) direct photodissociation; (b) predissociation; (c) coupled states photodissociation; (d) spontaneous radiative dissociation.*

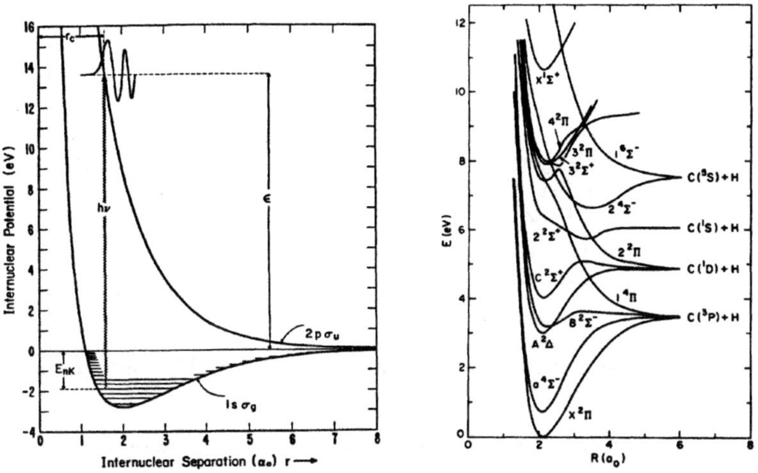

Figure 2. *Potential energy curves for H_2^+ (left) and CH (right) (from: Dunn 1968; van Dishoeck 1987b).*

polyatomic molecules, the processes are similar, but more complicated because of the multidimensionality of the potential energy surfaces.

The simplest process of direct photodissociation, illustrated in Figure 1*a*, dominates the photodissociation of a number of small astrophysically important molecules, such as H_2^+, CH^+, NH, and H_2O. The clearest example is provided by the H_2^+ ion for which the potential energy curves are reproduced in Figure 2*a*. A more complicated molecule such as CH (see Figure 2*b*) generally has several low-lying photodissociation channels. In this case, direct photodissociation can occur by absorption into, for example, the $2^2\Sigma^+$ state. The cross section for direct photodissociation is continuous as a function of wavelength, and its shape reflects that of the vibrational wave function for the lower state. The maximum usually occurs close to the vertical excitation energy, indicated by the arrow in Figure 1*a*. Note that the cross section is often negligibly small at threshold. Direct photodissociation can also occur by absorption into the repulsive part of a potential curve which exhibits a bound well at larger distances. A well-known example is provided by the Schumann-Runge continuum of O_2. In this case, the relative amounts of discrete and continuous absorption are very sensitive to small uncertainties in the relative positions of the upper and lower potential curves. An instructive example, illustrated in Figure 3, is provided by the photodissociation of the isovalent CH^+ and SiH^+ ions through the $A^1\Pi$ state. For CH^+, nearly 100% of the absorption cross section occurs into the discrete levels. For SiH^+, however, the curves are slightly displaced, and most of the absorption takes place into the continuum of the A state (Kirby and Singh 1983). Thus the photodissociation of SiH^+ through the A state will be very rapid, whereas for CH^+ it will be negligible.

In indirect photodissociation processes, the initial absorption occurs into a bound discrete excited state which subsequently interacts with the continuum of a final dissociating state. The process of predissociation, in which the bound potential curve is crossed by a repulsive state of a different symmetry, is illustrated in Figure 1*b*. The cross section consists in this case of a series of discrete peaks, broadened by the predissociation process.

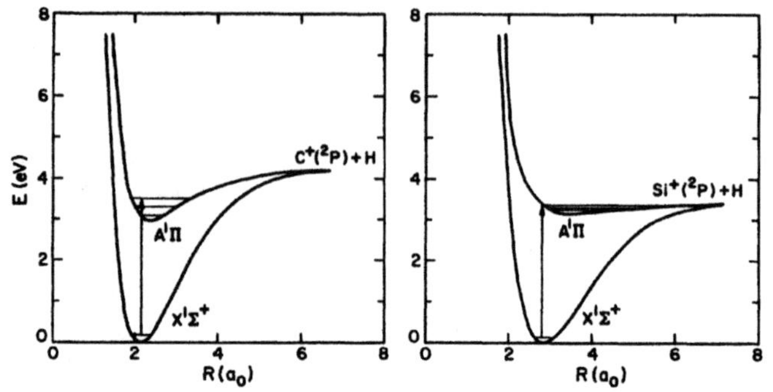

Figure 3. *Comparison of photodissociation of CH^+ and SiH^+ through the $A^1\Pi$ state (from: Kirby and van Dishoeck 1988).*

The photodissociation of the second most abundant interstellar molecule, CO, occurs by this mechanism, and will be discussed in more detail in §6.

The bound levels of an excited state can also couple with the continuum of a dissociative state of the same symmetry, which does not cross the bound state. This process can play an important role in molecules in which e.g., Rydberg–valence interactions cause avoided crossings between excited states. Examples are provided by OH and CH. The shape of the cross section depends on the details of the molecular structure, but it may consist of a continuous background with superposed a series of resonances (cf. Figure 1*c*).

Finally, the bound excited state can simply decay by spontaneous emission into lower-lying continuum states (see Figure 1*d*). For most molecules, this mechanism plays only a minor role, but it is the dominant photodissociation process of the most abundant interstellar molecule, H_2. The cross section consists in this case of a series of sharp, discrete peaks with widths determined by the natural radiative lifetimes of the levels.

2.2 Large molecules

For larger molecules with a significant number of vibrational degrees of freedom, the density of vibrational levels of the lower electronic states becomes so high that they form a quasi continuum with which the excited electronic states can couple non-radiatively. In this case of internal conversion, the probability is small that all the energy will be located in any specific mode leading to dissociation, and the formation of a highly-excited bound molecule is more likely. The excited molecule can subsequently relax by emission of infrared photons. Alternatively, the excited molecule can fluoresce down to the ground state, or it can undergo intersystem crossing to an electronic state with a different spin multiplicity from which it can phosphoresce down. Statistical arguments suggest that N–atom molecules with $N > 25$ are stable with respect to photodissociation (Duley 1986; Omont 1986; Léger and Puget 1984). Although a considerable amount of experimental data is available on smaller polyatomic molecules such as benzene (Leach 1987), virtually no gas-phase experiments exist on large species to test these theories. Large molecules typically have ionization potentials around 7 eV, so that they can easily be photoionized. Whether a small percentage of the absorptions above the ionization threshold will lead to dissociation, or whether the H atoms on the periphery of a large molecule can be removed efficiently by photodissociation, is still uncertain as well.

3. PHOTODISSOCIATION CROSS SECTIONS

3.1 *Theory*

The procedure to obtain photodissociation cross sections from theory has been outlined e.g. by Kirby and van Dishoeck (1988). Potential energy curves and transition dipole moments connecting the excited states with the ground electronic state can be obtained from ab initio quantum chemical calculations. As discussed in §2, the potential curves provide insight into the possible photodissociation processes. For direct photodissociation, the eigenvalue equations for the nuclear motion in the ground and excited states can then be solved by numerical integration, and cross sections can be obtained by integration of the product of the vibrational wave functions and the transition dipole moment function over the nuclear coordinate. For indirect processes, the transition moment function is used to compute oscillator strengths for absorption into the discrete upper level. If the coupling of the upper bound level with the final dissociative continuum is weak, a first order perturbation theory can be used to calculate the predissociation rates k^{pr} (in s^{-1}). The predissociation probability η_u is then obtained by comparing k^{pr} with the inverse radiative lifetime of the molecule A^{rad}, $\eta_u = k^{pr}/(k^{pr} + A^{rad})$. If the coupling between the bound and continuum excited states is strong, as often occurs for the case illustrated in Figure 1*c*, the coupled equations for the excited states have to be solved in order to compute the cross sections.

Theory can provide accurate cross sections—i.e., transition energies accurate to better than about 0.3 eV, and oscillator strengths and cross sections to better than 30%—for small molecules in which the number of (active) electrons is at most 30. In practice, reliable calculations are also limited to the lowest 4–5 electronic states per molecular symmetry. For most molecules of astrophysical interest, this range includes the dominant photodissociation channels. An important exception is the CO molecule discussed in §6.

3.2 *Experiment*

The information obtained from experiment is often complementary to that derived from theory. Laboratory measurements of absorption cross sections have been performed for chemically stable species, such as H_2O, CO_2, NH_3, CH_4, ... Most of these experiments are performed at rather low resolution, where the individual rotational lines are not resolved. Typical examples of measured cross sections are reproduced in Figure 4. For CH_4, various absorption continua are observed, with discrete features superposed at shorter wavelengths. The absorption of NO occurs mostly in discrete bands. Note that often the electronic states responsible for the absorptions have not yet been identified spectroscopically.

The absorption of a photon can result in reemission of another photon, dissociation or ionization of the molecule, and many experiments do not distinguish between these processes. If the photon energy is insufficient to ionize the molecule and if the absorption is continuous, it is likely that photodissociation is the dominant process. In other cases, additional information on the fluorescence and/or ionization cross sections is needed to infer the dissociation probabilities η_u. For some species, the fluorescence cross sections have been measured explicitly. Figure 4*b* compares the absorption and fluorescence results for NO. It appears that for some bands the fluorescence yield is large, whereas for other bands, photodissociation occurs with almost 100% probability. The absorption cross sections for continuous processes measured at low resolution are quite accurate. However, for discrete absorptions, high resolution and low pressures are essential to obtain reliable experimental results, because saturation effects can easily cause orders of magnitude errors. Another

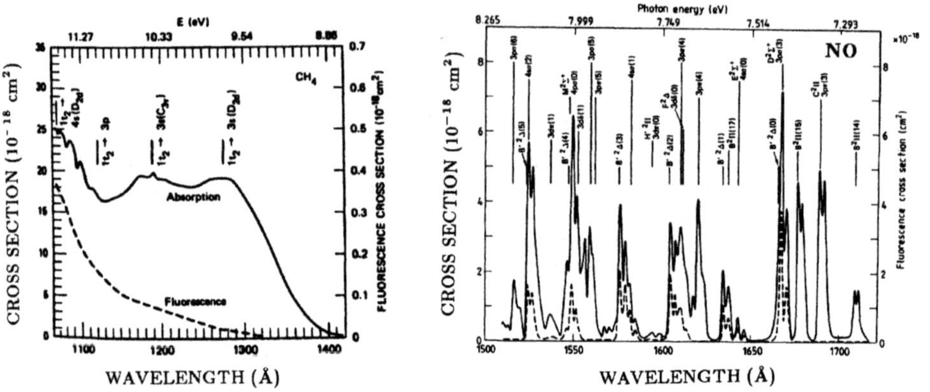

Figure 4. *Measured absorption (full lines) and fluorescence (dashed lines) cross sections for CH_4 (left) and NO (right) (from: Lee and Chiang 1983; Guest and Lee 1981).*

experimental problem arises from the lack of powerful continuous light sources at vacuum ultraviolet (VUV) wavelengths λ <1100 Å. Some beautiful experiments have been performed using VUV lasers, but these are usually limited to only a few specific wavelengths. Measurements using synchrotron radiation are possible, but only very few facilities have spectrographs with sufficiently high resolution (Yoshino *et al.* 1988). Therefore, even for many stable molecules, no reliable cross section data are available at the shortest wavelengths. Experimental data on reactive molecules, such as the radicals CH, OH, C_2H, ... are still extremely limited. Reviews of cross section measurements have been given by Hudson (1971), Robin (1975), Okabe (1978), Ashfold *et al.* (1979), Huebner and Carpenter (1979) and Lee (1984).

3.3 Photodissociation rates

The rate of photodissociation k_{pd} of a molecule by absorption from a lower level ℓ into a continuous upper channel u is given by

$$k_{pd}^{cont} = \int \sigma_{u\ell}(\lambda) x_\ell I(\lambda) d\lambda \ \ \mathrm{s}^{-1}, \tag{1}$$

where σ is the cross section for photodissociation in cm^2, x_ℓ is the fractional population in level ℓ, and I is the mean intensity of the radiation measured in photons cm^{-2} s^{-1} Å^{-1} as a function of wavelength λ in Å. For indirect processes, the rate of dissociation by absorption into a specific level of a bound upper state u from the lower level ℓ is

$$k_{pd}^{line} = \frac{\pi e^2}{mc^2} \lambda_{u\ell}^2 \ f_{u\ell} \ \eta_u \ x_\ell \ I(\lambda_{u\ell}) \ \ \mathrm{s}^{-1}, \tag{2}$$

where $f_{u\ell}$ is the line absorption oscillator strength and η_u is the dissociation efficiency of the upper state level which lies between zero and unity. The numerical value of the factor $\pi e^2/mc^2$ is 8.85×10^{-21} in the adopted units. The total photodissociation rate of

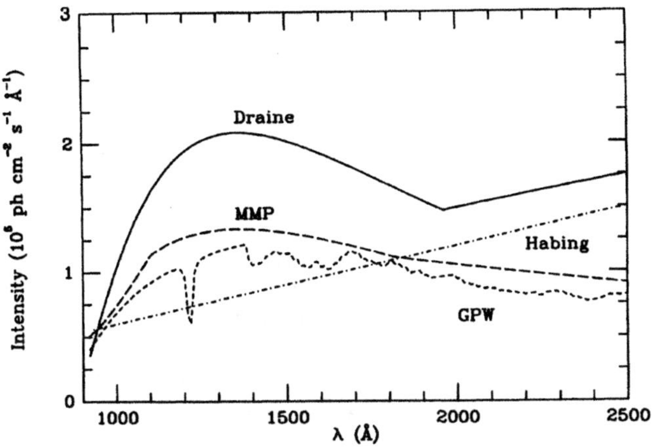

Figure 5. *The intensity of the interstellar radiation field as a function of wavelength cf. Draine (1978) (full line), Mathis et al. (1983) (long-dashed line), Gondhalekar et al. (1980) (short-dashed line) and Habing (1968) (dash-dotted line).*

a molecule is obtained by summation over all possible channels. The effectiveness of each channel depends on the characteristics of the radiation source. Photodissociation channels with large cross sections but at energies of low photon flux may be less significant than channels with smaller cross sections at the peak of the photon flux. In the following, only the interstellar case will be considered.

4. INTERSTELLAR RADIATION FIELDS

4.1 General background radiation field

Estimates of the intensity of the background interstellar radiation field in the solar neighbourhood due to the ensemble of stars in the Galaxy have been made by Habing (1968), Jura (1974), Draine (1978), Gondhalekar *et al.* (1980; GPW) and Mathis *et al.* (1983; MMP). The various radiation fields for $912 < \lambda < 2500$ Å—the important wavelength region for photodissociation and photoionization—are illustrated in Figure 5. The intensities vanish below 912 Å (13.6 eV) due to absorptions by atomic hydrogen in the vicinity of the stars where the photons originate. The various estimates differ typically by factors of two, so that it is important to refer all photodestruction rates to the same radiation field. In the following, the radiation field of Draine (1978)* will be used at $\lambda < 2000$ Å, augmented

* Unfortunately, discrepant versions of the Draine (1978) radiation field exist in the literature. These can be traced back to the formula specified by Roberge *et al.* (1981), who did not retain enough significant digits following conversion of units. Therefore, formula (3) in van Dishoeck (1987*a*) is inaccurate and gives intensities that are too large by factors of 2 at $\lambda < 1000$ Å compared with Draine's original formula. The numbers reported in Table 1 of van Dishoeck (1987*a*) were obtained, however, with the correct Draine formula.

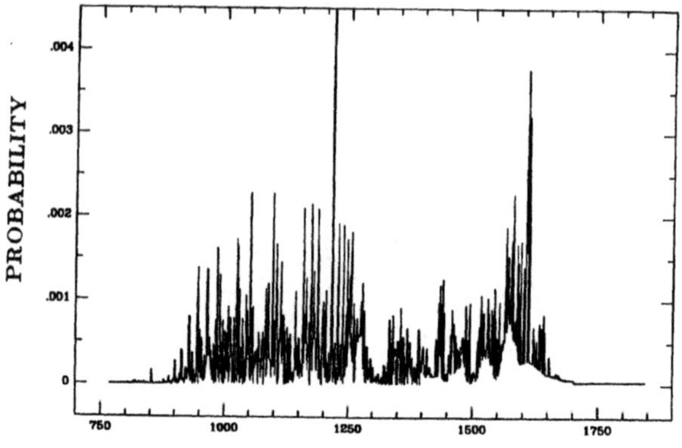

Figure 6. *The emission spectrum of H_2 following excitation by cosmic–ray induced secondary electrons (from: Gredel et al. 1987).*

by a simple formula (cf. van Dishoeck and Black 1982) for λ >2000 Å

$$\begin{aligned} I^o(\lambda) &= 3.2028 \times 10^{15}\lambda^{-3} - 5.1542 \times 10^{18}\lambda^{-4} + 2.0546 \times 10^{21}\lambda^{-5} \quad \lambda < 2000 \text{ Å} \\ &= 7.32 \times 10^2 \lambda^{0.7} \quad \lambda > 2000 \text{ Å}. \end{aligned} \tag{3}$$

Note that few direct measurements of the interstellar radiation field exist at λ <1100 Å, the important wavelength region for the H_2 and CO photodissociations and C photoionization. As discussed by van Dishoeck and Black (1988), all extrapolations of the measured intensities at longer wavelengths to these short wavelengths are highly uncertain, and the intensities given by formula (3) may well be too large by factors of a few. Thus the photodestruction rates of species which have their dominant channels at short wavelengths are particularly uncertain.

4.2 Lyman α radiation

In interstellar regions subjected to fast shocks, the radiation field may have a very high intensity at 1215.6 Å, because of recombination and collisional excitation of atomic hydrogen leading to Lyman α radiation (Hollenbach and McKee 1979; Neufeld and Dalgarno 1988). It is thus important to know whether a molecule has a photodissociation or photoionization channel at this wavelength. Molecules such as H_2, CO and N_2 cannot be destroyed by Lyman α radiation, whereas other species like OH and H_2O have cross sections of a few times 10^{-18} cm^2 at 1215.6 Å and are thus easily dissociated. Another strong peak in the radiation field in shocks is provided by the C III resonance line at 977 Å.

4.3 Cosmic–ray induced photons

Most models of interstellar clouds assume that the ultraviolet photons of the interstellar radiation field cannot penetrate the inner parts of dense clouds. However, Prasad and Tarafdar (1983) have suggested that a dilute flux of ultraviolet photons can be produced following excitation of H_2 by energetic secondary electrons resulting from cosmic ray ionization of H and H_2. The cosmic ray induced spectrum has been calculated by Sternberg *et al.* (1987), and with more detail recently by Gredel *et al.* (1987). The latter spectrum is reproduced in Figure 6. It consists of a large number of discrete peaks, mostly due to H_2 Lyman and Werner band photons, with a weak underlying continuum. In order to estimate photodissociation rates, coincidences between the cosmic–ray induced photons and the photodestruction cross sections have to be found. A list of cosmic–ray induced photodissociation rates for use in dense cloud models has been given by Gredel (1987). Because of the complicated nature of the process and the lack of high resolution cross sections at short wavelengths, the rates may be quite uncertain, although the large number of lines reduces the sensitivity to some extent.

5. INTERSTELLAR PHOTODISSOCIATION RATES

5.1 Unattenuated photodissociation rates

The photodissociation rates for a number of important astrophysical molecules are summarized in Table 1 for the unattenuated interstellar radiation field given by Draine (1978; Eq. (3)). In order to calibrate the significance of a rate of 3×10^{-10} s^{-1}, say, it is useful to note that this rate corresponds to a lifetime of a molecule of 10^2 years, which is short in comparison with dynamical lifetimes of interstellar clouds (often estimated to be $10^6 - 10^8$ years), and with most other chemical timescales. In particular, destruction of a molecule by reaction with an ion in a cloud of density 10^3 cm^{-3} with a fractional ionization less than 10^{-4} will occur with a characteristic lifetime of 300 years or more.

The rates in Table 1 are based on cross sections which have been derived either from theory, indicated by T, or from experiment, denoted by E. References to the principal sources for the cross sections are indicated. More information can be found in van Dishoeck (1987*a*) and in §5.2. The table includes an indication of the reliability of the results with the notation A (k_{pd} known to better than 50%), B (k_{pd} uncertain up to a factor of two), C (k_{pd} uncertain up to an order of magnitude), and D (no information on cross sections; k_{pd} may be uncertain by orders of magnitude). For some species, Table 1 lists a lower limit to the photodissociation rate, which is obtained by summing the rates through all known channels. This lower limit is often quite accurate. However, higher lying channels may exist for which the cross sections are not yet known. For example, for rates based on experimental results, additional contributions may come from cross sections at λ <1100 Å, which have not been measured or for which the branching ratios to dissociation and ionization are not known. Rates based on theory may be increased by higher–lying channels below 13.6 eV for which no reliable calculations can be made. In those cases, an additional effective oscillator strength of about 0.1 at λ =1000 Å has been assumed for the unknown higher–lying channels, which increases the total rates by $(1 - 2) \times 10^{-10}$ s^{-1}.

Table 1 also includes the most likely products of the photodissociation processes. Note that different products may be obtained for different photodissociation channels. For example, Figure 2*b* illustrates that photodissociation of a simple diatomic molecule like CH may result in C in the ^{3}P, ^{1}D and ^{1}S states, with the two excited species formed predominantly. The relative importance of the photodissociation channels changes with depth into a cloud, so that different products may be dominant at different depths. The photodissociation of polyatomic molecules can result in various fragments, and each of the molecular fragments can be electronically, vibrationally and rotationally excited. Only very limited information is available on the product distributions of polyatomic species as functions of photon energy. If the symmetries of the dissociating states are known, some guesses of the most likely products can be made on the basis of correlation diagrams, but

TABLE 1. Interstellar molecular photodissociation rates (in s^{-1})

Species	Products	k^o_{pd} L.L.[a]	k^o_{pd} Draine[a]	k^o_{pd} MMP[b]	α_2[c]	β_2[c]	γ_2[d]	γ_3[d]	Acc[e]	Ref[f]	
H_2	H+H	...	5.0(-11)	4.0(-11)	*g*	*g*	*g*	*g*	A	1	T
H_2^+	$H+H^+$	...	5.7(-10)	3.8(-10)	3.913	4.106	1.85	0.72	A	2	E,T
H_3^+	...	...	<1(-12)	<1(-12)	...	...	...	...	A	3	T
HeH^+	$He^+ +H$	...	<1(-12)	<1(-12)	...	...	...	...	A	4	T
CH	C+H	8.0(-10)	9.5(-10)	6.6(-10)	2.815	3.089	1.15	0.82	A	5	T
CH^+	$C+H^+$	...	3.2(-10)	2.5(-10)	5.760	5.604	2.5	0.94	A	6	T
CH_2	CH+H	5.0(-10)	7.2(-10)	4.9(-10)	3.543	3.813	1.67	0.79	C	7	T
CH_3	CH_2+H	...	5.0(-10)	3.2(-10)	3.567	3.792	1.88	0.67	D	8	T
CH_4	CH_2+H_2/2H	...	1.2(-9)	8.1(-10)	4.233	4.387	2.15	0.76	A	9	E
C_2	C+C	1.8(-10)	2.3(-10)	1.7(-10)	4.658	4.789	2.09	0.75	B	10	T
C_2H	C_2+H	2.0(-10)	5.1(-10)	3.4(-10)	3.676	3.881	1.89	0.67	C	11	T
C_2H_2	C_2H+H	...	3.2(-9)	2.1(-9)	3.536	3.750	1.84	0.71	B	9	E
C_2H_4	$C_2H_2+H_2$/2H	...	3.0(-9)	2.0(-9)	3.479	3.730	1.67	0.73	C	12	E
C_3	C_2+C	...	3.8(-9)	2.5(-9)	2.950	3.186	1.80	0.60	C	13	T
c–C_3H_2	C_3H+H	...	1.9(-9)	1.2(-9)	2.920	3.157	1.79	0.67	C	14	T
OH	O+H	...	4.2(-10)	2.9(-10)	3.815	4.028	1.72	0.77	A	15	T
OH^+	$O^+ +H$	...	1.1(-11)	8.5(-12)	5.862	5.658	2.8	1.05	A	16	T
H_2O[h]	OH+H	...	8.0(-10)	5.5(-10)	3.895	4.120	1.70	0.76	A	9,17	E
O_2	O+O	...	7.9(-10)	5.3(-10)	3.286	3.534	1.76	0.66	A	18	E
HO_2	OH+O	4.7(-10)	6.7(-10)	4.4(-10)	3.808	4.075	1.76	0.77	C	19	E,T
H_2O_2	OH+OH	...	9.4(-10)	6.2(-10)	3.756	3.967	1.82	0.71	B	9	E
O_3	O_2+O	...	1.9(-9)	1.2(-9)	3.301	3.559	1.46	0.75	B	20	E
CO	C+O	...	2.0(-10)	1.8(-10)	*g*	*g*	*g*	*g*	B	21	E
CO_2	CO+O	...	8.7(-10)	6.5(-10)	5.333	5.291	2.5	0.90	C	22	E
HCO	CO+H	...	1.1(-9)	1.0(-9)	2.028	2.285	0.8	0.50	C	23	T
H_2CO	$CO+H_2$/2H	...	1.0(-9)	6.7(-10)	3.457	3.689	1.74	0.74	B	24	E
CH_3OH	H_2CO+H_2	...	1.4(-9)	9.4(-10)	3.758	3.975	1.76	0.71	C	25	E
NH	N+H	...	5.0(-10)	3.3(-10)	3.727	3.942	1.97	0.68	A	26	T
NH^+	$N^+ +H$	...	5.4(-11)	3.9(-11)	3.398	3.693	1.39	0.99	C	27	T
NH_2	NH+H	5.8(-10)	7.4(-10)	5.0(-10)	3.476	3.737	1.63	0.83	B	28	T
NH_3[j]	NH_2+H	...	1.1(-9)	7.8(-10)	3.842	4.055	1.79	0.74	A	9	E
N_2[k]	N+N	5.0(-11)	2.3(-10)	2.1(-10)	6.435	6.022	3.82	1.20	C	29	E
NO	N+O	...	4.7(-10)	3.3(-10)	3.838	4.090	1.71	0.72	B	9	E
NO_2	NO+O	...	1.4(-9)	9.1(-10)	3.473	3.715	1.70	0.73	B	9	E
N_2O	N_2+O	...	1.9(-9)	1.2(-9)	3.836	4.023	2.05	0.72	B	30	E
CN[k]	C+N	1.5(-10)	3.0(-10)	2.5(-10)	5.860	5.656	3.07	1.05	C	31	T
HCN	CN+H	...	1.5(-9)	1.1(-9)	4.450	4.565	2.08	0.79	B	9	E
HC_3N[i]	C_3N+H/C_2H+CN		5.6(-9)	3.6(-9)	3.157	3.383	1.84	0.64	B	32	E
CH_3CN	CH_3+CN	...	2.4(-9)	1.6(-9)	4.206	4.354	2.04	0.76	B	33	E
HCℓ	Cℓ+H	...	9.8(-10)	6.5(-10)	3.713	3.928	1.79	0.72	A	34	T,E
NaCℓ	Na+Cℓ	4.8(-10)	3.8(-9)	3.6(-9)	1.565	1.764	0.90	0.90	C	35	E,T

[a] *Unattenuated rates using the Draine (1978) radiation field; L.L. indicates lower limit;* [b] *As a, but with the Mathis et al. (1983) radiation field;* [c] *Exponential fit parameters α and β cf. Eqn. (5) for grain model 2 and $A_V^{tot}=1$ mag;* [d] *Exponential fit parameters γ cf. Eqn. (4) for grain models 2 and 3, respectively;* [e] *Estimated accuary of rate, see text;* [f] *References for cross sections; E=experiment, T=theory;* [g] *Depth dependence dominated by self–shielding and H, H_2 shielding, see text;* [h] *Branching ratio to O+H_2 is about 10%;* [i] *Products highly uncertain;* [j] *Branching ratio to NH+H_2 is about 30%;* [k] *Depth dependence affected by H, H_2 shielding, effective γ may be much higher.*

TABLE 1. Interstellar molecular photodissociation rates (cont'd)

Species	Products	k^o_{pd} L.L.[a]	k^o_{pd} Draine[a]	k^o_{pd} MMP[b]	α_2[c]	β_2[c]	γ_2[d]	γ_3[d]	Acc[e]	Ref[f]	
SH	S+H	2.8(-10)	9.7(-10)	6.5(-10)	3.281	3.536	1.42	0.67	C	36	T
SH^{+i}	S^++H	...	2.5(-10)	1.9(-10)	4.158	4.416	1.8	0.96	C	36	T
H_2S	SH+H	...	3.1(-9)	2.0(-9)	3.716	3.935	1.87	0.69	A	37	E
CS	C+S	2.5(-10)	9.7(-10)	6.3(-10)	3.637	3.828	2.03	0.71	C	36	E
CS_2	CS+S	...	6.1(-9)	4.2(-9)	3.447	3.715	1.74	0.74	B	9,38	E
COS	CO+S	...	3.7(-9)	2.4(-9)	3.186	3.434	1.69	0.75	B	9	E
SO	S+O	...	3.7(-9)	2.4(-9)	3.753	3.949	1.95	0.74	B	39	E
SO_2	SO+O	...	1.9(-9)	1.2(-9)	3.768	3.961	1.88	0.73	B	9	E
SiH	Si+H	1.5(-9)	2.8(-9)	1.9(-9)	2.802	3.095	1.10	1.03	B	28,40	T
SiH^+	Si^++H	2.0(-9)	2.6(-9)	2.5(-9)	1.824	2.050	0.86	0.84	A	41	T
PH	P+H	...	5.8(-10)	3.8(-10)	3.204	3.440	1.73	0.72	C	36	T
PH^{+i}	P^++H	...	1.4(-10)	1.1(-10)	2.923	3.206	1.2	0.84	C	36	T
LiH	Li+H	...	5.1(-9)	3.5(-9)	2.351	2.597	1.21	1.21	A	42	T
NaH	Na+H	...	7.3(-9)	5.2(-9)	2.206	2.442	1.13	0.94	A	42	T
MgH	Mg+H	3.3(-11)	5.0(-10)	3.3(-10)	2.914	3.160	1.45	0.61	C	43	T
AℓH	Aℓ+H	2.0(-9)	3.2(-9)	3.7(-9)	1.193	1.342	0.77	0.74	C	44	T

References: (1) Allison and Dalgarno 1970; Stephens and Dalgarno 1972 ; (2) Dunn 1968 ; (3) Kulander and Bottcher 1978 ; (4) Roberge and Dalgarno 1982; (5) van Dishoeck 1987b; (6) Kirby et al. 1980; (7) Römelt et al. 1981; (8) Herzberg 1961; Macpherson et al. 1985; (9) Lee 1984; (10) Pouilly et al. 1983; (11) Shih et al. 1977, 1979; (12) Zelikoff and Watanabe 1953; (13) Römelt et al. 1978; (14) van Hemert et al. 1988; (15) van Dishoeck and Dalgarno 1983, 1984; (16) Saxon and Liu 1986; (17) Watanabe and Zelikoff 1953; Watanabe and Jursa 1964; (18) Huebner et al. 1975; Yoshino et al. 1983; Smith et al. 1984; Gies et al. 1981; (19) McAdam et al. 1987; Lee 1982; Langhoff and Jaffe 1979; (20) Molina and Molina 1986; Tanaka et al. 1953; Ogawa and Cook 1958a; (21) see §6; (22) Lewis and Carver 1983; NaKata et al. 1965; (23) Bruna et al. 1976; (24) Suto et al. 1986; Mentall et al. 1971; (25) Salahub and Sandorfy 1971; Nee et al. 1985; Ogawa and Cook 1958b; (26) Kirby and Goldfield 1988; (27) van Dishoeck 1986; (28) Saxon et al. 1982; (29) Carter 1972; (30) Zelikoff et al. 1953; (31) van Dishoeck 1986; Lavendy et al. 1987; (32) Connors et al. 1974; (33) Nuth and Glicker 1982; Suto and Lee 1985; (34) van Dishoeck et al. 1982; Nee et al. 1986; (35) Silver et al. 1986; Zeiri and Balint-Kurti 1983; (36) see §5.2; (37) Lee et al. 1987; (38) Cook and Ogawa 1969a; (39) Nee and Lee 1986; Phillips 1981; (40) Lewerenz et al. 1983; (41) Kirby and Singh 1983; (42) Kirby and Dalgarno 1978; (43) Kirby et al. 1979; (44) Matos et al. 1987.

these estimates are usually limited to the best characterized lowest-lying channels, which may not dominate the photodissociation.

The list of molecules in Table 1 is far from complete, but contains most small species of astrophysical interest for which some information is available in the vast molecular and chemical physics literature. A list of photodissociation rates for larger molecules has been compiled by Herbst and Leung (1986). As emphasized by them, data for large species are sparse and incomplete. It appears from Table 1 that the unshielded rates for most polyatomic molecules are large, about 10^{-9}–10^{-8} s^{-1}. However, at the depth into a cloud where the abundances of the larger molecules become significant, the cosmic ray induced photodissociation rates are probably more relevant.

Compared with the photodissociation rates listed by Black and Dalgarno (1977), most of the rates listed in Table 1 are significantly larger. This is partly due to the fact that the radiation field of Draine (1978) is a factor of two more intense than that adopted by Black and Dalgarno in the ultraviolet (cf. Figure 5). Further increases are due either to contributions from channels which were not recognized before, or to improved experimental data. The photodissociation rates in the unattenuated radiation field specified by Mathis *et al.* (1983) are included in the table for comparison, and are typically smaller by 40%.

The photodissociation rates for isotopic species are expected to be similar to those of the principal species if the photodissociation proceeds mostly by the direct mechanism. However, if indirect processes dominate, differences may occur if the predissociation of the isotopic variety starts at a different energy level with a different oscillator strength than that of the main species. In addition, depth–dependent effects may occur if the predissociating lines of the principal molecule do not shield those of the isotopic species. Examples are provided by HD and the isotopic varieties of CO (cf. §6).

5.2 *Recent developments*

Molecules for which information on the photodissociation cross sections has recently become available and/or which were not yet discussed by van Dishoeck (1987*a*) will be reviewed briefly below. The important case of CO will be discussed separately in §6.

Potential energy curves and transition dipole moment functions for the NH molecule have been computed by Goldfield and Kirby (1987). The photodissociation cross sections into the excited $^3\Sigma^-$ and $^3\Pi$ states give rise to an unshielded rate of about 5×10^{-10} s^{-1} (Kirby and Goldfield 1988), which is comparable to that of OH, and about a factor of two smaller than that of CH (van Dishoeck 1987*b*). The photodissociation rate of NH^+ is based on cross sections calculated by van Dishoeck (1986). Although many of the excited electronic potentials are repulsive, most of them have vertical excitation energies larger than 13.6 eV, so that the destruction by interstellar radiation is not very rapid.

A large photodissociation rate had been suggested for the CN molecule by Lavendy *et al.* (1984). As pointed out by van Dishoeck (1986), the limited size of these computations may have led to an overestimate of the rate. More recent computations by Lavendy *et al.* (1987) indicate a rate of about 7×10^{-10} s^{-1}, whereas the results of van Dishoeck (1986) suggest $k_{pd} \approx 3 \times 10^{-10}$ s^{-1}. The photodissociation occurs primarily at short wavelengths, $\lambda < 1050$ Å, in harmony with the fact that no CN photodissociation is observed experimentally for $\lambda > 1060$ Å (Nee and Lee 1985).

The electronic structure of molecules involving second row atoms is often similar to that of the corresponding first row species. Nevertheless, subtle differences can occur in the photodissociation processes. For example, the photodissociation of SiH^+ through the lowest $A^1\Pi$ channel is much more effective than that of CH^+, as discussed in §2.1 and illustrated in Figure 3. As a consequence, the photodissociation rate of interstellar SiH^+ is about an order of magnitude larger than that of CH^+. In general, the potential energy curves of molecules involving second row atoms are shifted to lower energies compared with the first row species. Thus, photodissociation channels that are inaccessible at $\lambda > 912$ Å in the latter case may contribute significantly to the photodissociation of the second row molecules. For example, the excited repulsive $^3\Pi$ and $^3\Sigma^-$ states are expected to contribute significantly to the photodissociation of interstellar SH^+, whereas they cause negligible photodissociation in the case of OH^+. The transition moments are probably of similar magnitude in the two cases. Since the intensity of the interstellar radiation field also increases with wavelength, the photodissociation of second row molecules will usually proceed more rapidly than that of the first row species.

The SH photodissociation rate was estimated on the basis of potential curves computed by Bruna and Hirsch (1987), together with transition moments for similar states of OH obtained by van Dishoeck and Dalgarno (1983). Note that photodissociation of interstellar SH can also occur by absorption into the $A^2\Sigma^+$ (v'=0) level, which is strongly predissociated throughout (Friedl *et al.* 1983). The interstellar SH^+ photodissociation rate was obtained by shifting the cross sections computed by Saxon and Liu (1986) for OH^+ by 200 Å to longer wavelengths.

The photodissociation of interstellar CS, SiO and SiS probably proceeds by processes similar to those for CO, with the difference that the dissociation energies and ionization thresholds are smaller by 3–4 eV. High resolution spectra of CS at VUV wavelengths (Stark *et al.* 1987*a*) show many diffuse bands. In particular, the strong B–X (0,0) and C–X (0,0) bands of CS are rapidly predissociated, in contrast with CO. If the oscillator strengths for these bands in CS are similar to those of CO, the photodissociation rate of interstellar CS must be significantly larger than that of CO. The SiO and SiS photodissociation rates are probably similar to that of CS. Photoabsorption cross sections for SO have recently been measured by Nee and Lee (1986) and Phillips (1981). Photodissociation of interstellar S_2 is expected to be at least as rapid as that of O_2.

Detailed potential energy curves of lower electronic states of PH and PH^+ have been computed by Bruna *et al.* (1981). The rate estimated for PH assumes that the cross sections to the repulsive $^3\Sigma^-$ and $^3\Pi$ states are similar to those of NH, but shifted to longer wavelengths. Photodissociation of PH^+ can occur already through the low-lying $1^2\Sigma^-$ and $1^2\Sigma^+$ states, which are bound in the case of NH^+. The PH^+ photodissociation rate is therefore expected to be at least as large as that of NH^+.

The lower limit to the photodissociation rate of interstellar AlH is based on potential curves and transition moments computed by Matos *et al.* (1987). Additional contributions from higher-lying states are expected to further increase the rate. The photofragmentation of various alkali–halides, including the KCl and NaCl molecules recently discovered in circumstellar envelopes (Cernicharo and Guélin 1987), has been investigated theoretically by Zeiri and Balint–Kurti (1983). Measurements of photodissociation cross sections for NaCl have been performed by Silver *et al.* (1986).

The photodissociation of HO_2 in the 1700–2700 Å range has been investigated experimentally by McAdam *et al.* (1987) and Lee (1982), and theoretically by Langhoff and Jaffe (1979). Little is known about possible photodissociation processes at higher energies. The photodissociation rate of interstellar HCO is based on potential curves and oscillator strengths computed by Bruna *et al.* (1976), and is still highly uncertain.

The photodissociation rate of c–C_3H_2 is based on preliminary results of calculations by van Hemert *et al.* (1988), and assumes that the strong absorptions around 1500 Å lead to dissociation of the molecule. The photodissociation of CH_3 can occur through several of the excited states found by Herzberg (1961), but no information is available on cross sections for these transitions, except at 2164 Å (MacPherson *et al.* 1985). The ionization potentials of both species lie around 10 eV, so that photoionization may be effective. Note that the uncertainties in the branching ratios for dissociation vs. ionization affect the estimated photodissociation rates of many small hydrocarbons such as CH_2, CH_3 and C_2H, which have ionization potentials around 10–11 eV. In Table 1, it has been assumed that at least 50% of the absorptions above the ionization threshold lead to ionization of the molecules.

5.3 Depth–dependent photodissociation rates

The intensity of the interstellar radiation field—and consequently the photodissociation rates—decreases with depth into an interstellar cloud due to several factors. First, the grains present in the cloud will scatter and absorb the photons with a cross section that varies almost inversely with wavelength. The amount of continuum attenuation at each depth into the cloud depends on the extinction, the albedo and the scattering phase function of the grains (Sandell and Mattila 1975; Roberge, Dalgarno and Flannery 1981), only the first of which is sometimes known accurately at ultraviolet wavelengths. Current models of interstellar grains (Draine and Lee 1984; Chlewicki and Greenberg 1984) favor a relatively low albedo $\omega_\lambda \approx 0.4$ and a rather forward scattering asymmetry parameter $g_\lambda \approx 0.6$–0.9 at $\lambda < 1500$ Å. Second, the intensity of the radiation field at $\lambda < 1100$ Å may be further diminished by the opacity due to H_2 and C in the clouds. The rates for molecules like CO and CN for which the photodissociation occurs primarily at $\lambda < 1100$ Å will be particularly affected. Finally, if the photodissociation of a molecule is dominated by discrete absorptions and if its abundance is sufficiently high, the dissociating lines will become saturated and the molecule will shield itself against dissociation deep in the cloud.

It is usually assumed that the contributions of line and continuum attenuation are separable. The depth dependence of the photodestruction rates due to grain attenuation can be computed by solving the equations of radiative transfer for specified grain properties and geometry of the cloud. If the cloud is assumed to have a plane–parallel geometry with radiation incident on both sides of the cloud, the depth–dependent rates depend strongly on the total extent of the cloud. The results for clouds with a total visual extinction A_V^{tot} are usually represented by single exponential decays to the centers

$$k_{pd} = C \exp(-\gamma A_V) \ \ \mathrm{s}^{-1}. \tag{4}$$

For thicker clouds, $A_V^{tot} > 5$ mag, this representation is accurate to within a factor of a few at all depths. For typical diffuse clouds with A_V^{tot}=1 mag, more accurate fits may be provided by bi–exponential decays

$$k_{pd} = C \exp(-\alpha A_V + \beta A_V^2) \ \ \mathrm{s}^{-1}. \tag{5}$$

Depth–dependent photodissociation rates have been computed for most species listed in Table 1 adopting grain models 2 and 3 of Roberge *et al.* (1981) for clouds with total visual extinctions ranging from 1 to 30 mag. Of the two grain models, model 2 gives results that are closest to those obtained with the grain properties recommended by Chlewicki and Greenberg (1984). The results for the exponents α and β for a typical diffuse cloud with A_V^{tot}=1 mag illuminated from both sides are presented in Table 1 for grain model 2, together with the exponents γ for a dense cloud with A_V^{tot}=20 mag. The latter exponents are also appropriate for a cloud exposed to radiation from one side only. The exponents α and β listed in Table 1 should not be used beyond a depth of 0.5 mag. The single exponents γ range from 1.5 to 3.5 with grain model 2. The exponent is small for species like NaH and SiH^+ for which the photodissociation occurs principally at longer wavelengths, whereas it is large for molecules like CN and N_2, which only photodissociate at the shortest wavelengths. The effective exponent for the latter species will be even larger if the attenuation due to H_2 and C is taken into account. For the more forward scattering grain model 3, γ is decreased to about 0.7–1.2. Results for γ with grain model 3 are included in Table 1. Note that the depth dependence of the rate of each photodissociation channel varies with wavelength: lower–lying photodissociation channels become relatively more important than higher–lying channels deep inside a cloud. It may thus be that the photodissociation of a molecule is dominated by a high–lying channel at the edge, whereas deeper inside a low–lying channel is most important. In those cases, the single exponential fit is particularly bad, and differences of 0.5 or more in γ are found if the total thickness of the cloud is varied from 2 to 20 mag. The factor C in Eqs. (4) and (5) is about half the unattenuated photodissociation rate if the cloud is illuminated on both sides by half the intensity specified by Eq. (3), so that in the absence of the cloud the unattenuated rates are recovered.

Expressions for the attenuation due to line shielding have been given by Hollenbach, Werner and Salpeter (1971), Federman, Glassgold and Kwan (1979) and van Dishoeck and Black (1986). In practice, self–shielding is taken into account only for the H_2, HD

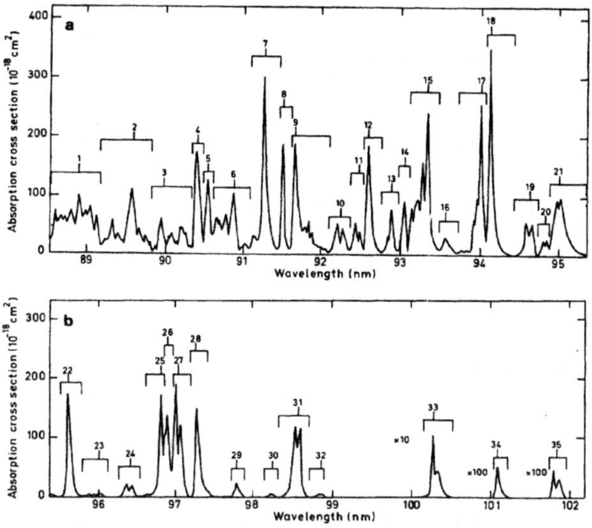

Figure 7. *CO absorption cross sections measured by Letzelter et al. (1987)*

and CO molecules. In applications in which the fine details of the depth dependences can be ignored, the shielding functions can be approximated by those of a representative single line with an appropriate scaling factor to account for the absolute value of the rate. Such simple expressions have been given by van Dishoeck (1987*a*) for H_2, and by Mamon, Glassgold and Huggins (1988) for CO. In addition, van Dishoeck and Black (1988) have presented a simple representation of the photodissociation rates of CO and its isotopes as functions of total CO and H_2 column densities.

6. PHOTODISSOCIATION OF INTERSTELLAR CO

The most important recent development with respect to photodissociation is that the photodissociation of CO is finally reasonably well understood. This has been accomplished mostly by the detailed laboratory measurements of photoabsorption cross sections and line positions by Letzelter *et al.* (1987) and Yoshino *et al.* (1988). The photodissociation processes of interstellar and circumstellar CO based on the new experimental data have been discussed in considerable detail by Letzelter *et al.* (1987), Viala *et al.* (1988), Mamon *et al.* (1988) and van Dishoeck and Black (1988), and only the most important aspects will be summarized here. Apart from being the second most abundant molecule, CO also illustrates well the difficulties in the determination of accurate cross sections at short wavelengths (cf. §3), as well as the many complications that arise in the calculation of depth-dependent rates (cf. §5).

The dissociation energy of CO is 11.09 eV, so that photodissociation can take place only at $\lambda < 1118$ Å. Although it has been clear for some time that the CO photodissociation is dominated by discrete absorptions, it was previously thought to occur mostly through the $C^1\Sigma^+-X^1\Sigma^+$ (1,0) and $E^1\Pi-X^1\Sigma^+$ (0,0) bands at 1063 and 1076 Å respectively (Bally

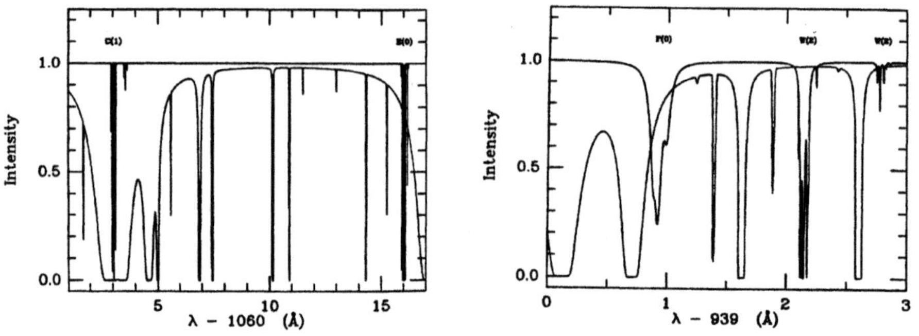

Figure 8. *Parts of the simulated high–resolution absorption spectrum of CO (full lines) and H and H_2 (dotted lines) at the center of the ς Oph cloud (from: van Dishoeck and Black 1988).*

and Langer 1982; Glassgold, Huggins and Langer 1985). Experiments by Fock, Gürtler and Koch (1980) indicated that many more excited states may contribute to photodissociation of the molecule at shorter wavelengths, but accurate determinations of the cross sections had to await the measurements by Letzelter *et al.* (1987). Part of the resulting absorption spectrum is reproduced in Figure 7, and cross sections for individual bands have been listed by Letzelter *et al.* They are significantly larger than those obtained in earlier experiments due to underestimates of the saturation effects in the older work. Note, however, that the resolution of 0.15 Å in the Letzelter *et al.* spectrum is not yet sufficient to resolve the rotational structure of the bands. Experiments at the 20 times higher resolution of 0.007 Å have recently been performed by Yoshino *et al.* (1988), and indicate that some of the cross sections given by Letzelter *et al.* may still be too low by a factor of about two. Since many of the absorptions involve high Rydberg states, no accurate theoretical calculations are possible. An additional uncertainty in the calculation of CO photodissociation rates is provided by the predissociation probabilities η_u for the upper states. The experiments of Letzelter *et al.* suggest that all η_u are close to unity, except for the C–X (1,0) and E–X (0,0) states, but most of the experiments refer to rather high rotational levels. The question remains whether the lower rotational levels are rapidly predissociated as well, as is assumed in all current models. With the new experimental data, the unshielded photodissociation rate of CO in the Draine (1978) radiation field is 2.0×10^{-10} s^{-1}. This rate is a factor of 40 larger than that adopted e.g. by Black and Dalgarno (1977) and Glassgold *et al.* (1985).

The depth–dependence of the absorption rate in each line is controlled by self–shielding, by shielding by coincident lines of H and H_2, and by dust attenuation. In order to account for the first two effects, van Dishoeck and Black (1988) have simulated the full absorption spectrum of CO at each depth into the cloud. Line positions were based on published analyses of the well–studied transitions and unpublished spectra by Stark *et al.* (1987*b*). Line widths were estimated from high–resolution spectra on the basis of the diffuseness of the bands. Since the predissociation rates are rapid, $A^{pr} \approx 10^{11} - 10^{12}$ s^{-1}, the lines are intrinsically very broad with line widths of 2–20 km s^{-1} in Doppler velocity units. Figures 8*a* and *b* contain two small portions of the absorption that would be produced by CO and by H and H_2 at the center of a diffuse cloud such as that toward ς Oph. It is clear that some bands like the C–X (1,0) band at 1063 Å are coincident with strong H or H_2 features, and are effectively blocked by them. Figure 8*b* also illustrates that the predissociation rates

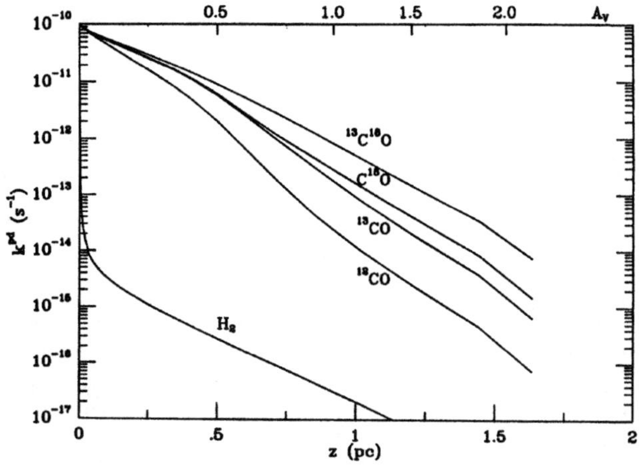

Figure 9. *Photodissociation rates of CO and its principal isotopes as functions of depth into a cloud with a total visual extinction of 5 mag (from: van Dishoeck and Black 1988).*

for some bands, such as that at 940 Å, are so large that the rotational structure becomes indistinct, even at a rotational excitation temperature of only 4 K.

The depth dependence of the CO photodissociation rate is shown in Figure 9 for a model cloud with a total visual extinction A_V^{tot}=5.2 mag, a total hydrogen density of 1000 cm^{-3} and a temperature of 15 K, exposed to the normal interstellar radiation field cf. Draine (1978). Grain model 2 of Roberge *et al.* (1981) was adopted. The CO photodissociation rate decreases rapidly with depth into the cloud due to self-shielding and shielding by H and H_2. The variation is, however, not as drastic as that of the H_2 photodissociation rate, as illustrated in the figure.

The calculation of the depth dependence of the photodissociation rate of the isotopic varieties ^{13}CO, $C^{18}O$ and $^{13}C^{18}O$ is complicated by the fact that lines in (0,0) bands can be effectively shielded by ^{12}CO inside the cloud, but not lines in (v',0) bands with $v' \geq 1$, owing to the larger isotope shift when a non-zero vibrational quantum number is involved. Unfortunately, the majority of the absorption bands at $\lambda < 1000$ Å shown in Figure 7 have not yet been identified spectroscopically, although reasonable guesses have been made by van Dishoeck and Black (1988). The depth-dependent rates of the isotopic species included in Figure 9 are therefore still uncertain, but appear to be larger than those of ^{12}CO inside a cloud by up to an order of magnitude. This isotope-selective photodissociation, originally recognized by Bally and Langer (1982), competes with ion-molecule exchange reactions in establishing the fractionation of the isotopic species, as discussed by Glassgold *et al.* (1985) and van Dishoeck and Black (1988).

From Figures 8*a* and *b*, it is clear that there are many coincidences between the CO photodissociation transitions and the broad lines of H_2 arising in the lowest rotational levels. In order to compute the cosmic ray induced CO photodissociation rate inside a dense cloud, coincidences between the CO absorption lines and the H_2 emission lines presented in Figure 6 have to be found. Note that these emission lines are much narrower than the broad absorption lines, so that the coincidences need to be nearly exact. Gredel

TABLE 2. Ionization potentials of selected species

Species	I.P. (eV)	Ref.	Species	I.P. (eV)	Ref.	Species	I.P. (eV)	Ref.
H_2	15.43	1	OH	13.0	1	SiH	8.0	1
CH	10.64	1	H_2O	12.62	2	SiO	11.43	1
CH_2	10.40	2	O_2	12.06	3	SH	10.43	1
CH_3	9.83	3	O_3	12.67	3	S_2	9.36	1
CH_4	12.55	3	HO_2	11.53	2	CS	11.34	1
C_2	12.15	1	NH	13.1	1	SO	10.29	1
C_2H	11.0	4	NH_2	11.4	2	H_2S	10.47	2
C_2H_2	11.41	2	NH_3	10.15	2	HCl	12.74	1
C_2H_4	10.50	2	N_2	15.58	1	PH	10.2	6
C_3	12.1	3	NO	9.25	1	PN	11.85	1
CO	14.01	1	NO_2	9.78	2	NaH	6.9	7
CO_2	13.79	5	N_2O	12.89	2	CH_3CN	12.19	3
HCO	9.88	2	CN	14.17	1	CH_3OH	10.83	3
H_2CO	10.88	2	HCN	13.59	3			

References: (1) Huber and Herzberg 1979; (2) Herzberg 1966; (3) Rosenstock et al. 1977; (4) Shih et al. 1977; (5) Cossart–Magos et al. 1987; (6) Bruna et al. 1981; (7) Langhoff 1985.

et al. (1987) have found about 60 close coincidences, resulting in a fairly large cosmic ray induced photodissociation rate of CO.

7. PHOTOIONIZATION PROCESSES AND CROSS SECTIONS

Absorption of a photon with an energy above the lowest ionization threshold can result both in photoionization and in photodissociation of a molecule. The photoionization can produce either a stable molecular ion in its ground or excited electronic state, or it can result in dissociation of the ion into a charged and a neutral fragment. Photoionization can also occur by absorption into excited Rydberg states lying above the first ionization potential, followed by autoionization:

$$\begin{aligned} XY + h\nu &\rightarrow XY^+ + e && \text{photoionization} \\ &\rightarrow X^+ + Y + e && \text{dissociative photoionization} \\ &\rightarrow XY^{**} \rightarrow XY^+ + e && \text{autoionization} \end{aligned} \tag{6}$$

The photoionization cross section is finite at threshold and generally consists of a smooth background with superposed resonances due to interactions with Rydberg states of the neutral molecule. Information about the cross sections can be obtained both from theory and experiment. On the theoretical side, the calculation is complicated by the description of the wave function of the electron in the field of the molecular ion XY^+. The various theoretical methods have recently been summarized e.g. by Langhoff (1985) and Lucchese and McKoy (1983). Extensive calculations have been performed mostly for common species such as H_2, N_2, CO and NO, for which direct comparison with experiments is possible. Although the computations give reasonable results at higher energies, they often fail to reproduce the detailed structure near threshold, the region of astrophysical interest.

Many different experimental techniques have been applied to obtain total and partial photoionization cross sections (see e.g. Hudson 1971; Sandorfy, Ausloos and Robin 1974;

TABLE 3. Interstellar atomic photoionization rates

Species	λ_{ion} (Å)	k^o_{pi} (s^{-1}) Draine[a]	k^o_{pi} (s^{-1}) MMP[b]	α_2[c]	β_2[c]	γ_2[d]	γ_3[d]	Ref.
H	911.75	...	...	...	...	...	...	
H^-	16420	1.1(-7)	1.4(-7)	0.302	0.329	0.17	0.17	1
He	504.3	...	...	...	...	...	...	
Li	2299	3.5(-10)	2.3(-10)	3.557	3.820	1.70	0.71	2
C	1101	3.0(-10)	2.2(-10)	5.426	5.336	2.96	0.99	3
N	853	...	...	...	...	...	...	
O	910.5	...	...	...	...	...	...	
Na	2413	1.5(-11)	1.0(-11)	3.769	3.996	1.78	0.72	4
Mg	1622	7.9(-11)	5.1(-11)	3.091	3.335	1.67	0.70	5
Al	2071	4.7(-9)	3.2(-9)	3.456	3.740	1.68	0.76	6
Si	1521	3.1(-9)	2.0(-9)	3.488	3.704	1.91	0.67	7
P	1182	1.0(-9)	7.0(-10)	4.786	4.832	2.65	0.88	8,9
S	1197	5.9(-10)	4.4(-10)	5.193	5.180	2.58	0.90	10
Cℓ	956	9.5(-11)	9.6(-11)	6.695	6.192	4.01	1.25	11
K	2856	2.9(-11)	1.9(-11)	3.429	3.684	1.65	0.75	12
Ca	2028	3.4(-10)	2.3(-10)	3.353	3.627	1.68	0.82	13
Ca^+	1044	2.3(-12)	1.8(-12)	5.895	5.665	3.33	1.09	14
Ti	1818	2.3(-10)	1.5(-10)	3.943	4.132	1.95	0.72	9
Cr	1832	1.5(-9)	9.8(-10)	3.282	3.535	1.70	0.76	9
Mn	1668	3.3(-11)	2.2(-11)	2.971	3.228	1.66	0.75	8,9
Fe	1575	2.8(-10)	1.8(-10)	3.204	3.427	1.86	0.67	15
Co	1577	5.4(-11)	3.5(-11)	3.039	3.278	1.73	0.73	9
Ni	1624	9.8(-11)	6.4(-11)	3.027	3.270	1.70	0.74	9
Zn	1320	4.4(-10)	3.0(-10)	5.116	5.081	2.43	0.94	16
Rb	2986	2.7(-11)	1.8(-11)	3.396	3.656	1.42	0.74	17

[a–d] *See footnotes to Table 1.*

References for cross sections: (1) Wishart 1979; (2) Caves and Dalgarno 1972; Serrao 1982; (3) Cantù et al. 1981; Burke and Taylor 1979; Hoffmann et al. 1983; (4) Butler and Mendoza 1983; (5) Bates and Altick 1973; Preses et al. 1984; Mendoza and Zeippen 1987; (6) LeDourneuf et al. 1975; Kohl and Parkinson 1973; Roig 1975; (7) Chapman and Henry 1972b; (8) McGuire 1968; (9) Reilman and Manson 1979; (10) Chapman and Henry 1972a; Dill et al. 1975; (11) Brown et al. 1980; Ruščić and Berkowitz 1983; (12) Weisheit 1972; Sandner et al. 1981; (13) Scott et al. 1983; McIlrath et al. 1972; Carter et al. 1971; (14) Black et al. 1972; (15) Kelly 1972; Lombardi et al. 1978; Tondello 1975; Hansen et al. 1977; (16) Marr and Austin 1969; (17) Weisheit 1972; Suemitsu and Samson 1983.

Samson 1976; Berkowitz 1979; Brion and Thomson 1984; Gallagher *et al.* 1988 and references cited), although still only few high resolution measurements exist around threshold. Information on the branching ratios to photoionization and photodissociation is still incomplete at low energies, and is based mostly on the older high resolution measurements. The experiments for species such as H_2, CO and O_2 suggest that the quantum yield for photoionization is more than 50% around threshold if no major change in configuration occurs during the ionization. If the transition is highly non–vertical, as is the case for CH_4 and NO_2, the photoionization yield can be very small around threshold. At higher energies –typically about 20 eV for molecules of astrophysical interest— the ionization efficiency approaches unity. The percentage of absorptions leading to dissociative photoionization is usually low at threshold, <5–10 %, but increases with increasing photon energy. The probability for formation of an ion–pair $X^+ + Y^-$ is also usually small at low energies,

TABLE 4. Interstellar molecular photoionization rates

Molecule	λ_{ion} (Å)	k^{o}_{pi} (s^{-1}) Draine[a]	k^{o}_{pi} (s^{-1}) MMP[b]	α_2[c]	β_2[c]	γ_2[d]	γ_3[d]	Ref.
CH	1164	7.6(-10)	5.5(-10)	5.311	5.247	2.84	0.97	1
CH_4	980	6.8(-12)	6.9(-12)	6.733	6.217	4.04	1.26	2,3
C_2	1020	4.1(-10)	3.4(-10)	6.171	5.853	3.54	1.14	4
C_2H_2 ...	1087	3.3(-10)	2.5(-10)	5.713	5.547	3.12	1.04	2,3
C_2H_4 ...	1181	4.1(-10)	3.1(-10)	5.447	5.370	2.77	0.95	2,3
C_2H_6 ...	1064	2.0(-10)	1.6(-10)	6.013	5.751	3.28	1.10	2,3
O_2	1027	7.7(-11)	6.8(-11)	6.366	5.978	3.73	1.18	5
H_2O	984	3.3(-11)	3.1(-11)	6.531	6.088	3.85	1.21	6,7
NH_3	1220	2.8(-10)	2.0(-10)	5.134	5.117	2.70	0.93	8,9
NO	1345	2.6(-10)	1.9(-10)	5.092	5.107	2.32	0.86	10
NO_2	1268	1.5(-10)	1.1(-10)	5.585	5.464	2.77	0.99	11
N_2O	962	1.9(-10)	1.9(-10)	6.688	6.188	4.00	1.25	12
H_2S	1185	7.1(-10)	5.1(-10)	5.023	5.027	2.69	0.91	7,13
CS_2	1232	1.7(-9)	1.2(-9)	5.239	5.198	2.62	0.93	14,15
COS	1109	7.0(-10)	5.4(-10)	5.592	5.471	2.96	1.00	16
HCl	973	8.7(-11)	8.2(-11)	6.535	6.087	3.89	1.22	17
H_2CO ..	1142	4.7(-10)	3.5(-10)	5.369	5.304	2.82	0.96	18
P1[e]	950	1.8(-11)	1.8(-11)	6.707	6.200	4.05	1.25	
P2	1000	5.9(-11)	5.1(-11)	6.339	5.962	3.70	1.18	
P3	1100	1.9(-10)	1.4(-10)	5.503	5.393	3.00	1.00	
P4	1200	3.7(-10)	2.6(-10)	4.893	4.930	2.58	0.88	
P5	1300	5.7(-10)	3.9(-10)	4.413	4.537	2.26	0.79	
P6	1500	9.8(-10)	6.6(-10)	3.791	3.995	1.95	0.69	

[a–d] *See footnotes to Table 1.* [e] *Rates obtained assuming a constant cross section of* 10^{-17} cm^2 *and an ionization threshold given by* λ_{ion}.

References to cross sections: (1) Barsuhn and Nesbet 1978; (2) Hudson 1971; (3) Metzger and Cook 1964; (4) Padial et al. 1985; (5) Matsunaga and Watanabe 1967; (6) Haddad and Samson 1986; (7) Watanabe and Jursa 1964; (8) Samson et al. 1987; (9) Watanabe and Sood 1965; (10) Watanabe et al. 1967; (11) Nakayama et al. 1959; (12) Cook et al. 1968; (13) Lee et al. 1987; (14) Cook and Ogawa 1969a; (15) Carnovale et al. 1981; (16) Cook and Ogawa 1969b; (17) Faegri and Kelly 1982; (18) Mentall et al. 1971.

although e.g. $NO^+ + O^-$ can be produced from NO_2 already at 1145 Å.

Reviews of atomic ionization cross sections have been given by Burke (1976), Roberge et al. (1981), Mendoza (1983), Keenan (1984) and Péquinot and Aldrovandi (1986).

8. INTERSTELLAR PHOTOIONIZATION RATES

Photoionization of atoms and molecules by the interstellar radiation field can occur only if their lowest ionization potential is less than 13.6 eV. The ionization potentials of a number of molecules of astrophysical interest are collected in Table 2. The table lists the lowest ionization threshold, wherever data are available. Note that this threshold may be substantially lower than the vertical ionization potential of the molecule. For example, for CH_4, the ionization threshold is about 12.5 eV, whereas the vertical ionization potential is 14.2 eV. Thus, the photoionization cross section of CH_4 is extremely small at threshold, and becomes substantial only at energies larger than 14.2 eV, resulting in a negligible interstellar photoionization rate. Most simple molecules have their ionization threshold below 13.6 eV. On the other hand, important molecules such as $H_2(v=0)$, CO, CN and N_2 cannot be photoionized in interstellar clouds. Photoionization of H_2 out of vibrationally excited levels $v \geq 4$ can take place at wavelengths longer than 912 Å (Ford *et al.* 1975), and may be important in some astrophysical environments (Black and van Dishoeck 1987). In planetary nebulae, where the radiation field extends beyond 13.6 eV, photoionization of H_2 $(v=0)$, CO and N_2 becomes possible (Black 1983).

Tables 3 and 4 list the resulting photoionization rates in the unattenuated interstellar radiation field of Draine (1978; cf. Eq. (3)) for a number of important atomic and molecular species. Although many of the atomic cross sections and rates are accurately known, the rates for most molecules are uncertain by at least a factor of two. An order of magnitude estimate of the photoionization rates of species not listed in Table 4 can be obtained by assuming a constant cross section of 10^{-17} cm^2 above the (vertical) ionization threshold. Some prototypical calculations for different thresholds are included at the bottom of Table 4. Comparison with rates for species for which the cross sections are known shows that the assumption of $\sigma_{\rm ion} = 10^{-17}$ cm^2 most likely results in an underestimate of the rate by a factor of a few. The variations of the photoionization rates with depth for the cases discussed in §5.3 are included as well. It is expected that virtually all of the molecular ionizations lead to the formation of a stable molecular ion.

9. CONCLUDING REMARKS

Substantial progress has been made in the last decade in our understanding of the photodissociation and ionization processes of small molecules. In particular, the experiments by Letzelter *et al.* (1987) and Yoshino *et al.* (1988) have finally removed most of the uncertainties in the photodissociation of one of the dominant interstellar molecules, CO. Nevertheless, there are still many simple molecules of astrophysical interest such as the radicals CH_2, C_2H, CH_3, HCO ... and the ions HCO^+, CH_2^+, CH_3^+, N_2H^+, ... for which only little is known about the photodissociation or ionization. Information on the photodissociation of larger molecules is still sparse and incomplete, not only with respect to cross sections, but also regarding the possible photodissociation products. Detailed theoretical calculations on smaller species and experiments on larger molecules are needed to remove these uncertainties. In addition, the stability of very large molecules with respect to photodissociation urgently needs to be investigated experimentally.

ACKNOWLEDGMENTS

It is a pleasure to thank J.H. Black for useful discussions and for a compilation of recent atomic photoionization cross sections. The author is grateful to W.G. Roberge for the use of his continuum radiative transfer code, to E. Herbst and R. Gredel for sharing their notes on the photodissociation of more complicated species, and to K. Kirby for communicating her results on the photodissociation of SiH^+ and NH prior to publication. This work was supported by NSF grant RII 86–20342 to Princeton University. The hospitality of the Institute for Advanced Study is appreciated.

REFERENCES

Allison, A.C. and Dalgarno, A. 1970, *Atomic Data*, **1**, 289.
Ashfold, M.N.R., Macpherson, M.T., and Simons, J.P. 1979, *Topics Curr. Chem.*, **86**, 1.
Bally, J. and Langer, W.D. 1982, *Ap. J.*, **255**, 143; erratum: **261**, 747.
Barsuhn, J. and Nesbet, R.K. 1978, *J. Chem. Phys.*, **68**, 2783.

Bates, D.R. and Spitzer, L. 1951, *Ap. J.*, **113**, 441.
Bates, G.N. and Altick, P.L. 1973, *J. Phys. B*, **6**, 653.
Berkowitz, J. 1979, *Photoabsorption, Photoionization and Photoelectron Spectroscopy* (Academic Press, New York).
Black, J.H. 1983, in *Planetary Nebulae*, ed. D.R. Flower (Reidel, Dordrecht) p. 91.
Black, J.H. and Dalgarno, A. 1977, *Ap. J. Suppl.*, **34**, 405.
Black, J.H. and van Dishoeck, E.F. 1987, *Ap. J.*, **322**, 412.
Black, J.H., Weisheit, J.C., and Laviana, E. 1972, *Ap. J.*, **177**, 567.
Brion, C.E. and Thompson, J.P. 1984, *J. Electron. Spectr. Rel. Phenom.*, **33**, 301.
Brown, E.R., Carter, S.L., and Kelly, H.P. 1980, *Phys. Rev. A*, **21**, 1237.
Bruna, P.J. and Hirsch, G. 1987, *Mol. Phys.*, **61**, 1359.
Bruna, P.J., Buenker, R.J., and Peyerimhoff, S.D. 1976, *J. Mol. Struct.*, **32**, 217.
Bruna, P.J., Hirsch, G., Peyerimhoff, S.D., and Buenker, R.J. 1981, *Mol. Phys.*, **42**, 875.
Burke, P.G. 1976, in *Atomic Processes and Applications*, eds. P.G. Burke and B.L. Moiseiwitsch (North Holland, Amsterdam), p. 200.
Burke, P.G. and Taylor, K.T. 1979, *J. Phys. B*, **9**, L353.
Butler, K. and Mendoza, C. 1983, *J. Phys. B*, **16**, L707.
Cantù, A.M., Mazzani, M., Pettini, M., and Tozzi, G.P. 1981, *Phys. Rev. A*, **23**, 1223.
Carnovale, F., White, M.G., and Brion, C.E. 1981, *J. Electron. Spectr. Rel. Phenom.*, **24**, 63.
Carter, V.L. 1972, *J. Chem. Phys.*, **56**, 4195.
Carter, V.L., Hudson, R.D., and Breig, E.L. 1971, *Phys. Rev. A*, **4**, 821.
Caves, T.C. and Dalgarno, A. 1972, *J. Quant. Spectrosc. Rad. Transf.*, **12**, 1539.
Cernicharo, J. and Guélin, M. 1987, *Astr. Ap.*, **183**, L10.
Chapman, R.D. and Henry, R.J.W. 1972*a*, *Ap. J.*, **168**, 169.
Chapman, R.D. and Henry, R.J.W. 1972*b*, *Ap. J.*, **173**, 243.
Chlewicki, G. and Greenberg, J.M. 1984, *M. N. R. A. S.*, **210**, 791; **211**, 719.
Connors, R.E., Roebbel, J.L. and Weiss, K. 1974, *J. Chem. Phys.*, **60**, 5011.
Cook, G.R., Metzger, P.H., and Ogawa, M. 1968, *J. Opt. Soc. Am.*, **58**, 129.
Cook, G.R. and Ogawa, M. 1969*a*, *J. Chem. Phys.*, **51**, 2419.
Cook, G.R. and Ogawa, M. 1969*b*, *J. Chem. Phys.*, **51**, 647.
Cossart-Magos, C., Jungen, M. and Launay, F. 1987, *Mol. Phys.*, **61**, 1077.
Dill, D., Starace, A.F., and Manson, S.T. 1975, *Phys. Rev. A*, **11**, 1596.
Draine, B.T. 1978, *Ap. J. Suppl.*, **36**, 595.
Draine, B.T. and Lee, H.M. 1984, *Ap. J.*, **285**, 89.
Duley, W.W. 1986, *Quart. J. R. Astr. Soc.*, **27**, 403.
Dunn, G.H. 1968, *Phys. Rev.*, **172**, 1.
Eddington, A.S. 1926, *Proc. Roy. Soc. A*, **111**, 424.
Faegri, K. and Kelly, H.P. 1982, *Chem. Phys. Lett.*, **85**, 472.
Federman, S.R., Glassgold, A.E. and Kwan, J. 1979, *Ap. J.*, **227**, 466.
Fock, J.-H., Gürtler, P. and Koch, E.E. 1980, *Chem. Phys.*, **47**, 87.
Ford, A.L., Docken, K.K., and Dalgarno, A. 1975, *Ap. J.*, **200**, 788.
Friedl, R.R., Brune, Wm.H., and Anderson, J.G. 1983, *J. Chem. Phys.*, **79**, 4227.
Gallagher, J.W., Brion, C.E., Samson, J.A.R., and Langhoff, P.W. 1988, *J. Phys. Chem. Ref. Data*, in press.
Gies, H.P.F., Gibson, S.T., McCoy, D.G., Blake, A.J., and Lewis, B.R. 1981, *J. Quant. Spectrosc. Rad. Transf.*, **26**, 469.
Glassgold, A.E., Huggins, P.J. and Langer, W.D. 1985, *Ap. J.*, **290**, 615.
Goldfield, E. and Kirby, K. 1987, *J. Chem. Phys.*, **87**, 3986.
Goldreich, P. and Scoville, N. 1976, *Ap. J.*, **205**, 144.
Gondhalekar, P.M., Phillips, A.P., and Wilson, R. 1980, *Astr. Ap.*, **85**, 272.
Gredel, R. 1987, Ph. D. Thesis, University of Heidelberg.
Gredel, R., Lepp, S., and Dalgarno, A. 1987, *Ap. J. (Letters)*, **323**, L137.
Guest, J.A. and Lee, L.C. 1981, *J. Phys. B*, **14**, 3401.
Habing, H.J. 1968, *Bull. Astr. Inst. Neth.*, **19**, 421.
Haddad, G.N. and Samson, J.A.R. 1986, *J. Chem. Phys.*, **84**, 6623.
Hansen, J.E., Ziegenbein, B., Lincke, R., and Kelly, H.P. 1977, *J. Phys. B*, **10**, 37.
Herbst, E. and Leung, C.M. 1986, *M. N. R. A. S.*, **222**, 689.
Herzberg, G. 1961, *Proc. Roy. Soc. A*, **262**, 291.
Herzberg, G. 1966, *Electronic Spectra of Polyatomic Molecules* (van Nostrand, Princeton).
Hofmann, H., Saha, H.P., and Trefftz, E. 1983, *Astr. Ap.*, **126**, 415.
Hollenbach, D.J. and McKee, C.F. 1979, *Ap. J. Suppl.*, **41**, 555
Hollenbach, D.J., Werner, M.W. and Salpeter, E.E. 1971, *Ap. J.*, **163**, 165.
Huber, K.P. and Herzberg, G. 1979, *Constants of Diatomic Molecules* (van Nostrand, Princeton).

Hudson, R.D. 1971, *Rev. of Geophys. and Space Physics*, **9**, 305.
Huebner, R.H., Celotta, R.J., Mielczarek, S.R., and Kuyatt, C.E. 1975, *J. Chem. Phys.*, **63**, 241.
Huebner, W.F. and Carpenter, C.W. 1979, *Los Alamos National Laboratory Informal Report* LA-8085-MS.
Jura, M. 1974, *Ap. J.*, **191**, 375.
Keenan, F.P. 1984, *M. N. R. A. S.*, **206**, 449.
Kelly, H.P. 1972, *Phys. Rev. A*, **6**, 1048.
Kirby, K. and Dalgarno, A. 1978, *Ap. J.*, **224**, 444.
Kirby, K. and Goldfield, E. 1988, in preparation.
Kirby, K., Roberge, W.G., Saxon, R.P. and Liu, B. 1980, *Ap. J.*, **239**, 855.
Kirby, K., Saxon, R.P. and Liu, B. 1979, *Ap. J.*, **231**, 637.
Kirby, K. and Singh, P.D. 1983, unpublished results.
Kirby, K. and van Dishoeck, E.F. 1988, *Adv. Atom. Mol. Phys.*, in press.
Kohl, J.L. and Parkinson, W.H. 1973, *Ap. J.*, **184**, 641.
Kulander, K. and Bottcher, C. 1978, *Chem. Phys.*, **29**, 141.
Langhoff, P.W. 1985, in *Molecular Astrophysics*, eds. G.H.F. Diercksen *et al.*, NATO ASI Series **157** (Reidel, Dordrecht), p. 551.
Langhoff, S.R. and Jaffe, R.L. 1979, *J. Chem. Phys.*, **71**, 1475.
Lavendy, H., Gandara, G. and Robbe, J.M. 1984, *J. Mol. Spectrosc.*, **106**, 395.
Lavendy, H., Robbe, J.M., and Gandara, G. 1987, *J. Phys. B*, **20**, 3067.
Leach, S. 1987, in *Polycyclic Aromatic Hydrocarbons in Astrophysics*, eds. A. Léger *et al.* Nato ASI Series **191** (Reidel, Dordrecht), p. 99.
LeDourneuf, M., Vo Ky Lan, Burke, P.G., and Taylor, K.T. 1975, *J. Phys. B*, **8**, 2640.
Lee, L.C. 1982, *J. Chem. Phys.*, **76**, 4909.
Lee, L.C. 1984, *Ap. J.*, **282**, 172.
Lee, L.C. and Chiang, C.C. 1983, *J. Chem. Phys.*, **78**, 688.
Lee, L.C., Wang, X., and Suto, M. 1987, *J. Chem. Phys.*, **86**, 4353.
Léger, A. and Puget, J.L. 1984, *Astr. Ap.*, **137**, L5.
Letzelter, C., Eidelsberg, M., Rostas, F., Breton, J., and Thieblemont, B. 1987, *Chem. Phys.*, **114**, 273.
Lewerenz, M., Bruna, P.J., Peyerimhoff, S.D., and Buenker, R.J. 1983, *Mol. Phys.*, **49**, 1.
Lewis, B.R. and Carver, J.H. 1983, *J. Quant. Spectrosc. Rad. Transf.*, **30**, 297.
Lombardi, G.G., Smith, P.L., and Parkinson, W.H. 1978, *Phys. Rev. A*, **18**, 2131.
Lucchese, R.R. and McKoy, B.V. 1983, in *Electron-Molecule Collisions and Photoionization Processes*, eds. V. McKoy *et al.* (Verlag Chemie), p. 13.
Macpherson, M.T., Pilling, M.J., and Smith, M.J.C. 1985, *J. Phys. Chem.*, **89**, 2268.
Mamon, G.A., Glassgold, A.E., and Huggins, P.J. 1988, *Ap. J.*, in press.
Marr, G.V. and Austin, J.M. 1969, *J. Phys. B*, **2**, 107.
Mathis, J.S., Mezger, P.S., and Panagia, N. 1983, *Astr. Ap.*, **128**, 212.
Matos, J.M.O., Malmqvist, P.Å., and Roos, B.O. 1987, *J. Chem. Phys.*, **86**, 5032.
Matsunaga, F.M. and Watanabe, K. 1967, *Sci. Light*, **16**, 31.
McAdam, K., Veyret, B., and Lesclaux, R. 1987, *Chem. Phys. Lett.*, **133**, 39.
McGuire, E.J. 1968, *Phys. Rev.*, **175**, 20.
McIlrath, T.J. and Sandeman, R.J. 1972, *J. Phys. B*, **5**, L217.
Mendoza, C. 1983, in *Planetary Nebulae*, ed. D.R. Flower (Reidel, Dordrecht), p. 143.
Mendoza, C. and Zeippen, C.J. 1987, *Astr. Ap.*, **179**, 346.
Mentall, J.E., Gentieu, E.P., Krauss, M., and Neumann, D. 1971, *J. Chem. Phys.*, **55**, 5471.
Metzger, P.H. and Cook, G.R. 1964, *J. Chem. Phys.*, **41**, 642.
Molina, L.T. and Molina, M.J. 1986, *J. Geophys. Res.*, **91**, 14501.
Nakata, R.S., Watanabe, K., and Matsunaga, F.M. 1965, *Sci. Light*, **14**, 54.
Nakayama, T., Kitamura, M.Y. and Watanabe, K. 1959, *J. Chem. Phys.*, **30**, 1180.
Nee, J.B. and Lee, L.C. 1985, *Ap. J.*, **291**, 202.
Nee, J.B. and Lee, L.C. 1986, *J. Chem. Phys.*, **84**, 5303.
Nee, J.B., Suto, M., and Lee, L.C. 1985, *Chem. Phys.*, **98**, 147.
Nee, J.B., Suto, M., and Lee, L.C. 1986, *J. Chem. Phys.*, **85**, 719.
Neufeld, D.A. and Dalgarno, A. 1988, in preparation.
Nuth, J.A. and Glicker, S. 1982, *J. Quant. Spectrosc. Rad. Transf.*, **28**, 223.
Ogawa, M. and Cook, G.R. 1958*a*, *J. Chem. Phys.*, **28**, 173.
Ogawa, M. and Cook, G.R. 1958*b*, *J. Chem. Phys.*, **28**, 747.
Okabe, H. 1978, *Photochemistry of small molecules* (Wiley, New York).
Omont, A. 1986, *Astr. Ap.*, **164**, 159.
Omont, A. 1987, in IAU Symposium **120**, *Astrochemistry*, eds. M.S. Vardya and S.P. Tarafdar (Reidel, Dordrecht), p. 357.
Padial, N.T., Collins, L.A., and Schneider, B.I. 1985, *Ap. J.*, **298**, 369.

Péquignot, D. and Aldrovandi, S.M.V. 1986, *Astr. Ap.*, **161**, 169.
Phillips, L.F. 1981, *J. Phys. Chem.*, **85**, 3994.
Pouilly, B., Robbe, J.M., Schamps, J. and Roueff, E. 1983, *J. Phys. B*, **16**, 437.
Prasad, S.S. and Tarafdar, S.P. 1983, *Ap. J.*, **267**, 603.
Preses, J.M., Bukhardt, C.E., Garver, W.P., and Leventhal, J.J. 1984, *Phys. Rev. A*, **29**, 985.
Reilman, R.F. and Manson, S.T. 1979, *Ap. J. Suppl.*, **40**, 815; errata: **46**, 115; **62**, 939.
Roberge, W.G. and Dalgarno, A. 1982, *Ap. J.*, **255**, 489.
Roberge, W.G., Dalgarno, A. and Flannery, B.P. 1981, *Ap. J.*, **243**, 817.
Robin, M.B. 1975, *Higher Excited States of Polyatomic Molecules* (Academic Press, New York).
Roig, R.R. 1975, *J. Phys. B*, **8**, 2939.
Römelt, J., Peyerimhoff, S.D., and Buenker, R.J. 1978, *Chem. Phys. Lett.*, **58**, 1.
Römelt, J., Peyerimhoff, S.D., and Buenker, R.J. 1981, *Chem. Phys.*, **54**, 147.
Rosenstock, H.M., Draxl, K., Steiner, B.W. and Herron, J.T. 1977, *J. Phys. Chem. Ref. Data*, **6**, sup. 1.
Ruščić, B. and Berkowitz, J. 1983, *Phys. Rev. Letters*, **50**, 675.
Salahub, D.R. and Sandorfy, C. 1971, *Chem. Phys. Lett.*, **8**, 71.
Samson, J.A.R. 1976, *Phys. Rep.*, **28**, 304.
Samson, J.A.R., Haddad, G.N., and Kilcoyne, L.D. 1987, *J. Chem. Phys.*, **87**, 6416.
Sandell, G. and Mattila, K. 1975, *Astr. Ap.*, **42**, 357.
Sandner, W., Gallagher, T.F., Safinya, K.A., and Gounand, F. 1981, *Phys. Rev. A*, **23**, 2732.
Sandorfy, C., Ausloos, P.J., and Robin, M.B., eds. 1974, *Chemical Spectroscopy and Photochemistry in the Vacuum-Ultraviolet* (Reidel, Dordrecht).
Saxon, R.P. and Liu, B. 1986, *J. Chem. Phys.*, **85**, 2099.
Saxon, R.P., Lengsfield, B.H. and Liu, B. 1982, *J. Chem. Phys.*, **78**, 312.
Scott, P., Kingston, A.E., and Hibbert, A. 1983, *J. Phys. B*, **16**, 3945.
Serrao, J.M.P. 1982, *J. Phys. B*, **15**, 2009.
Shih, S.-K., Peyerimhoff, S.D. and Buenker, R.J. 1977, *J. Mol. Spectrosc.*, **64**, 167.
Shih, S.-K., Peyerimhoff, S.D. and Buenker, R.J. 1979, *J. Mol. Spectrosc.*, **74**, 124.
Silver, J.A., Worsnop, D.R., Freedman, A., and Kolb, C.E. 1986, *J. Chem. Phys.*, **84**, 4378.
Smith, P.L., Griesinger, H.E., Black, J.H., Yoshino, K. and Freeman, D.E. 1984, *Ap. J.*, **277**, 569.
Stark, G., Yoshino, K., and Smith, P.L. 1987*a*, *J. Mol. Spectrosc.*, **124**, 420.
Stark, G., Smith, P.L., Yoshino, K., and Parkinson, W.H. 1987*b*, private communication.
Stephens, T.L. and Dalgarno, A. 1972, *J. Quant. Spectrosc. Rad. Transf.*, **12**, 569.
Sternberg, A., Dalgarno, A., and Lepp, S. 1987, *Ap. J.*, **320**, 676.
Suemitsu, H. and Samson, J.A.R. 1983, *Phys. Rev. A*, **28**, 2752.
Suto, M. and Lee, L.C. 1985, *J. Geophys. Res.*, **90**, 13037.
Suto, M., Wang, X., and Lee, L.C. 1986, *J. Chem. Phys.*, **85**, 4228.
Tanaka, Y., Inn, E.C.Y., and Watanabe, K. 1953, *J. Chem. Phys.*, **21**, 1651.
Tondello, G. 1975, *Mem. Soc. Astron. Ital.* **46**, 113.
van Dishoeck, E.F. 1986, unpublished results.
van Dishoeck, E.F. 1987*a*, in IAU Symposium **120**, *Astrochemistry*, eds. M.S. Vardya and S.P. Tarafdar (Reidel, Dordrecht), p. 51.
van Dishoeck, E.F. 1987*b*, *J. Chem. Phys.*, **86**, 196.
van Dishoeck, E.F. and Black, J.H. 1982, *Ap. J.*, **258**, 533.
van Dishoeck, E.F. and Black, J.H. 1986, *Ap. J. Suppl.*, **62**, 109.
van Dishoeck, E.F. and Black, J.H. 1988, *Ap. J.*, submitted.
van Dishoeck, E.F. and Dalgarno, A. 1983, *J. Chem. Phys.*, **79**, 873.
van Dishoeck, E.F. and Dalgarno, A. 1984, *Ap. J.*, **277**, 576.
van Dishoeck, E.F., van Hemert, M.C., and Dalgarno, A. 1982, *J. Chem. Phys.*, **77**, 3693.
van Hemert, M.C., Stehouwer, A., and van Dishoeck, E.F. 1988, in preparation.
Viala, Y.P., Letzelter, C., Eidelsberg, M., and Rostas, F. 1988, *Astr. Ap.*, in press.
Watanabe, K. and Jursa, A.S. 1964, *J. Chem. Phys.*, **41**, 1650.
Watanabe, K., Matsunaga, F.M., and Sakai, H. 1967, *Appl. Optics*, **6**, 391.
Watanabe, K. and Sood, S.P. 1965, *Sci. Light*, **14**, 36.
Watanabe, K. and Zelikoff, M. 1953, *J. Opt. Soc. Am.*, **43**, 753.
Weisheit, J.C. 1972, *Phys. Rev. A*, **5**, 1621.
Wishart, A.W. 1979, *M. N. R. A. S.*, **187**, 59P.
Wurm, K. 1934, *Zs. für Astrophys.*, **8**, 281; 1935, *ibid.* **9**, 62.
Yoshino, K., Freeman, D.E., Esmond, J.R. and Parkinson, W.H. 1983, *Planet. Space Sci.*, **31**, 339.
Yoshino, K., Stark, G., Smith, P.L., and Parkinson, W.H. 1988, *J. de Physique*, in press; and in prep.
Zeiri, Y. and Balint-Kurti, G.G. 1983, *J. Mol. Spectrosc.*, **99**, 1.
Zelikoff, M. and Watanabe, K. 1953, *J. Opt. Soc. Am.*, **43**, 756.
Zelikoff, M., Watanabe, K., and Inn, E.C.Y. 1953, *J. Chem. Phys.*, **21**, 1643.

RADIATIVE PUMPING AND COLLISIONAL EXCITATION OF MOLECULES IN DIFFUSE INTERSTELLAR CLOUDS.

E. Roueff*, H. Abgrall*, J. le Bourlot*, Y. Viala+
* DAMAp, + DEMIRM
Observatoire de Paris
F-92195 Meudon Principal Cedex
France

ABSTRACT. Recent work on radiative processes and collisional excitation in molecular Hydrogen and its deuterated isotopic substitute and in molecular Carbon is reviewed. Particular attention is drawn to non-adiabatic coupling effects on the intensities of Lyman and Werner band systems of the vacuum ultraviolet spectrum of H_2 and to the role of nuclear spin on ortho-para transitions in H_2 due to H^+ collisions. The inter-relation between those processes and state to state chemistry is stressed out. We discuss the implications of these new data in a recent comprehensive model of diffuse interstellar clouds (Viala et al., 1987).

1. INTRODUCTION

In diffuse interstellar clouds transparent to the UV galactic radiation field, most observational results come from narrow interstellar absorption transitions from which column densities can be obtained. Particularly interesting are the molecular absorption lines arising from different rotational levels belonging to the ground electronic and vibrational state. In this case, the individual rotational levels column densities should help to determine the physical conditions prevailing in these regions such as the temperature and the density. Relevant high spectral resolution observations are available for H_2 up to J = 7 and HD up to J = 1 which were obtained by the Copernicus satellite (Spitzer et al, 1974 ; Spitzer & Morton, 1976 ; Savage et al, 1977 ; Wright & Morton, 1979) and for molecular carbon for which satellite and ground based observations give column densities up to J = 14. (Hobbs, 1981 ; van Dishoeck & Black, 1986 ; Danks & Lambert, 1983 ; Hobbs & Campbell, 1982 ; Chaffee et al, 1980 ; van Dishoeck & de Zeeuw, 1984).

However, a single excitation temperature cannot explain the various observations so that a detailed analysis of the different excitation processes has to be performed for each considered molecule. Black & Dalgarno (1977) have analysed the case of the Hydrogen molecule and have shown that radiative pumping followed by UV fluorescence in the Lyman and Werner band systems and subsequent radiative cascades have to be considered as well as collisional excitation by H, H_2, H^+, e^- of the ground vibrational rotational states. Van Dishoeck & Black (1982) have analysed the various processes arising for the interstellar C_2 molecule and pointed out, that in addition to the mechanisms mentioned above for H_2, one has to include intercombination transitions between the nearby X $^1\Sigma_g^+$ ground state and a $^3\Pi_u$ excited state.

The equilibrium propulations n_{oJ} in the ground vibrational state for a molecule submitted to radiation pumping followed by fluorescence and subsequent radiative cascades inside the ground state <u>and</u> collisions are expressed in the following expression :

T. J. Millar and D. A. Williams (eds.), Rate Coefficients in Astrochemistry, 73–85.

$$n_{oJ}\left[\sum_{J''}\left(\sum_{P} K^{P}_{oJ,oJ''}\, n_P\right) + A_{oJ,oJ''} + \sum_{E} \zeta_{oJ,v'J'}\right]$$

$$= \sum_{J''} n_{oJ''}\left[\left(\sum_{P} K^{P}_{oJ'',oJ}\, n_P\right) + A_{oJ'',oJ} + \sum_{E} \zeta_{oJ'',v'J'}\, D_{v'J',oJ}\right] \qquad (1)$$

The sum over the ground rotational levels J" extends to a certain maximum value J''_{max} . $K^{P}_{oJ,oJ''}$ represents rotational collision rate constants due to perturbers P, A is the cascading emission probability within the ground vibrational state. ζ represents the radiative pumping probability and $D_{v'J',oJ}$ is a weighting factor which represents the probability that a vibration-rotational level (v'J') of the upper electronic state (E) decays to the level v = 0,J of the ground electronic state either directly or through radiative cascade inside this state.

This general formulation has been first introduced by Black & Dalgarno (1976) for H_2, where electric quadrupole cascades take place, enlarged to HD by Viala et al (1987) where faint dipole transitions can occur (Abgrall et al, 1982) and recently extended to C_2 by Le Bourlot et al (1987) where the effect of intercombination transitions has been introduced in the $D_{v'J',oJ}$ factor.

We note that this expression involves on the one hand pure molecular data such as the A, D factors which can be calculated once and on the other hand physical conditions dependent quantities such as $K^P\, n_P$ (temperature and density dependent) and ζ which involves the galactic UV radiation field.

In the present review, we discuss in section 2 the effects of rotational coupling on the intensities of Lyman and Werner band systems of the vacuum ultraviolet emission spectrum of H_2 towards discrete as well as continuous states of the ground $X\ ^1\Sigma_g^+$ level of H_2. In section 3 we present theoretical calculations by D. Gerlich on the reactive scattering of $H^+ + H_2$ (leading to ortho-para transitions in H_2) and its deuterated analog $D^+ + H_2 \rightarrow HD + H^+$. Section 4 is devoted to recent theoretical calculations on collisional excitation of C_2 due to parahydrogen. We finally discuss in section 5 the implications of these new data in a model of diffuse interstellar cloud and derive our conclusions in section 6.

2. NON ADIABATIC EFFECTS ON THE INTENSITIES OF LYMAN AND WERNER BAND SYSTEMS OF THE VACUUM ULTRAVIOLET SPECTRUM OF H_2.

2.1 Discrete emission spectrum.

The value of the rotational coupling between the $B\ ^1\Sigma_u^+$ and $C\ ^1\Pi_u$ states which are the two upper states arising in the Lyman and Werner band systems of H_2 has been calculated by Ford (1974). He subsequently calculated the effect of this coupling on the rotational line strengths of the first Lyman (v'_{max} = 25) and Werner (v'_{max} = 9) band systems for J" = 0 to 5, by using a matrix method described in Julienne (1973). The electro-rotational (non-adiabatic) coupling splits the doubly degenerate $C\ ^1\Pi_u$ state into two components which have a specific + or - character relative to the inversion of the electronic coordinates (Λ doubling). When introducing the total parity (Brown et al, 1975), one obtains e and f character total wave functions which behave respectively as $(-)^{J'}$ or $(-)^{J'+1}$ relative to inversion of electronic and nuclear coordinates. The f character total wave function involves only $C\ ^1\Pi_u^-$ whereas the e character total wave function has a contribution from the $B\ ^1\Sigma_u^+$ and $C\ ^1\Pi_u^+$ states.

One important consequence is that the $C\ ^1\Pi_u^-$ is not affected by this non-adiabatic effect. The f levels can then only be reached through Q transitions ($\Delta J = 0$) through electric dipole transitions since all levels of the ground $X\ ^1\Sigma_g^+$ state have e character. This is a very important point when one compares theoretical calculations on level positions and on transition probabilities with experiments since one can test the quality of the potential functions (diagonal term of the interaction including adiabatic relativistic corrections) on the unperturbed level positions.

However, the P and R transitions ($\Delta J = \mp 1$) which involve e type levels will be subjected to the coupling arising between $B\ ^1\Sigma_u^+$ and $C\ ^1\Pi_u^+$ when $J > 0$ [(1)]. We have reinvestigated recently (Abgrall et al,1987) this problem in connection with high spectral resolution emission spectra obtained by means of the National 10m VUV spectograph of the Meudon Observatory. A scanning photomultiplier or channeltron whose response is linear allows to measure absolute intensities of the different lines. Using most recent values of the involved electronic quantities (see table 1), we have reproduced previous calculations performed by Ford (1975). Furthermore, we have performed a direct integration to calculate the eigenvalues and the wave functions of the coupled system describing the nuclear radial functions related to $B\ ^1\Sigma_u^+$ and $C\ ^1\Pi_u^+$ state by using the renormalized Numerov algorithm as described by Johnson (1978). This has the advantage over the matrix method used by Ford (1975) and Julienne (1973) to include automatically the effect of the vibrational continuum and the results are then reliable up to the dissociation limit.

Table 1

References of the electronic quantities used in the theoretical calculations.

potentials	$X\ ^1\Sigma_g^+$	Bishop and Shih (1976)
	$B\ ^1\Sigma_u^+$ $C\ ^1\Pi_u$	} Dressler and Wolniewicz (1986)
non-adiabatic coupling		Ford (1974)
Electronic transition moments	B-X C-X	Dressler and Wolniewicz (1985)

The effect of the rotational coupling is readily seen on the ratio of the emission intensities of P and R lines connected to the same v'J' upper level, since in this case one eliminates the density of the upper level and one can compare directly theoretical results to the intensity measurements. Table 2 reproduces a part of Table VI of Abgrall et al (1987) in order to show the quantitative importance of this effect. The HL column refers to the ratios of the Hönl-London factors which appear when one neglects coupling effects and the role of

---

(1) The $C\ ^1\Pi_u^+$ state does not possess a $J' = 0$ level so that coupling effects take place only when $J' > 0$.

the centrifugal potential term $[J(J+1) - \Lambda^2]/R^2$ in the one dimension differential equation of the nuclear radial wave function. NC refers to the results obtained when one neglects the coupling and takes into account the centrifugal term. M and D columns refer to the values obtained through the matrix and direct resolution method, respectively. The last column gives the experimental ratios. Analysis of table 2 shows that the introduction of the coupling improves the comparison between experimental and theoretical values. For low vibrational states, the results obtained with the matrix or direct method are identical whereas some discrepancies appear for higher excited levels. For purpose of interstellar studies we have recalculated all transition probabilities up to $J'' = 9$ in the various Lyman and Werner absorption and emission transitions (Abgrall & Roueff, 1987). Previous systematic calculations of the Lyman and Werner band systems by Allison & Dalgarno (1970) were obtained for $J' = 0$, i.e. without the centrifugal term and without coupling.

Table 2

Ratios of emission probabilities of some P and R lines.

	v'	J'	v"	HL	NC	M	D	Expt
B	27	2	9	1.500	1.492	1.611	1.473	1.5
B	27	1	1	2.000	1.893	1.045	0.996	1.0
B	24	3	1	1.333	1.221	1.130	1.052	1.0
B	14	2	3	1.500	1.483	0.776	0.777	0.7
B	14	1	3	2.000	1.984	3.733	3.736	3.4
B	12	4	2	1.250	0.836	0.003	0.003	0.0
B	12	3	1	1.333	1.788	0.415	0.412	0.4
C	12	2	2	0.667	0.685	1.226	1.004	1.0
C	11	4	2	0.800	0.877	1.252	0.998	1.0
C	5	2	6	0.667	1.398	2.410	2.420	2.7

2.2 Continuous emission spectrum.

Absorption of molecular hydrogen followed by fluorescence in the continuum of the ground $X\ ^1\Sigma_g^+$ state is the photodissociation mechanism in diffuse interstellar clouds where the radiation field vanishes shortwards of the ionization limit of atomic Hydrogen (912 Å). This spontaneous radiative dissociation process was first suggested by Solomon (1966), then confirmed by Stecher & Williams (1967). Theoretical calculations of this mechanism have been performed by Stephens & Dalgarno (1972) for $J' = 0$ (no centrifugal barrier and no-coupling). A beautiful experimental confirmation of this structured continuous fluorescence was obtained on photographic plates by Dalgarno et al (1970). We have performed again such calculations by using the previously determined excited wave functions when non-adiabatic effects as well as centrifugal term are taken into account. The final wave function is now the continuum wave function inside the ground $X\ ^1\Sigma_g^+$ electronic state. This leads to a bound to free state emission probability which has the units T^{-1} energy^{-1}. On the other hand, Noll & Schmoranzer (1987 a) have studied the structured continuous vacuum ultraviolet emission from selectively excited rotation-vibrational levels of H_2 in the $B\ ^1\Sigma_u^+$ and $C\ ^1\Pi_u^+$ electronic states by using monochromatized synchrotron radiation. They obtained significant differences in the position of the peaks and also in their intensities when comparing the structured continuous fluorescence of B $v' = 13$, $J' = 0$ and 4 which they attributed to centrifugal distortion effects. It is also interesting to search the effect of rotational coupling using the C state. We find in our calculations that the B $v' = 13$, $J' = 4$ state has a 0.004 fraction of C character so that, indeed, centrifugal distortion is the main effect explaining the discrepancy. Figure 1 shows a comparison between experiment and our close-coupled calculations convoluted with instrumental resolution for B $v' = 10$, $J = 4$. The agreement is excellent but the percentage of C character is also weak in this case (4 %). Nevertheless, stronger mixing can occur. We have found cases with percentage close to 0.5 where the profiles are strongly modified. Such effects are presently investigated both theoretically and experimentally (Abgrall et al, 1987).

Figure 1

Comparison between experimental results (Noll and Schmoranzer, 1987 b), and close-coupled calculations relative to continuous fluorescence of the $v' = 10$, $J' = 4$ B $^1\Sigma_u^+$ state.

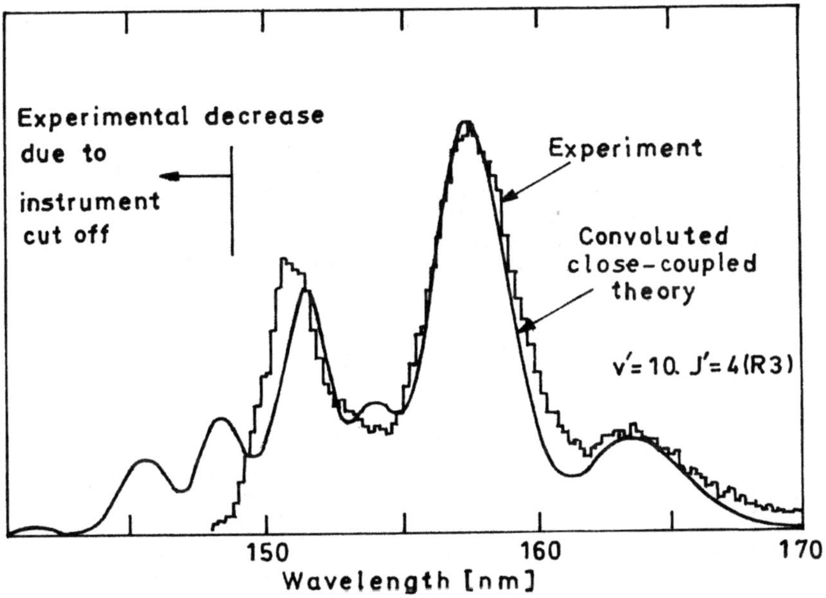

3. RADIATIVE SCATTERING IN $H^+ + H_2$ (J) AND ITS ISOTOPIC SUBSTITUTE $D^+ + H_2$ (J) COLLISIONS.

Radiative processes and rotational excitation collisions preserve the para or ortho character of the H_2 molecule due to the conservation of the total parity of the system (ΔJ even selection rule). However, at low collision energies, H_2 (J) collides with H^+ in a reactive process which proceeds through the strongly coupled intermediate complex H_3^+ which leads to a redistribution of the rotational states of H_2 without any selection rule on J and induces ortho-para transitions. This is the most simple reactive collision system for which the corresponding electronic potential surface is known (Giese & Gentry, 1974). A most dynamically biased (MDB) statistical theory has been developed (Schlier, 1980) which has been tested to be rather accurate in comparison with the available experiments dealing with isotopic substitutes of $H^+ + H_2$, i.e. $D^+ + H_2$, $H^+ + D_2$ (Gerlich et al, 1980 ; Teloy, 1978). Particular attention has been drawn by Gerlich (1977, 1982) to the influence of nuclear spin and symmetry in the $H^+ + H_2$ (J) reaction cross sections. The H_3^+ system involving three fermions should indeed have an A_2 character symmetry property in the S_3 substitution group (Quack, 1977). This implies a block diagonal form of the corresponding S matrix. An approximate conservation law of the total nuclear spin I (3/2 or 1/2 with statistical weight 4 or 2, respectively) can be inferred based on the very plausible assumption that the coupling of the nuclear spin to the other degrees of freedom is negligible. I = 3/2 can only be formed and decay from and to orthohydrogen while for the decay of the I = 1/2 complex the probability of forming para-hydrogen can be calculated simply by determining the statistical weight of the energetically accessible rotational states

$$P\left(J : \text{even} \leftarrow I = \frac{1}{2}\right) = \frac{\sum_{\text{even } J} (2J+1)}{\sum_{\text{all } J} (2J+1)} \tag{2}$$

This simple argument leads to the following consequence that at low energies below the J = 2 threshold the conversion rate of J = 1 to J = 0 is 1/12 of the Langevin rate (4/6 x 0 + 2/6 x 1/4). D. Gerlich (1987) has calculated the state to state reaction rate coefficients involving $\Delta J = 1$ which are given in Table 3 in function of temperatures. The value of the $1 \rightarrow 0$ rate coefficient is approximately equal to 2 10^{-10} cm^3 s^{-1} which compares relatively well with the value of 3 10^{-10} cm^3 s^{-1} assumed by Dalgarno et al (1973).

Table 3

$H^+ + H_2 (J) \rightarrow H^+ + H_2 (J')$

MDB calculations (Gerlich, 1987)

Reaction rate coefficients in 10^{-10} cm^3 s^{-1}								
T	1→0	2→1	3→2	4→3	5→4	6→5	7→6	8→7
10	2.41	10.7	3.16	8.69	2.56			
30	2.30	10.3	3.0	8.37	2.44	6.52	1.96	5.24
50	2.24	10.1	2.92	8.22	2.38	6.42	1.92	5.16
70	2.16	9.92	2.86	8.12	2.33	6.36	1.89	5.10
90	2.11	9.80	2.81	8.06	2.30	6.32	1.87	5.06
110	2.08	9.73	2.79	8.02	2.29	6.29	1.85	5.03
130	2.05	9.65	2.77	8.00	2.27	6.28	1.84	5.01
160	2.01	9.53	2.75	7.97	2.26	6.27	1.84	4.99
160	1.99	9.47	2.74	7.96	2.25	6.26	1.83	4.98
190	1.98	9.41	2.73	7.95	2.25	6.26	1.83	4.97
210	1.94	9.28	2.71	7.93	2.24	6.26	1.82	4.96
230	1.90	9.13	2.69	7.90	2.23	6.25	1.82	4.95
250	1.89	9.10	2.69	7.89	2.23	6.25	1.82	4.94
270	1.88	9.00	2.69	7.88	2.23	6.26	1.82	4.94

However, the 2 → 1 reaction rate coefficients differ markedly from the value deduced by the relation given by Black and Dalgarno (1977) :

$$k_{J+2 \to J+1} = \frac{g\ (J+1)}{g\ (J+2)}\ k_{J+1 \to J} \tag{3}$$

where g(J) is the total (rotational x nuclear) degeneracy of level J.

The $D^+ + H_2 \to H^+ + HD$ reaction proceeds through the same electronic potential surface with a 460 K exoergicity. MDB statistical calculations have been performed (Gerlich, 1987) which show deviations from the thermal equilibrium distribution for the populations of the rotational states of the HD product formed at low energies and also a significant difference in the rate coefficients whenever molecular hydrogen is in its J = 0 or J = 1 rotational state. Comparison with available experiments (Henchman et al, 1981) is satisfactory especially for the 295 K temperature (see table 4).

Table 4

Population of the rotational states of the HD product formed at low energies in the exothermic reaction $D^+ + H_2 \to H^+ + HD + 460$ K

a : Boltzmann equilibrium distribution (T = 230 K) ; b : MDB statistical calculations (Gerlich, 1987)

J		0	1	2	3	4	5
a		0.2532	0.4349	0.2385	0.0635	0.0091	0.0007
b	H_2 J=1	0.1135	0.3355	0.5508			
	H_2 J=0	0.1714	0.3404	0.4881			

Rate coefficients for the reaction $D^+ + H_2 \to H^+ + HD$ (units : 10^{-10} $cm^3\ s^{-1}$)

a : thermal rotational population of H_2 ; b : H_2 J=0 only ; c : H_2 J=1.

T	a J therm	b J = 0	c J = 1	exp (Henchmann et al, 1981)
10	25.1	25.1	15.9	
50	21.2	23.0	15.6	
90	18.4	22.2	15.8	
110	17.8	21.9	15.9	
130	17.5	21.7	16.0	
201				22.
230	17.0	20.8	16.4	
295				17.
330	16.8	20.1	16.7	

4. RADIATIVE AND COLLISIONAL PROCESSES IN C_2.

The C_2 molecule is difficult to study experimentally through its singlet band systems and has recently received much attention from different theoretical groups. Radiative pumping in diffuse interstellar clouds occurs through the Phillips X $^1\Sigma_g^+ \rightarrow$ A $^1\Pi_u$ (in the red and infrared) and Mulliken X $^1\Sigma_g^+ \rightarrow$ D $^1\Sigma_u^+$ (ultraviolet) band systems which were studied in extensive calculations by Chabalowsky et al (1983), van Dishoeck (1983), and in a less extent by Pouilly et al (1983) which derived also the photodissociation cross sections. Radiative cascades inside the ground state take place through very slow quadrupole transitions calculated by van Dishoeck & Black (1982). Le Bourlot & Roueff (1986) have shown that for excited vibrational levels of the X $^1\Sigma_g^+$ state (which are higher than the ground vibrational state of the metastable a $^3\Pi_u$ state), intercombination transitions were the main decaying radiative channels. They have calculated all individual intercombination transition probabilities, and have shown that a unique Re^2 value for the a $^3\Pi_u \rightarrow$ X $^1\Sigma_g^+$ is a very misleading approximation as it can be seen on figure 2. Collision rates with molecular hydrogen have recently been calculated by Chambaud et al (1987) in an Infinite Order Sudden (IOS) approximation. The molecular potential energy surface is obtained with C_2 and H_2 treated as rigid rotors held at their internuclear equilibrium separations and developed as a sum of terms product of a function of the intermolecular separation and angular factors.

Figure 2

Re^2 transition moment obtained by inversing the emission transition probability A in intercombination transitions between a $^3\Pi_u$, $\Omega = 1$, $J' = 1 \rightarrow$ X $^1\Sigma_g^+$, $J'' = 0$

$Re^2 = C^{te}\ A_{v'J',v''J''} / \sigma^3 q_{v'v''}$

σ = wave number ; $q_{v'v''}$ = Franck-Condon factors taken from van Dishoeck & Black (1982)

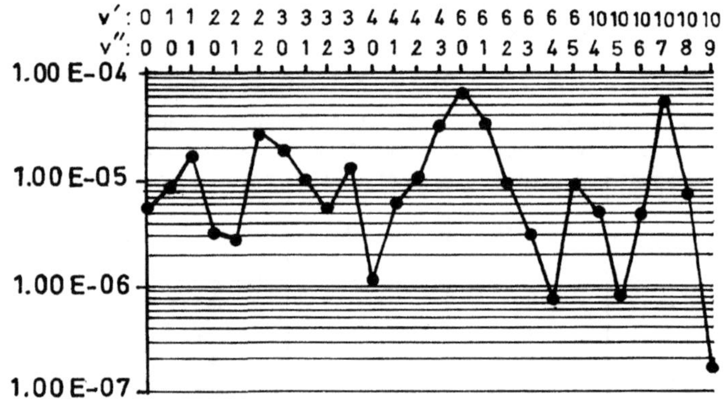

5. ASTROPHYSICAL IMPLICATIONS FOR DIFFUSE INTERSTELLAR CLOUDS.

We have developed recently a comprehensive model of the ζ Oph diffuse interstellar cloud (Viala et al, 1987) where the populations of the first ten rotational levels of H_2 and HD have been calculated by taking into account UV pumping to excited electronic states followed by fluorescence and radiative decay inside the ground state, collisional processes, and chemical reactions.

In addition, a chemical network of about a thousand reactions relating eighty five species has been set up in order to reproduce most column densities of the observed atomic and molecular species and to search for still unobserved molecules and led us to a two-shell structure for the ζ Oph cloud as Black & Dalgarno (1977), with different physical parameters.

Model 6 of table 6 in Viala et al gives the best fit to the available observations and we study now in table 5 the changes arising from the new data reported in § 3. Comparison between VRA and column **a** shows that the new data on $\Delta J = 1$ transitions in $H^+ + H_2(J)$ collisions by Gerlich (1987) modify considerably the $J = 0$ and $J = 1$ column densities. Higher rotational levels which are mainly populated through radiative pumping are almost not affected by the new data. In columns **b** and **c**, we study the effect of the $D^+ + H_2(J)$ reaction rate coefficients. The new data lead to a small decrease of the HD molecule column density in the non thermal and thermal distribution hypotheses of the HD product with no difference between both results. Study of H_2 and HD results reported in table 5 from the standard radiation field of Mathis et al (1983) with the recent accurate molecular data shows that the physical conditions derived previously have to be changed in order to reproduce satisfactorily the observations. In particular, too much H_2 ($J = 1$) and HD ($J = 1$) is obtained. These column densities can be reduced either by decreasing the temperature of the envelope or by extending the core at the expense of the envelope.

Table 5

Column densities in log (cm^{-2}) computed for a two shell structure of the ζ Oph cloud. Observations are reported in Viala et al (1988).

VRA reports results obtained with the physical conditions of model 6 table 6 of Viala et al (1988) using as standard radiation field the Mathis et al (1983) expression (column M83 of table 8) ; these calculations are performed with $k[H^+ + H_2(J{=}1) \rightarrow H^+ + H_2(J{=}0)] = 3\ 10^{-10}\ cm\ s^{-1}$ and relation (3),
and $k[D^+ + H_2(J) \rightarrow H^+ + HD(J)] = 3.3\ 10^{-9}\ (1 - 1.7\ 10^{-3}\ T)$ (linear parametrization of experimental results Henchman et al (1981)), and assuming a thermal distribution of the HD(J) products corresponding to T = 230 K.

a : same as VRA but with $k[H^+ + H_2(J) \rightarrow H^+ + H_2(J')]$ given in table 1

b : same as a with $k[D^+ + H_2(J)] \rightarrow H^+ + HD(J)$ obtained from table 2 with non thermal distribution of HD

c : same as b assuming thermal distribution of HD.

Model		Observations	VRA	a	b	c
H		20.72 ± 0.03	20.64	20.64	20.64	20.64
H_2		20.05 ± 0.08	20.69	20.68	20.68	20.68
H_2	J = 0	20.51 ± 0.08	20.52	20.52	20.51	20.51
	J = 1	20.10 ± 0.08	20.19	20.18	20.18	20.18
	J = 2	18.56 + 0.10, - 0.19	18.52	18.52	18.52	18.52
	J = 3	17.07 + 0.30, - 0.39	16.98	16.98	16.98	16.98
	J = 4	15.68 + 0.10, - 0.19	14.96	14.96	14.96	14.96
	J = 5	14.63 ± 0.05	14.22	14.22	14.22	14.22
	J = 6	13.69 ± 0.05	13.43	13.43	13.43	13.43
	J = 7	13.55 ± 0.05	13.13	13.14	13.14	13.14
HD		14.37 ± 0.03	14.95	14.95	14.89	14.89
HD	J = 0	14.26 ± 0.03	14.93	14.95	14.87	14.87
	J = 1	13.44 ± 0.04	13.55	13.55	13.48	13.48

One of the conclusions reached in Viala et al (1987) was that the constraints on the density in the core were quite floppy from the observations of the various rotational levels of H_2 and that the rotational excitation of C_2 should be very critical for the physical conditions in the core.

Le Bourlot et al (1987) have indeed reconsidered the rotational excitation of C_2 by solving the equilibrium populations equations (eq. 1) with inclusion of the intercombination transitions in the formalism with the recent molecular data reported in § 4. They found two different classes of models describing satisfactorily the observations of the first rotational levels of H_2 and the various rotational levels of C_2 (figure 3), depending on the value of the interstellar radiation field, a very critical parameter.

Different standard radiation fields are found in the literature. We chose to take the values given by Mathis et al (1983) which extend from the ultraviolet to the far infrared. The chemical equilibrium and the populations of the excited rotational states of H_2 and HD are sensitive to the vacuum ultraviolet part of the radiation field (912 Å $\leq \lambda \leq$ 2000 Å) which is badly determined, whereas we have verified that the rotational excitation of C_2 depends mainly on the red and infrared range corresponding to the radiative pumping in the Phillips band system. This red and infrared part of the radiation field is well constrained by photometric data and cannot be increased arbitrarily.

Present calculations proceed as follows : A multiplying factor f_{UV} is applied to the radiation field of Mathis et al (1983). It controls the chemical equilibrium of the different species, the atomic to molecular Hydrogen ratio, and the rotational excitation of molecular Hydrogen. The rotational excitation of C_2 is subsequently studied from the resolution of equations (1) with the same physical conditions as those obtained for chemical equilibrium except for the infrared part of the radiation field which is scaled by a multiplying factor f_{IR}. The corresponding results are shown in figure 3. Model A is relative to a two shell structure with a relatively small and not very dense cold core ($n = 600\ cm^{-3}$, $T = 30$ K), and a very diffuse envelope ($n = 80\ cm^{-3}$) at $T = 100$ K with a moderately large UV and IR radiation field ($f_{UV} = 5$, $f_{IR} = 1.5$).

Model B is very near to the one proposed by Viala et al (1987) involving a high value of the UV and IR radiation field ($f_{UV} = 7.5$, $f_{IR} = 3.75$), a relatively dense, cold and large core ($n = 1800\ cm^{-3}$, $T = 40$ K) and a diffuse envelope of $100\ cm^{-3}$ at $T = 120$ K. Such density contrasts may be discriminated by high spatial resolution observations of the CO emission towards the line of sight of the star, which are planned at the 30 m IRAM millimetertelescope. Figure 3 shows a relatively satisfactory agreement between the observations (Danks & Lambert, 1983) and both models except for the $J = 4$ relative column density. Error bars do not exist for the present observations and it would be highly desirable to confirm or invalidate the observational peak at $J = 4$.

6. CONCLUSIONS.

Limiting our discussion to the excitation of H_2, HD and C_2, we can stress the following points :

* As far as gas phase processes are concerned, we have at present for molecular Hydrogen a thorough knowledge of the various radiative and collision processes with precise experimental checks. We should then exploit the discrepancies between the observations and the models in order to reinvestigate the physical conditions prevailing in the various interstellar clouds. These should be tested as well as the remaining unknowns involving mainly surface processes, i.e. the influence of the sticking coefficient for the formation of H_2 on grains (Hollenbach et al, 1971) and its subsequent vibrational distribution. In this work, we have assumed an equipartition of the energy released together with a thermal distri

Figure 3

Relative column densities of the C_2 molecule in function of the rotational quantum number J.

Obs : observational results of Danks & Lambert (1983)

A : 1.5 x standard radiation field of Mathis et al (1983) from 6000 Å to far infrared

B : 3.75 x standard radiation field of Mathis et al (1983) from 6000 Å to far infrared

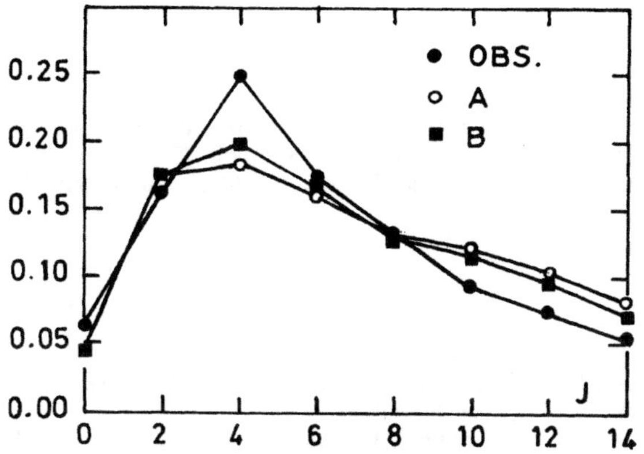

bution in the internal modes whereas Duley & Williams (1986) predict that H_2 forms vibrationally hot but rotationally cold, at least, on amorphous silicates. We have also confirmed the important role of the radiative transfer in H_2 for which shielding effects are critical for the determination of the atomic to molecular ratio as well as for the rotational excitation.

* For the deuterated substitute HD molecule, gas phase processes are predominent and well known in the same context as for H_2. Leaving aside the HD (J = 0) observations which have been questioned by Crutcher & Watson (1981), we are now in the position to modify the astrophysical parameters (especially the temperature and/or the size of the envelope) in order to find an agreement between the observations and the models including the recent molecular data.

* The study of the excitation of C_2 was essentially motivated by the loose constraints on the core from the H_2 excitation study and by the recent determination of the various molecular data relative to C_2 based mainly on ab initio calculations. Experiments involving the singlet band systems of this molecule are indeed very difficult (but welcome !). We have shown that two classes of models can reproduce within some conditions, the various excited rotational states of C_2 and H_2 depending mainly on the strength of the interstellar radiation field. A moderate field (as seen from earth) favours a lower density in the core. This leads to different density contrasts between the envelope and the core (in the frame of the idealized two-shell model) which may be discriminated by high spatial resolution observations.

We have learned a lot with the excitation of H_2 and C_2. From this study, we find rather complementary constraints on the physical parameters prevailing in the ζ Oph diffuse interstellar cloud, at the opposite of van Dishoeck & Black (1986), since we are rather confident in the relevant molecular data. However, we do not discuss here the abundances of the various species although some of them are very badly reproduced in our models. This should of course be studied at the same time and we are presently considering the implications in our model of recent data on reactive processes such as the negative result of the N + H_3^+ reaction (Herbst et al, 1987) and the photodissociation measurements on CO (Letzelter et al, 1987).

The rapid increase of knowledge on up to now poorly known processes will certainly improve our understanding of diffuse interstellar clouds although some recent values challenge our present view.

7. ACKNOWLEDGEMENTS.

All computations have been done at the 'Centre Inter Régional de Calcul Electronique' in Orsay (France). We are indebted to D. Gerlich, T. Noll and H. Schmoranzer for communicating results before publication. We thank J. Black and E. van Dishoeck for fruitful discussions.

8. REFERENCES.

Abgrall, H., Roueff, E., Viala, Y. : 1982, *Astron. Astrophys. Suppl. Ser.*, **50**, 505.
Abgrall, H., Launay, F., Roueff, E., Roncin, J.Y. : 1987, *J. Chem. Phys*, **87**, 20.
Abgrall, H., Roueff, E. : 1987, to be submitted.
Abgrall, H., Bieniek, R., Noll, T., Roueff, E., Schmoranzer, H. : 1987, work in progress.
Allison, A.C., Dalgarno, A. : 1970, *At. Data*, **1**, 289.
Bishop, D.M., Shih, S.K. : 1976, *J. Chem. Phys.*, **64**, 162.
Black, J.H., Dalgarno, A. : 1976, *Astrophys. J.*, **203**, 132.
Black, J.H., Dalgarno, A. : 1977, *Astrophys. J. Suppl. Ser.*,**34**, 405.
Brown, J.M., Hougen, J.T., Huber, K.P., Johns, J.W.C., Kopp, I., Lefebvre Brion, H., Merer, A.J., Ramsay, D.A., Rostas, J., Zare, R.N. : 1975, *J. Molec. Spectr.*, **55**, 500.
Chabalowsky, C.F., Peyerimhoff, S.D., Buenker, R.J. : 1983, *Chem. Phys*, **81**, 57.
Chaffee, F.H., Lutz, B.L., Black, J.H., van den Bout, P.A., Snell, R.L. : 1980, *Astrophys. J.*, **236**, 474.
Chambaud, G., Lavendy, H., Levy, B., Robbe, J.M., Roueff, E. : 1987, submitted to *Chem. Phys.*
Crutcher, R.M., Watson, W.D. : 1981, *Astrophys. J.*, **244**, 855.
Dalgarno, A., Herzberg, G., Stephens, : 1970, *Astrophys. J.*, **162**, L4.
Dalgarno, A., Black, J.H., Weisheit, J.C. : 1973, *Astrophys. Lett.*, **14**, 77.
Danks, A.C., Lambert, D.L. : 1983, *Astron. Astrophys.*, **124**, 188.
Dressler, K., Wolniewicz, L. : 1985, *J. Chem. Phys.*, **82**, 4720.
Dressler, K., Wolniewicz, L. : 1986, *J. Chem. Phys.*, **85**, 2821.
Duley, W.W., Williams, D.A. : 1986, *Mon. Not. R. Astr. Soc.*, **223**, 177.
Ford, A.L. : 1974, *J. Molec. Spectr.*, **53**, 364.
Ford, A.L. : 1975, *J. Molec. Spectr.*, **56**, 251.
Gerlich, D. : 1977, Dissertation, Freiburg.
Gerlich, D., Nowotny, U., Schlier, C., Teloy, E. : 1980, *Chem. Phys.*, **47**, 245.
Gerlich, D., Bohli, H.J. : 1981, *E.C.A.P.*, Kowalski, J., Zu Putlitz, G., & Weber, H.G., eds., 930.
Gerlich, D. : 1982, *S.A.S.P.*, **3**, Lindinger, W., Howorka, F., Mäsk, T.D., & Eggen, F., eds., 304.
Gerlich, D. : 1987, private communication.
Giese, G.F., Gentry, W.R. : 1974, *Phys. Rev. A* , **10**, 2156.

Henchman, M.J., Adams, N.G., Smith, D. : 1981, *J. Chem. Phys.*, **75**, 1201.
Herbst, E., de Frees, D.J., Mc Lean, A.D. : 1987, *Astrophys. J.*, **321**, 898.
Hobbs, L.M. : 1981, *Astrophys. J.*, **243**, 485.
Hobbs, L.M., Campbell, B. : 1982, *Astrophys. J.*, **271**, L95.
Hollenbach, D., Werner, M.W., Salpeter, E.E. : 1971, *Astrophys. J.*, **163**, 165.
Johnson, B.R. : 1978, *J. Chem. Phys.*, **69**, 4679.
Julienne, P.S. : 1973, *J. Molec. Spectr.*, **48**, 508.
Le Bourlot, J., Roueff, E. : 1986, *J. Molec. Spectr.*, **120**, 157.
Le Bourlot, J., Roueff, E., Viala, Y.P. : 1987, *Astron. Astrophys.*, in press.
Letzelter, C., Eidelsberg, M., Rostas, F., Breton, J., Thieblemont, B. : 1987, *J. Chem. Phys.*, **114**, 273.
Noll, T., Schmoranzer, H. : 1987 a, *Phys. Scripta*, **36**, 129.
Noll, T., Schmoranzer, H. : 1987 b, private communication.
Pouilly, B., Robbe, J.M., Schamps, J., Roueff, E. : 1983, *J. Atom. Mol. Phys*, **16**, 437.
Quack, M. : 1977, *Mol. Phys.*, **34**, 477.
Savage, B.D., Bohlin, R.C., Drake, J.F., Budich, W. : 1977, *Astrophys. J.*, **216**, 291.
Schlier, C.G. : 1980, in 'Energy storage and redistribution in Molecules', ed. by J. Hinze, Plenum Press, 585.
Solomon, P., as cited by Field G.B., Somerville, W.B., Dressler, K. : 1966, *Annual Rev. Astron. Astrophys.*, **4**, 207.
Spitzer, L., Cochran, W.D., Hirschfeld, A. : 1974, *Astrophys. J. Suppl.*, **28**, 373.
Spitzer, L., Morton, W.A. : 1976, *Astrophys. J.*, **204**, 731.
Stecher, T.P., Williams, D.A. : 1967, *Astrophys. J. (Letters)*, **149**, L29.
Stephens, T.L., Dalgarno, A. : 1972, *J. Quant. Spectr. Rad. Transf.*, **12**, 569.
Teloy, E. : 1978, X ICPEAC, G. Watel, ed., North Holland, Amsterdam, 591.
van Dishoeck, E.F. : 1983, *Chem. Phys.*, **77**, 277.
van Dishoeck, E.F., de Zeeuw, T. : 1984, *Mon. Not. R. Astr. Soc.*, **206**, 383.
van Dishoeck, E.F., Black, J.H. : 1986 a, *Astrophys. J. Suppl.*, **62**, 116.
van Dishoeck, E.F., Black, J.H. : 1986 b, *Astrophys. J.*, **307**, 332.
Viala, Y.P., Roueff, E., Abgrall, A. : 1987, *Astron. Astrophys.*, in press.
Wright, E.L., Morton, D.C. : 1979, *Astrophys. J.*, **227**, 483.

CHARGE TRANSFER IN ASTROPHYSICAL PLASMAS

Ronald McCarroll
Laboratoire de Physique et Optique Corpusculaires
Université Pierre et Marie Curie
4, place Jussieu, 75252 Paris Cedex 05

ABSTRACT. A review is presented of charge transfer reactions of multiply charged ions with atomic hydrogen and helium in astrophysical plasmas. The general features of the collision process are analyzed within the framework of the molecular model and a brief discussion is given of the theoretical methods used to calculate the cross-sections. A comparison of recent theoretical results with the available experimental data at laboratory energies would indicate that the theoretical model should be reliable for predicting cross-sections at astrophysical energies.

1. INTRODUCTION

In coronal type astrophysical plasmas, multiply charged ions can be created by strong UV or X radiation fields. Such is the case, for example, in gaseous and planetary nebulae under the action of the UV radiation from a bright star and in Seyfert galaxies under the action of X-rays produced by active galactic nuclei (Péquignot et al., 1978 ; Péquignot, 1980 a, b). The subsequent recombination of these ions leads to an emission spectrum which can serve as a diagnostic probe of the physical conditions - temperature, density and chemical abundances.

For the typical densities of astrophysical plasmas, where multiply charged ions are observed, 3-body reactions are of negligible importance and recombination can take place only by three distinct modes :

- Direct radiative recombination

$$A^{+q} + e \rightarrow A^{+q-1}(n) + h\nu \qquad 1.1$$

- Dielectronic recombination

$$A^{+q} + e \rightarrow (A^{+q-1})^{**} \rightarrow A^{+q-1}(n) + h\nu \qquad 1.2$$

where recombination takes place via the formation of an autoionizing state of A^{+q-1}. Stabilization occurs by the subsequent emission of a photon.

T. J. Millar and D. A. Williams (eds.), Rate Coefficients in Astrochemistry, 87–102.

- Charge transfer

$$A^{+q} + X \rightarrow A^{+q-1}(n) + X^{+} \qquad 1.3$$

where X designates a cosmic abundant species such as H or He.

Adopting typical values of the corresponding rate coefficients for reactions (1.1), (1.2) and (1.3) under favourable circumstances, it is easily seen that provided the neutral densities are of the order of 10^{-3} n_e or greater (n_e is the electron density) charge transfer can become the dominant mechanism for recombination. Such is the case in most astrophysical plasmas for temperatures not exceeding 10^5 K.

Direct confirmation of the influence of charge transfer has recently been provided by the observations (Clegg et al., 1986) of the NeIII and OIII spectra in several planetary nebulae. The evidence is particularly convincing for the quintet transitions at 260 nm in NeIII, for which charge transfer would appear to be the only possible excitation mechanism.

It should be remarked that since the charge transfer reaction (at least for $q \geqslant 3$) selectively populates a small number of excited states, the recombination spectra resulting from (1.3) differs considerably from that produced by radiative or dielectronic recombination. There is thus a great need for reliable cross-sections in low energy collisions (thermal to 10 eV).

Some mention should also be made of the inverse of reaction (1.3), leading to ionization. This can be particularly important when charge transfer recombination occurs directly via the ground state of A^{+q-1} (often the case for q = 2 and X = H, He or for q = 3 with X = He). Charge transfer ionization turns out to be critical for the fractional abundances of certain ions of Si and Fe (Baliunas and Butler, 1980 ; Gargaud et al., 1982 ; Neufeld and Dalgarno, 1987).

The progress accomplished in recent years in our understanding of the detailed mechanisms for charge transfer has been largely stimulated by the great advance in experimental techniques. Electron cyclotron resonance ion sources (ECRIS) now make it possible to produce ions with charge q up to 8 or more in the energy range $\sim$ 100 -5000 eV/amu. Low energy ions (down to about 5 eV/amu) can subsequently be produced in recoil ion sources. Laser induced ion sources are also capable of generating directly low eV ions. New detection techniques enable the measurement of state selective electron capture cross-sections either by the analysis of the emitted photons or by the translational energy spectra of the scattered (or recoil) ions. Of course each technique has its own specific limitations and it should be borne in mind that experimental measurements of charge transfer cross-sections are still not possible in the 0.1 - few eV energy range, required for most astrophysical plasmas. Recoil ion sources appear to offer the most promising approach for low energy experiments but so far measurements have been confined to a few selected ions. Thus, in spite of the considerable technological progress of the past decade, theoretical predictions of the charge transfer rate coefficients are still required for most astrophysical applications. On the other hand, the wealth of experimental data in the 50-5000 eV/amu energy range allows us to test the theoretical models in a way which was not hitherto possible.

2. PHYSICAL MODEL OF CHARGE TRANSFER

At the low energies of astrophysical interest (and indeed for energies up to about 1 keV/amu), when the collision velocity is small compared with the velocity of an electron in a classical Bohr orbit, the charge transfer process is most simply described in terms of the molecular model.

The entry channel ($A^{+q} + X$) is attractive at large internuclear distances, due to the polarization potential $-\alpha q^2/2R^4$ where α is the polarizability of X and R is the internuclear distance. As a consequence, there is no activation barrier and, if conditions are favourable, the cross-section may become very large at thermal energies.

For $q \geqslant 2$, the exit channels ($A^{+q-1}(n) + X^+$) are dominated by the repulsive Coulomb potential $(q-1)/R$. A curve crossing of the adiabatic curves of the $(AX)^{+q}$ molecular ion may be expected to occur at a value of R given by

$$\frac{(q-1)}{R} + \frac{q^2\alpha}{2R^4} = E_i - E_f \qquad 2.1$$

where E_i, E_f are the internal energies of the initial and final reaction products. (This formula is reasonably accurate for $R \gtrsim 5\,a_0$.) Of course, because of the Wigner non-crossing rule, most of the crossings become avoided crossings and, in general, a network of avoided crossings may be expected. A typical illustration is provided by the N^{5+}/He system (Fig. 1).

An electronic transition leading to charge transfer is only possible in the vicinity of a curve crossing, since it is only when the electronic energies are quasi-degenerate that a breakdown of the Born-Oppenheimer approximation is likely.

An approximate estimate of the collision cross-sections can be deduced from the minimum energy separation at the avoided crossings by the

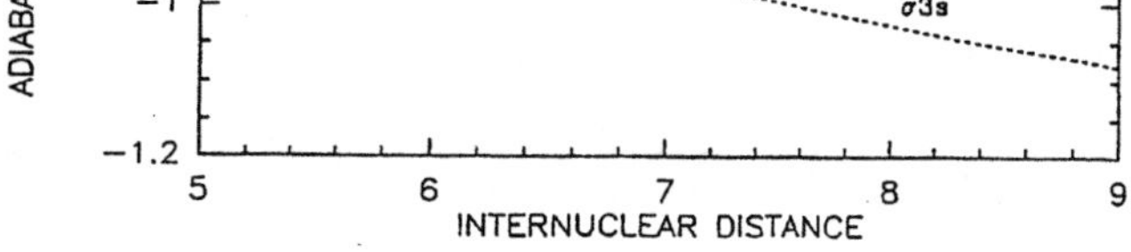

Figure 1. *Adiabatic energies of the Σ states of N^{5+}/He, illustrating the avoided crossings responsible for charge transfer.*

Landau-Zener formula (Child, 1974). Assuming the crossing to be characterized by R_X, the position of the crossing and Δ_X the minimum energy separation the cross-section Q is given by

$$Q = 2\pi \int_o^{b_m} 2P\,(1-P)\,bdb \qquad 2.2$$

where b_m is the maximum value of impact parameter b for which R_X is classically accessible and

$$P = \exp\left[-\frac{\pi\,\Delta_X^2\,R_X^2}{2(q-1)v}\right] \qquad 2.3$$

where v is the radial velocity at R_X and all quantities are expressed in atomic units. For energies upwards of a few eV it is valid to assume straight-line trajectories. For thermal energies, account must be taken of possible orbiting (Langevin) effects (Gargaud et al., 1981).

A transition will occur only if Δ_X is neither too small nor too large. If

$\Delta_X \gtrsim$ few eV $\qquad P \to 0 \implies Q \to 0$
adiabatic limit

$\Delta_X \lesssim 10^{-2}$ eV $\qquad P \to 1 \implies Q \Rightarrow 0$
diabatic limit

In practice, therefore, charge transfer is only likely if

$$10^{-2} < \Delta_X < 2-3 \text{ eV}$$

Such values of Δ_X imply that R_X must be in the range

$$12\ a_o \gtrsim R_X \gtrsim 5\ a_o$$

Although for $q \gtrsim 3$, there is usually at least one reaction channel for which some suitable R_X exists, the number of favourable channels is small, thereby leading to a highly selective population of excited states.

It is also convenient to treat separately the two basic types of charge transfer (Butler and Dalgarno, 1980).

- Type I, in which electron capture takes place with no change in the orbital configuration of the ionic core, for example

$$C^{3+}\ (2s)^2S + H \to C^{2+}\ (2s,3s)^3S + H^+ \qquad 2.4$$

- Type II, in which electron capture is accompanied by a rearrangement of the ionic core orbitals, for example

$$C^{3+}\ (2s)^2S + H \to C^{+}\ (2p^2)^1D + H^+ \qquad 2.5$$

At low eV energies, it is easily seen from (2.3) that the cross-sections depend critically on Δ_X. Accurate adiabatic potential energy curves are therefore required in the vicinity of the crossings. Unfortunately ab-initio molecular structure calculations are rarely satisfactory (except for very simple systems) since the entry and exit channels have different electronic orbital configurations. For, unless the absolute energy can be guaranteed to a precision of about 0.1 eV, the crossing radius R_X (which depends on the energy difference of the entry and exit channels) may be subject to a sizeable error, especially for large R_X. Since Δ_X depends exponentially on R_X, large errors in Δ_X can arise. For that reason, model or pseudo-potential methods which guarantee accurate asymptotic energies are often more satisfactory when long-range crossings are involved.

When the cross-sections are large (of the order of $10^{-15}\,cm^2$ or greater), the Landau-Zener model is usually adequate. But when the cross-sections are small ($\lesssim 5 \times 10^{-16}\,cm^2$), the Landau-Zener model can become unreliable especially as the diabatic limit is approached. Besides, when capture occurs into a state of non-zero angular momentum, the influence of rotational (Coriolis) coupling may become appreciable. To obtain reliable quantitative cross-sections, it is therefore necessary to treat the collision dynamics taking account of all of the non-adiabatic couplings over the complete range of internuclear distances. At energies of a few eV or less, where large trajectory effects arise, a quantal method is to be preferred ; at higher energies either a semi-classical (impact parameter) or quantal method is satisfactory.

However, irrespective of the method used to treat the collision dynamics, there remains a fundamental difficulty with the molecular model, arising from the incapacity of a finite adiabatic (or diabatic) basis set to represent correctly the asymptotic conditions. In practice, the effect (the so-called translation effect) is not serious at low eV energies when the transition is well localized at the curve crossing. But at higher energies, it may become appreciable if the transition takes place over a range of internuclear distances around the crossing. This is the case for many long distance crossings as the diabatic limit is approached. In the semi-classical method, these spurious translation effects canbe remedied by introducing an electron translation factor (Bates and McCarroll, 1958). A particularly convenient form of translation factor was proposed by Schneidermann and Russek (1969). A variant of this latter by Errea et al. (1984) has proved quite satisfactory at least for energies less than $\sim$ 1 keV/amu. A brief review of the translation factor problem has been given recently by McCarroll, 1987.

In the quantal method, the explicit introduction of electron translation factors is inconvenient and it has proved more appropriate to introduce some suitable system of reaction co-ordinates (Mittleman, 1969 ; Thorson and Delos, 1978 ; Delos, 1981 ; Soloviev and Vinitsky, 1985) which modifies the definition of the adiabatic states so that the asymptotic conditions of the collision are automatically accounted for. Since the method is not yet standard, a description of its essential features will be presented in section 3.

3. QUANTUM MECHANICAL FORMULATION WITH APPROPRIATE REACTION CO-ORDINATES

Let us recall the 3 Jacobi co-ordinate systems for the 3-body problem, which allow us to express the kinetic energy operator T in separable form.

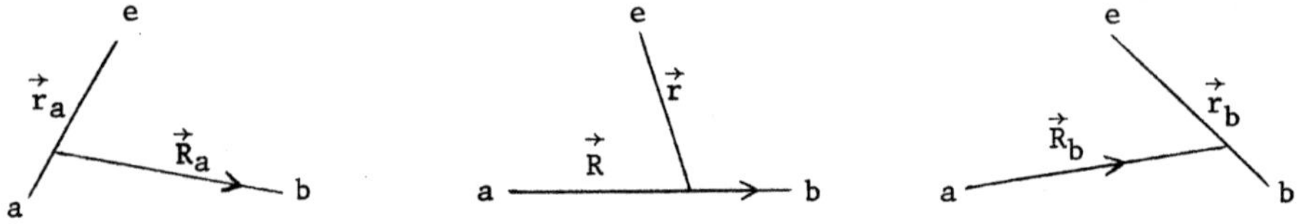

(a + e) + b	(a + b + e)	a + (b + e)
$\vec{r}_a$, $\vec{R}_a$	$\vec{r}$, $\vec{R}$	$\vec{r}_b$, $\vec{R}_b$
initial (atomic)	interaction (molecular)	final (atomic)

$$T = -\frac{1}{2\mu}\nabla_R^2 - \frac{1}{2m}\nabla_r^2 = -\frac{1}{2\mu_a}\nabla_{R_a}^2 - \frac{1}{2m_a}\nabla_{r_a}^2 = -\frac{1}{2\mu_b}\nabla_{R_b}^2 - \frac{1}{2m_b}\nabla_{r_b}^2 \qquad 3.1$$

$$\mu = M_a M_b/(M_a + M_b) \qquad \mu_a = M_b(M_a + 1)/(M_a + M_b + 1)$$

$$\mu_b = M_a(M_b + 1)/(M_a + M_b + 1)$$

$$m = (M_a + M_b)/(M_a + M_b + 1) \qquad m_a = M_a/(M_a + 1) \qquad m_b = M_b/(M_b + 1) \qquad 3.2$$

The wave-function of the initial state $(a + e)_i$ may then be expressed as

$$\exp(i\vec{k}_i \cdot \vec{R}_a)\, \phi_i(\vec{r}_a) \qquad 3.3$$

while that of the final state $a + (b + e)_f$ has the form

$$\exp(i\vec{k}_f \cdot \vec{R}_b)\, \phi_f(\vec{r}_b) \qquad 3.4$$

where $\vec{k}_i$, $\vec{k}_f$ are the initial and final wave vectors.

The co-ordinates $(\vec{r}_a, \vec{R}_a)$ are required to describe the initial state whereas $(\vec{r}_b, \vec{R}_b)$ are required for the final state, no one single Jacobi co-ordinate system being adequate to describe the entire collision process. For example, if we use (as is frequently done) the $(\vec{r}, \vec{R})$ co-ordinate system and expand the total wave-function Ψ in the form

$$\Psi(\vec{r}, \vec{R}) = \sum \vec{F}_n(\vec{R})\, \chi_n(\vec{r}\,; R) \qquad 3.5$$

we obtain the following set of coupled differential equations (in matrix notation)

$$-\frac{1}{2\mu}\left[\nabla_R^2 + 2i\,\underline{\vec{P}}\cdot\vec{\nabla}_R + i\,\vec{\nabla}_R\cdot\underline{\vec{P}} + \underline{P}^2\right]\underline{F}(\vec{R}) = E\underline{F}(\vec{R}) \qquad 3.6$$

where

$$\vec{P}_{mn} = <\chi_m | -i \vec{\nabla}_R | \chi_n > \qquad 3.7$$

Note that $\vec{\nabla}_R$ must be carried out with respect to $\vec{r}$ fixed. As in the semi-classical formulation, the matrix elements $\vec{P}_{mn}$ may be non-vanishing in the asymptotic limit. Similar problems occur had we used $(\vec{r}_a, \vec{R}_a)$ or $(\vec{r}_b, \vec{R}_b)$ for our co-ordinate system. To overcome the problem, Thorson and Delos (1978) introduce a co-ordinate system depending on the position of the electron. For example, introducing the variables $(\vec{u}, \vec{r})$, proposed by Gargaud et al. (1987)

$$\vec{u} = \vec{R} + \frac{1}{\mu}\vec{s} \qquad 3.8$$

$$\vec{s} = \frac{\vec{r}.\vec{R}}{R^2}\left(\vec{r} - \frac{1}{2}\frac{\vec{r}.\vec{R}}{R^2}\vec{R}\right) \qquad 3.9$$

it is easy to verify that $\vec{u}$ tends to $\vec{R}_a$ or $\vec{R}_b$ depending on whether $R/r_a \to \infty$ or $R/r_b \to \infty$. So the co-ordinates $(\vec{u}, \vec{r})$ can be used to describe both the initial and final asymptotic conditions. Of course the choice of $\vec{s}$ given by (3.9) is not unique. It is the simplest choice which eliminates artificial isotopic effects (Galilean invariance in semi-classical language) and guarantees the correct asymptotic form.

The full wave function is now expanded as

$$\psi(\vec{u}, \vec{r}) = \sum T_n(\vec{u})\, \chi_n(\vec{r}\ ;\ u) \qquad 3.10$$

where the adiabatic states are defined with respect to $\vec{u}$ constant. The justification is that the adiabatic functions will not change appreciably if we replace R by u whatever the value of $\vec{r}$ (consequence of the small mass ratio $1/\mu$). A straightforward (but lengthy) calculation yields

$$-\frac{1}{2\mu}\left[\nabla_u^2 + 2i(\underline{\vec{P}} + \underline{\vec{A}}).\vec{\nabla}_u + i\,[\vec{\nabla}_u \cdot (\underline{\vec{P}} + \underline{\vec{A}})] + (\underline{\vec{P}} + \underline{\vec{A}})^2\right]\underline{T}(\vec{u}) = E\underline{T}(\vec{u}) \qquad 3.11$$

where

$$\vec{A}_{mn} = i\ (\varepsilon_m - \varepsilon_n)\ <\chi_m | \vec{s}\, | \chi_n> \qquad 3.12$$

and $\vec{P}_{mn}$ is as defined by (3.7). It is easy to verify that the matrix elements of $(\underline{\vec{P}} + \underline{\vec{A}})$ vanish in the asymptotic limit.

A partial wave decomposition can be carried out in the usual way (bearing in mind that the adiabatic wave-functions are defined with respect to body-fixed co-ordinates). We obtain two types of coupling matrix elements :

- radial coupling elements of the form

$$<\chi_m \left| \frac{\partial}{\partial R} + \frac{z}{R}\frac{\partial}{\partial z} \right| \chi_n> \qquad 3.13$$

- rotational coupling elements of the form

$$<\chi_m \left| i\, L_y + (x\frac{\partial}{\partial z} + z\frac{\partial}{\partial x}) \right| \chi_n> \qquad 3.14$$

where the z-axis is in the direction of $\vec{R}$ and the y-axis is perpendicular to the classical collision plane.

These are identical to the matrix elements arising in the common translation factor semi-classical formula of Errea et al.

4. SOLUTION OF COUPLED EQUATIONS IN THE QUANTUM MECHANICAL FORMALISM AND INTRODUCTION OF DIABATIC STATES

To solve (3.11), we expand $T_n(\vec{u})$ on a basis set of symmetric top functions $D^K_{\Lambda_n M}$ where K is the total angular momentum, M its projection on the z-axis and Λ_n the projection of the angular momentum of molecular state n on the internuclear axis.

$$T_n(\vec{u}) = \sum_{KM} (-)^K \{(2K+1)/4\pi\}^{1/2} D^K_{\Lambda_n M} f_n^{(K)}(u) \frac{1}{u} \tag{4.1}$$

which yields

$$\frac{d^2}{du^2} \underline{f}^{(K)} + 2\underline{B} \frac{d}{du} \underline{f}^{(K)} + \underline{W}\,\underline{f}^{(K)} = 0 \tag{4.2}$$

where
$$B_{ij} = \langle \chi_i | \frac{\partial}{\partial R} | \chi_j \rangle \, \delta(\Lambda_i, \Lambda_j) \tag{4.3}$$

$$W_{ij} = V_{ij} + C_{ij} + \frac{2}{u^2} K L_{ij} \, \delta(\Lambda_i, \Lambda_j \pm 1) \tag{4.4}$$

$$V_{ij} = \left[2\mu(E - \varepsilon_i) - \frac{K(K+1)}{R^2} \right] \delta(i, j) \tag{4.5}$$

$$C_{ij} = \langle \chi_i | \frac{\partial^2}{\partial R^2} | \chi_j \rangle \qquad L_{ij} = \langle \chi_i | i\,L_y | \chi_j \rangle \tag{4.6}$$

To solve (4.2) it is convenient to remove the first order derivative by the transformation

$$\underline{g}^{(K)} = \underline{D}\,\underline{f}^{(K)} \tag{4.7}$$

where
$$\frac{d}{dR} \underline{D} + \underline{B}\,\underline{D} = 0 \qquad \underline{D}(\infty) = \underline{I} \tag{4.8}$$

giving

$$\frac{d^2}{dR^2} \underline{g}^{(K)} - 2\mu \underline{V}^d \underline{g}^{(K)} + \{ 2\mu E - \frac{K(K+1)}{R^2} \} \underline{g}^{(K)} = 0 \tag{4.9}$$

where
$$\underline{V}^d = \underline{D}^{-1} \left[\underline{\varepsilon} - \frac{1}{2\mu u^2} K \underline{L} \right] \underline{D} \tag{4.10}$$

is termed the diabatic matrix (as originally defined by Smith). Although the diagonal elements of $\underline{V}^d$ usually lie close to the adiabatic energies with the avoided crossings replaced by real crossings, a direct physical interpretation of $\underline{V}^d$ can be misleading if the non-diagonal terms become large. As a general rule, it is safest to consider the introduction of $\underline{V}^d$ as a useful mathematical transformation to solve the coupled equations.

5. APPLICATIONS AND EXPERIMENTAL TESTS

In view of the selectivity of charge transfer processes, each specific system tends to have a unique character. However, a qualitative classification can be made according to the kind of avoided crossings which contribute effectively to the total cross-section. In this review, three representative systems are chosen - CIV/H, AℓIV/H, ArVII/He - which provide an excellent testing ground for the theoretical model and which illustrate the precautions which must be followed to obtain accurate results.

In the applications which follow, the molecular structure parameters - adiabatic energies, non-adiabatic couplings - have been determined using model potential techniques. The collision dynamics have been treated either by the methods described in sections 3 and 4 or by their semi-classical equivalent.

5.1. System CIV/H

This system, which is typical of many astrophysical reactions, furnishes an interesting case, where charge transfer takes place primarily via a very long distance avoided crossing. It has been the subject of conside-

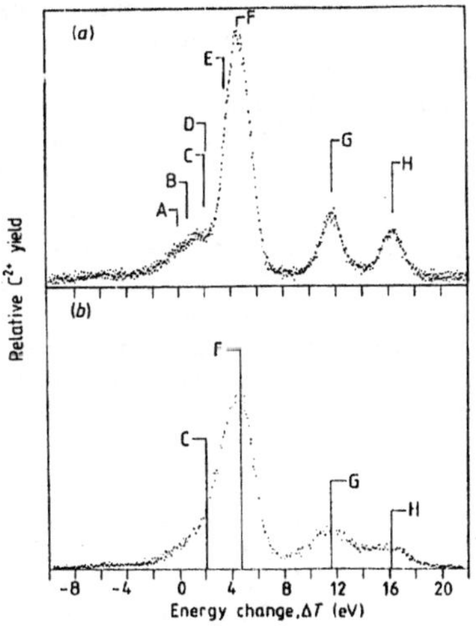

Figure 2. *Energy change spectra for electron capture by 15 keV C^{3+} in H using (a) a position sensitive detector and (b) a particle multiplier (Wilkie et al., 1986).*

rable experimental investigation - translation energy spectroscopy (Wilkie et al., 1986) and photon emission spectroscopy (Ćirić et al., 1985). A typical translation spectra is shown in Fig. 2.

It is observed that electron capture takes place principally into the $(2s3s)^3S$ (type I) and the $(2p^2)^1S, ^1D$ (type II) states of CIII, in conformity with the predictions of the molecular model.

To test the quantitative predictions of the theory, we shall confine our attention to the type I transition involving capture to the ^3S state via the avoided crossing at 11.4 a_o (Fig. 3).

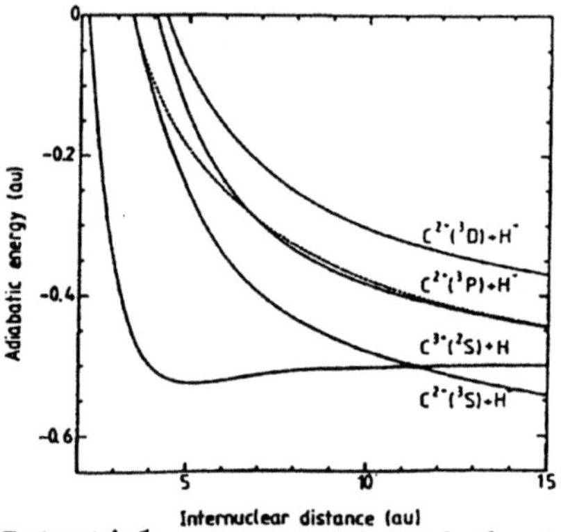

Figure 3. *Potential energy curves of the triplet states of CH^{3+}.*

The crossing tends to be diabatic for energies upward of 100 eV/amu. The Landau-Zener model is unsatisfactory under these conditions and detailed semi-classical or quantal calculations are necessary for a realistic comparison of theory with experiment.

Calculations with and without making allowance for translation effects show that these latter are quite considerable in the experimental energy range. The results of a 2-state semi-classical calculation (Opradolce et al., 1988) are tabulated below.

TABLE 1. Cross-sections (in units of $10^{-16}\,cm^2$) for capture into the ^3S state of CIII in CIV/H collisions. Columns (a) and (b) refer to calculations neglecting translation effects and with the origin of electron co-ordinates centred respectively on the H and C nuclei. Column (c) refers to calculations taking account of translation effects.

E (eV/amu)	(a)	(b)	(c)
30	4.47	4.97	4.70
75	3.26	4.12	3.65
125	2.85	4.10	3.39
250	2.70	4.97	3.60
500	2.88	6.55	4.23
1000	2.95	9.44	5.09
2000	3.24	13.15	6.26

More exhaustive calculations taking account of coupling with other secondary channels do not modify these results to any great extent. The cross-section for ^3S cpature is compared with the absolute measurements of Ćirić et al. in Fig. 4.

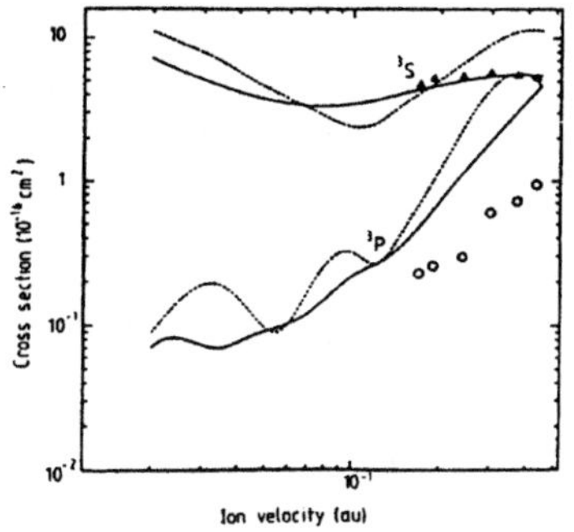

Figure 4. *Cross-sections (in units of 10^{-16} cm^2) for capture to the ^3S and ^3P states of CIII. The full curves refer to the calculations of Opradolce et al. (1988), the dotted curves to those of Bienstock et al. The experimental results of Ćirić et al. are designated by open triangles and circles.*

The excellent agreement with experiment suggests that both the molecular parameters and the translation effects are accurately described by the theoretical model. It may be noted that the calculations of Bienstock et al. (1982) neglected translation effects.

5.2. System AℓIV/H

This system provides a simple example of a 3-state process (Fig 5) where, in addition to the normal radial coupling, rotational coupling also is operative in the vicinity of an avoided crossing. Electron capture takes

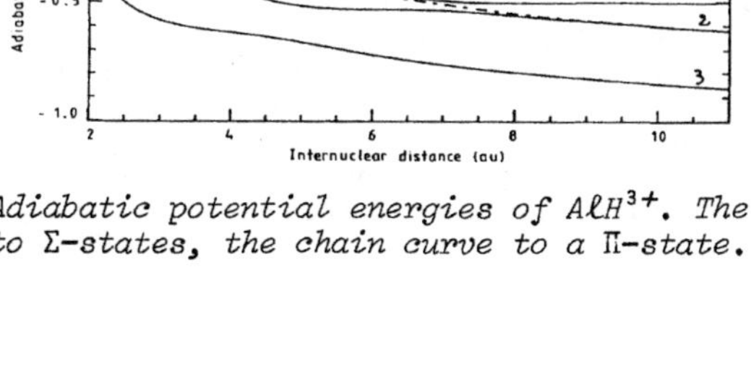

Figure 5. *Adiabatic potential energies of $A\ell H^{3+}$. The full curves refer to Σ-states, the chain curve to a Π-state.*

place mainly by capture into the $(2p^6 3p)^2P$ state of AℓIII via the avoided Σ-Σ crossing located at 7.2 a_0. But it may be observed that the Σ-Π energy gap narrows considerably around 6.4 a_0.

The Σ-Σ minimum energy separation being relatively large (∿ 1.1 eV), the collision process tends to be adiabatic at low eV energies. The transition is therefore well localized at the avoided crossing and translation effects are fairly small (at least up to energies of 1 keV/amu). On the other hand, the effect of rotational coupling leads to a considerable enhancement of the cross-section at low energies. Particularly striking is the sensitivity to the energy of the Π state. In Fig. 6 an illustration is given of the dependence of the cross-section on the diabatic Σ-Π energy splitting (see Gargaud et al., 1988, for details). The maximum in the cross-section accurs when the diabatic splitting is ∿ 0.8 eV, which corresponds to a tangential degeneracy of the adiabatic energies at 6.4 a_0. It is thus apparent that a small error in the Σ-Π energy splitting may lead to a large error in the cross-section.

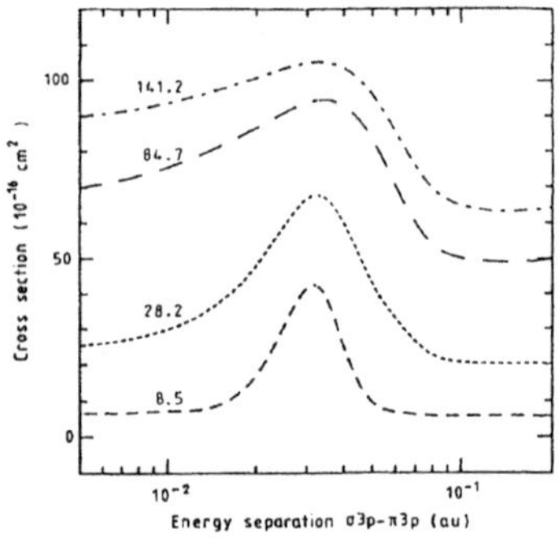

Figure 6. *Results of the 3-state calculation for electron capture into the 3p state of AℓIII, plotted as a function of the diabatic Σ-Π energy splitting. The cross-section is in units of 10^{-16} cm^2, the energy separation in atomic units. The numbers on the curves refer to the collision energies in units of eV/amu.*

The AℓIV/H system therefore enables us to make a critical assessment of the theoretical model and of the molecular structure calculations. Although, there is still insufficient experimental data to draw a form conclusion, the perfect agreement of the theory with the experiments of Phaneuf et al. (1985) at 37 eV/amu confirms the importance of rotational coupling.

Combining the findings of Gargaud et al., 1982, 1987, on such analogous systems as CV/H and SiIII/H with those on AℓIV/H, a general tendancy can be deduced. Rotational coupling tends to be important mainly in the adiabatic regime, whereas in the diabatic regime its influence is usually fairly small. For example in CV/H collisions at eV energies, rotational coupling considerably enhances capture into the 3p state of CIV but is negligible for capture into the 3d state.

5.3. System ArVII/He

Experimentally, it is much easier to study charge transfer with chemically stable targets than with reactive species such as H. It may therefore be expected that experimental data on charge transfer with He to be more reliable than with atomic H. The ArVII/He system is of particular interest since it has been studied experimentally both at keV energies by conventional ion sources (Panov, 1981) and at low energies by recoil ion techniques (Hvelplund and Pedersen, 1987 ; McCullough et al., 1987).

The system exhibits typical avoided crossings at large internuclear distances. Fortunately, the ArVII ion core being fairly diffuse, the sub-ℓ levels of ArVI are well separated, so that, at least for low energies, the charge transfer cross-section may be interpreted in terms of a series of non-overlapping avoided crossings. The main electron capture channels involve the 4d, 4p, 4s and (at higher energies) 3d states of ArVI. The avoided crossings leading to 4p and 4d capture are diabatic in character at low energies, so that we do not expect rotational coupling to play an appreciable role for the 4d, 4p and 4s channels. Moreover, in the energy range of interest, translation effects are small. Therefore, in spite of its apparent complexity, the ArVII/He system is a rare example where a direct test of the molecular structure calculations is possible, without having to allow for the complication of translational and rotational effects.

In the work of Benmeuraiem et al. (1987), ArVII/He is treated as an effective 2-electron system, with the ArVII ionic core represented by an appropriate model potential. A variant of the method of Grice and Herschbach (1974) is then used to obtain the minimum energy separation using molecular orbitals generated by a model potential, which guarantees the correct dissociation energies of all the reaction products (Benmeuraiem, 1987). The results are tabulated below.

TABLE 2. Curve crossing parameters for ArVII/He.

$n\ell$	R_x (a_o)	Δ_x (eV)
4d	13.4	0.005
4p	7.4	0.52
4s	5.8	1.88
3d	3.6	6.6

Following, the findings of Opradolce et al. (1983) that the Landau-Zener model should be reasonably accurate in the range 5-1000 eV/amu, it is scarcely worthwhile, at least for the present discussion to go beyond the simple Landau-Zener model.

The results, illustrated in Fig. 7, agree remarkably well with the experimental data, both that of Panov at energies above 100 eV/amu and that of Hvelplund et al. at energies in the eV range. Such a satisfactory agreement between theory and experiment indicates that our computed values of Δ_x must be accurate to better than 20 %. These results strongly suggest that model potential techniques can be successfully adapted to

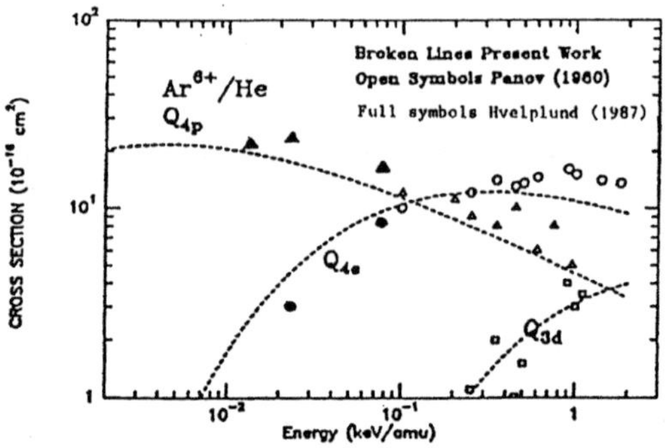

Figure 7. *Electron capture cross-sections (units of 10^{-16} cm^2) in ArVII/He collisions as a function of energy (keV/amu).*

compute accurately the molecular structure parameters at those internuclear distances required for the charge transfer process.

6. CONCLUSION

The examples chosen in section 5 illustrate the diversity of situations which arise in calculating the molecular model to calculate charge transfer rate coefficients at low energies. Extrapolation of experimental data down to energies of astrophysical interest can be misleading, unless adequate care is taken to ensure that translation and rotation effects are fully accounted for. On the other hand, neither translational nor rotational effects are likely to be of critical importance for a direct calculation at low eV energies if the cross-section for the dominant channel is large ($\gtrsim 5 \times 10^{-16}$ cm^2). In that case, the cross-sections depend mainly on the minimum energy separation at the avoided crossings. Comparison of theory with experiment suggests that model potential techniques can be used to compute accurate molecular parameters for the cross-section calculation even for relatively complex systems.

Although, more detailed experimental data would be welcome at low energies for a few key systems, the indications are strong that the theoretical model is capable of estimating reliable charge transfer rate coefficients at astrophysical energies, provided due account is taken of the specific nature of each system.

Acknowledgements. I wish to thank Larbi Benmeuraiem, Muriel Gargaud, Liliana Opradolce and Pierre Valiron for their contribution to some of the results cited in section 4. I am also indebted to Bob McCullough, Preben Hvelplund and Anders Bárány for communicating unpublished data.

REFERENCES

Baliunas, S.L. and Butler, S.E., 1980, *Astrophys. J. Lett.* **235**, L45
Bates, D.R. and McCarroll, R., 1958, *Proc. Roy. Soc. A* **245**, 175
Benmeuraiem, L., McCarroll, R. and Opradolce, L., 1987, XV ICPEAC, Brighton, Book of Abstracts, 557
Benmeuraiem, L., 1987, Thèse de doctorat, Université de Bordeaux
Bienstock, S., Heil, T.G., Bottcher, C. and Dalgarno, A., 1982, *Phys. Rev. A* **25**, 2850
Butler, S.E. and Dalgarno, A., 1980, *Astrophys. J.* **241**, 838
Child, M.S., 1974, *Molecular Collision Theory*, Academic Press, London
Ćirić, D., Brazuk, A., Dijkkamp, D., De Heer, F.J. and Winter, H., 1985, *J. Phys. B : Atom. Mol. Phys.* **18**, 3629
Clegg, R.E.S., Harrington, J.P. and Storey, P.J., 1986, *Monthly Notices Roy. Astron. Soc.* **221**, 61 P
Delos, J.B., 1981, *Rev. Mod. Phys.* **53**, 287
Errea, L.F., Méndez, L. and Riera, A., 1984, *J. Phys. B : Atom. Mol. Phys.* **17**, 3271
Gargaud, M., Hanssen, J., McCarroll, R. and Valiron, P., 1981, *J. Phys. B : Atom. Mol. Phys.* **14**, 2259
Gargaud, M., McCarroll, R. and Valiron, P., 1982, *Astron. Astrophys.* **106**, 197
Gargaud, M., McCarroll, R. and Valiron, P., 1987, *J. Phys. B : Atom. Mol. Phys.* **20**, 1555
Gargaud, M., McCarroll, R. and Opradolce, L., 1988, *J. Phys. B : Atom. Mol. Phys.*, to appear
Grice, R. and Herschbach, D.R., 1974, *Mol. Phys.* **27**, 159
Hvelplund, P. and Pedersen, J.O.K., 1987, XV ICPEAC, Brighton, Book of Abstracts, 545
Hvelplund, P., 1987, Private communication
McCarroll, R., 1987, *Recent Studies in Atomic and Molecular Processes*, ed. by Kingston, A.E., Plenum, New York
McCullough, R.W., Wilson, S.M. and Gilbody, H.B., 1987, *J. Phys. B : Atom. Mol. Phys.* **20**, 2031
Mittleman, M.H., 1969, *Phys. Rev.* **188**, 221
Neufeld, D.A. and Dalgarno, A., 1987, *Phys. Rev. A* **35**, 3142
Opradolce, L., Valiron, P. and McCarroll, R., 1983, *J. Phys. B : Atom. Mol. Phys.* **16**, 2017
Opradolce, L., Benmeuraiem, L., McCarroll, R. and Piacentini, R.D., 1988, *J. Phys. B : Atom. Mol. Phys.*, to appear
Panov, M.N., 1981, XI ICPEAC, Kyoto, Invited Lectures, ed. K. Oda and L. Takayanagi, Amsterdam, North Holland, 437
Péquignot, D., 1980 a, *Astron. Astrophys.* **81**, 356
Péquignot, D., 1980 b, *Astron. Astrophys.* **83**, 52
Péquignot, D., Aldrovandi, S.M.V. and Stasinska, G., 1978, *Astron. Astrophys.* **63**, 313
Phaneuf, R.A., Kimura, M., Sato, H. and Olson, R.E., 1985, *Phys. Rev. A* **31**, 2914
Schneidermann, S.V. and Russek, A., 1969, *Phys. Rev.* **181**, 311
Smith, F.T., 1969, *Phys. Rev.* **179**, 111

Soloviev, E.A. and Vinitsky, S.I., 1985, *J. Phys. B : Atom. Mol. Phys.* **18**, L557
Thorson, W.R. and Delos, J.B., 1978, *Phys. Rev. A* **18**, 135
Wilkie, F.G., McCullough, R.W. and Gilbody, H.B., 1986, *J. Phys. B : Atom. Mol. Phys.* **19**, 239

EXPERIMENTAL MEASUREMENTS OF THE RATE CONSTANTS FOR NEUTRAL-NEUTRAL REACTIONS

Ian W.M. Smith
Department of Chemistry
University of Birmingham
P.O. Box 363
Birmingham B15 2TT
England

ABSTRACT. This paper gives a brief review of experimental methods used to obtain rate data for neutral-neutral reactions involving free radicals. The application of these methods to reactions of two types - radical-molecule and radical-radical reactions - is illustrated by reference to some recent studies in the author's laboratory. As most radical-molecule reactions have significant activation energies they will be insignificantly slow at the temperatures of interstellar clouds. However, most radical-radical reactions are rapid and their rates increase as the temperature is lowered. The experimental data base for these reactions, which are the only neutral-neutral reactions likely to be important in most astrochemical environments, is still very limited and, in particular, measurements at low temperatures are badly needed.

1. INTRODUCTION

The important role in astrochemistry of homogeneous reactions involving charged species (e.g., ion-molecule reactions and ion-electron recombinations) is reflected in the proceedings of this meeting. However, it has been recognized for some time [1] that neutral-neutral reactions may also play some role in astrochemistry, even at the exceedingly low temperatures of interstellar clouds.

My main goal in this paper is to review the present state of experimental measurement and our current knowledge of neutral-neutral reaction rates, with special emphasis placed on those reactions involving free radicals which may be rapid even at very low temperatures. Within this overall context, I shall develop two main themes:

(i) a description of what can now be achieved by modern methods in regard to the range (of species and of temperature) and accuracy of rate constant measurements;

(ii) a comparison of the magnitude of rate constants and the variation of those rate constants with temperature for different types of neutral-neutral bimolecular reactions.

T. J. Millar and D. A. Williams (eds.), Rate Coefficients in Astrochemistry, 103–116.

It will be helpful to begin with some definitions and with a classification of bimolecular reactions according to the kinds of reagents involved. I shall take a radical to be any species, atomic or molecular, with one or more unpaired electrons. A molecule contains two or more atoms and no unpaired electrons: in a saturated molecule, only single bonds link the atoms; whereas an unsaturated molecule contains one or more double or triple bonds.

TABLE 1. Types of Neutral-Neutral, Bimolecular Reactions

Reaction type	$(E_{act}/R)/K$	V(x)	Ref.
Molecule + Molecule e.g. $H_2 + D_2 \rightarrow 2HD$	$>3.5 \times 10^4$		[2]
Radical + Saturated Molecule e.g. $OH + H_2 \rightarrow H_2O + H$	2100		[3]
Radical + Unsaturated Molecule e.g. $OH + CO \rightarrow CO_2 + H$	~ 0		[3]
Radical + Radical e.g. $O + OH \rightarrow O_2 + H$	-100		[3]

Table 1 lists four classes of bimolecular reaction and an example of each type. The reactions are listed in order of increasing electronic complexity: as one proceeds down the list the number of bonds broken, formed, or significantly changed decreases. The degree of electronic re-organisation might be classified in terms of the number of bonds undergoing a major change. Thus, for a molecule-molecule reaction, of the kind exemplified by $H_2 + D_2 \rightarrow 2HD$, this change is four: two new single bonds are formed, the two single bonds originally present are lost. In the transition state for such a reaction, there is presumably four partial bonds connecting the four atomic 'centres' which participate directly in the reaction. By contrast, in a radical-radical reaction, such as $O + OH \rightarrow O_2 + H$, one bond is broken and one formed but the transition state can involve simply the partial formation of a bond between the two oxygen atoms, the overall reaction proceeding by way of the electronic ground state of the well-known HO_2 radical.

The changes in the extent of electronic reorganisation for different reactions are reflected in the way in which the electronic potential changes along the reaction path. These profiles of potential energy are depicted in the third column of Table 1. For most four-centre molecule-molecule reactions there is a high barrier to reaction. Indeed, for $H_2 + D_2$, it is not clear that reaction occurs via a true bimolecular reaction [2], it may proceed via the initial formation of H or D atoms. In any case, such reactions are very slow, even at room temperature, and will have insignificant rates under the conditions prevalent in interstellar clouds (although they may have some role to play in shocked interstellar gas [4]).

The height of potential barriers to reaction are usually deduced from experimental measurements of activation energies. The activation energy is a convenient way to express the temperature dependence for a reaction when the rate is measured under properly thermalised conditions. It is strictly defined by the equation:

$$E_{act} = -d\ell nk/d(1/RT)$$

a definition which correctly allows for the possibility that E_{act} is itself a function of temperature: i.e., that a plot of ℓnk versus $(1/RT)$ is not strictly linear.

Radical-molecule reactions have been more extensively studied than any other class of bimolecular reaction between neutral species. Accurate rate constants have been measured for many reactions in the limited temperature range between ca 220 and 500 K [3], and for a fair fraction of these reactions rate data exist for $T > 500$ K. Positive activation energies are the rule, with values usually between 0 and 50 kJ mol^{-1}. Rate measurements below 200 K are very rare. Extrapolation of rate constants to the very low temperature of interstellar clouds is unlikely to be accurate but in virtually all cases it seems safe to assume that the rates will be too small for the reactions to be significant.

The interaction of two free radicals gives rise to at least two potential energy surfaces [5,6]. Usually, one or more of these surfaces are monotonically attractive, as electrons from each radical 'pair up' to form a band. Often, the resultant adduct contains a bond which is weaker than that formed initially, so that the adduct can dissociate to products via an exothermic process. When this is impossible, a chemical reaction can result only if the adduct is stabilised, which requires the removal of energy either by collision with a third-body or by emission of a photon. When the rate of a radical-radical reaction is controlled by the long-range attraction between the radical reagents, it is likely to increase as the temperature is lowered. Although this negative temperature dependence over a limited range is often represented by a negative activation energy, there is really no theoretical justification for a rate law of the Arrhenius form for such reactions. Theoretical efforts to calculate the rates of reactions in which long-range attractive forces dominate are described elsewhere in this volume by Clary [7].

Reactions which proceed by initial addition of a radical to a double or triple bond possess characteristics which are typically intermediate between those of radical-radical reactions and those of reactions between a radical and a saturated molecule. Activation energies are usually positive but small. There is also evidence that even if there is no significant barrier obstructing formation of the adduct, the transition state may be appreciably 'tighter' than those for radical-radical reactions.

In the remainder of this paper, I shall concentrate on reactions of radicals with saturated molecules and those between free radicals.

2. EXPERIMENTAL METHODS

The selected ion flow-tube [8] is an enormously versatile apparatus for measuring the rate constants of ion-molecule reactions. Its success is based on the sensitivity and universality of mass spectrometric detection of charged species. No comparable, all embracing, technique is available for reactions involving only neutral species. The most direct and generally accurate experiments are based either on the flowtube, with radicals being generated directly or indirectly via a microwave discharge, or on the creation of radicals by pulsed photolysis of a molecular precursor. Changes in concentration of the radicals which are generated (or, less commonly, of one of the products of the primary reaction) are generally observed using some form of spectroscopy.

The aim in such kinetic measurements is to isolate the primary reaction: that is, to ensure that the changes in concentration which are being observed are due, essentially entirely, to the reaction one wishes to study. To achieve this goal, the method must combine good time resolution with high detection sensitivity [9]. Then the first-order rate of the primary reaction can be made much greater than the sum of the rates of all the side and secondary reactions which might also remove the radical reagent. The existence or otherwise of a sensitive spectroscopic technique for a particular radical is the most important factor in determining whether its reactions can be isolated for kinetic study.

2.1 The Discharge-Flowtube Method [10,11]

The principle of the discharge-flowtube technique for studying neutral-neutral reactions is the same as that of the SIFT. However, the conditions are usually rather different, in that the walls of the flowtube are treated to minimise heterogeneous reactions, particularly the removal of radicals, the flowrates are less, and conditions of 'plug-flow' are established [10,11]. Reagents are mixed in gas flowing down a cylindrical pipe and species are detected close to the downstream end and at one or more distances d from the point where the reagents are mixed. Knowing the gas flowrate, distances can be converted to reaction times and changes in concentration can be followed as d and/or the concentration of the added reagent is systematically changed. The range of first-order decay constants that can be measured is ca 5-500 s^{-1}.

One advantage of the flowtube technique is that it can be used in conjunction with a wide variety of detection methods: electron spin resonance, laser magnetic resonance, mass spectrometry, optical absorption and fluorescence spectroscopies. The method has been applied extensively to the reactions of a number of small radicals, especially atoms: e.g., H, O, N, F, Cl, OH, ClO and HO_2. Usually atoms are generated by dissociation of the parent diatomic in the microwave discharge. If required, they can be converted to molecular radicals by rapid reactions such as $H + NO_2 \rightarrow OH + NO$. The flowtube method has been applied less to reactions of larger radicals, but successful rate and yield measurements have been carried out on, for example, reactions of the methoxy (CH_3O) radical. In general, the detection sensitivity is high since (relative) radical concentrations are measured in a steady-state mode. Consequently, although the inherent time resolution is not very high, in many systems removal of the radicals by homogeneous side and secondary reactions can be eliminated.

There are two major disadvantages of flowtube experiments. The first is that one cannot always be sure that heterogeneous reactions have been eliminated. Secondly, the technique is restricted to a small range of total pressures, ca 1-10 Torr, which is an important limitation in the study of association reactions. Although these limitations make pulsed photolysis experiments preferable, if they can be performed, the discharge-flowtube method has produced a wealth of accurate and reliable rate data for elementary reactions involving neutral species.

2.2 Pulsed Photolysis Experiments [11]

In kinetics experiments based on pulsed photolysis, radicals are created by photodissociation of a suitable parent molecule, using either a 'conventional' flashlamp or, increasingly, a pulsed laser. The ultimate time resolution which can be attained depends on which type of source is available: using a flashlamp, the maximum first-order rate constant that can be measured is ca 10^5 s^{-1}, but this limit is increased by at least two orders-of-magnitude if a pulsed ultraviolet laser is used to generate the radicals.

In most experiments, ultraviolet or infrared absorption, resonance fluorescence, or laser-induced fluorescence (LIF) is used to follow how transient concentrations change after the photolysis pulse. These optical techniques vary considerably in their sensitivity and hence to the extent to which they isolate the primary reaction. LIF is extremely sensitive, enabling one to follow decays of concentrations from an initial value of ca 10^{10} cm^{-3}, but its use is restricted to species with a bound-bound electronic transition within the range of tunable dye lasers. LIF has been used to follow the kinetics of reactions of, inter alia, the radicals OH [12-14], CN [15] and CH_3O [16,17]. It is more difficult to apply to radical atoms which usually have allowed electronic transitions only in the vacuum ultraviolet. Some LIF measurements utilising two-photon excitation of atoms have been reported [18].

In principle, there is no limit to the total pressures which can be used in pulsed photolysis experiments. Indeed, measurements have been carried out in gases at pressures up to 200 atm. [19], as well as in solutions. However, the sensitivity of detection may be lowered: for example, by fluorescence quenching [14]. The possibility of working at high pressures and with rapid decays ensures that heterogeneous reactions can be eliminated. One difficulty is the lack of 'clean' photochemical sources for some radicals, though the range of pulsed photochemical studies has been extended in recent years by the use of pulsed CO_2 lasers to generate radicals by infrared multiphoton dissociation [20]. A second problem can arise if the reaction under investigation can proceed by parallel pathways. Although progress is being made [17], it is difficult in photolysis experiments to observe reagent and product concentrations simultaneously and even more difficult to relate them quantitatively.

2.3 State-to-State Kinetics and Reaction Dynamics

The chief purpose of this paper is to review the experimental determination of thermal rate constants, emphasising those types of reaction which might be important in astrochemistry. However, there have been scarcely any direct kinetic measurements below 200 K, so the evaluation of rate data for most astrochemical modelling requires a long extrapolation to lower temperatures, aided, if possible, by theoretical calculations which necessarily include some degree of approximation. Experiments which provide dynamical information (e.g., reaction cross-sections or state-to-state rate coefficients) are important in this context since they provide a further, and often more searching, test of theoretical models.

I do not intend to dwell here on the techniques, usually involving molecular beams and/or tunable lasers, which are needed in experiments that yield dynamical data. However, it is worth emphasising the additional complexities which arise in both kinetics and dynamics experiments when the reagents are both reactive free radicals. Consequently, experiments that involve the 'stable' free radicals NO, NO_2 and O_2 have provided especially important information about radical-radical reactions.

3. RADICAL-MOLECULE REACTIONS

The overwhelming majority of reactions between small free radicals and molecules involve transfer of a single atom. Such reactions are very important in atmospheric chemistry and in combustion, and experimental results are regularly evaluated and recommendations made in respect of rate constants and their temperature dependence [3]. The literature on these reactions is vast. Here, I shall only allude to some recent and current experiments in my group as examples of what can now be done. This work relies on the production of OH(OD) or CN radicals by pulsed photolysis and the determination of kinetic decays of the radicals in either their vibrational ground state, v=0, or in v=1 by time-resolved LIF. I shall concentrate on reactions of CN.

In our experiments [15], a frequency-doubled Nd:YAG laser, or a dye laser operating at a wavelength below 585 nm, is used to dissociate NCNO:

$$NCNO + h\nu \rightarrow CN + NO \qquad (1)$$

Others have used ultraviolet photolysis of ICN or C_2N_2 to generate these radicals. Process (1) produces a small fraction of CN in v=1, large enough to observe but too small to affect the kinetics of v=0 if vibrational relaxation occurs at a similar rate to chemical reaction. By tuning the detection dye laser to lines in the appropriate bands [the (1,0) band to follow CN(v=0) and (0,1) to follow CN(v=1)] in the $B^2\Sigma^+ - X^2\Sigma^+$ system, the decay rates of the unexcited and excited radicals can be measured independently.

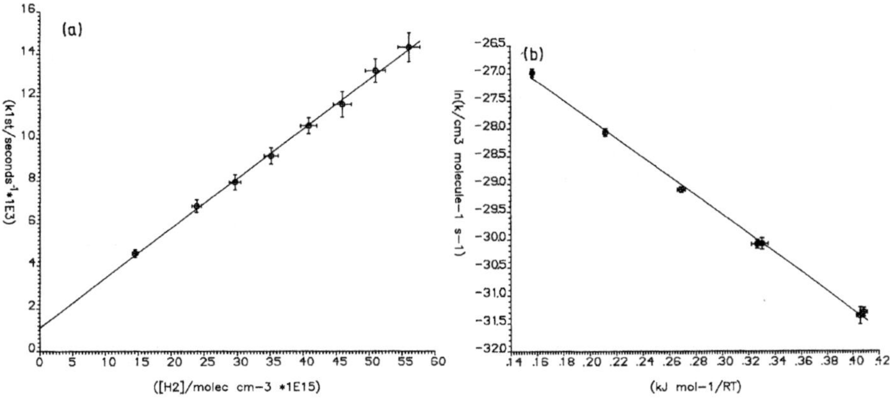

Fig. 1 Kinetic data for the reaction of CN(v=0) with H_2: (a) a plot of first-order decay constants versus concentration of H_2 at 447 K, and (b) an Arrhenius plot of the reaction rate constants for 295<T<761 K.

We now have rather extensive data on the rates of CN(v=0,1) + H_2. A plot of k_{1st} against [H_2] at 447 K and an Arrhenius plot for CN(v=0) + H_2 are given in Fig. 1. They indicate the quality of data obtainable in these experiments. Between 295 and 761 K, the rate constants are described by the expressions:

$$k(v=0) = (2.6+0.1) \times 10^{-11} \exp(-2070\pm120/T) \text{ cm}^3 \text{ molecule}^{-1}\text{s}^{-1}$$
$$k(v=1) = (2.6\pm0.2) \times 10^{-11} \exp(-2050\pm260/T) \text{ cm}^3 \text{ molecule}^{-1}\text{s}^{-1}$$

i.e., there is insignificant difference between the loss rates for the excited and unexcited radicals. In practice, rate constants derived for removal of species in vibrationally excited states will be the sum of those for reaction and non-reactive relaxation; in this example

$$CN(v=1) + H_2 \rightarrow HCN + H$$
$$CN(v=1) + H_2 \rightarrow CN(v=0) + H_2 .$$

The absence of any observable acceleration demonstrates that the addition of 24.7 kJ mol^{-1} to the CN vibration does nothing to enhance the rate of reaction despite the existence of a barrier of the order of 17.2 kJ mol^{-1}. This is undoubtedly because the CN vibration is almost completely uncoupled from motion along the reaction coordinate in this system. Consequently, the reaction is adiabatic with respect to this vibration and the vibronic curves associated with different CN vibrational states are parallel, at least up to the transition state region [21]. We have found a similar result for OH(OD) with HCl and HBr: again excitation in the attacking radical has virtually no effect on the reaction rates.

The reactions of CN with H_2 and OH with H_2 [3] have far too high activation energies to be significant in interstellar clouds. This is undoubtedly the rule in regard to radical-molecule reactions, although there may be exceptions. We have ourselves been surprised to find recently that the reaction between CN and NH_3 has rate parameters (between 295 and 761 K) that are more characteristic of a radical-radical reaction:

$$k(v=0) = (1.5\pm0.2) \times 10^{-11}\exp(+180\pm70/T) \text{ cm}^3 \text{ molecule}^{-1}\text{s}^{-1}$$

The reason for the rapid rate and negative activation energy are not known.

4. RADICAL-RADICAL REACTIONS

Reactions between two different free radicals are difficult to study experimentally except when one of the reagents is NO, NO_2 or O_2. The problem is that both radicals have to be formed in the reactor and the absolute concentration of at least one of the radicals has to be known if a rate constant is to be determined.

Most reactions between radicals belong to one of two general categories. In both, the first act is usually the formation of a new chemical bond as an electron from each of the radicals form a bonding pair. However, it has to be appreciated that the interaction between any pair of radicals give rise to more than one potential energy surface and that not all collisions will occur on a surface which is monotonically attractive at long range and which has a deep well corresponding to the formation of the new bond. In the simplest case (atomic radical(s) in ^2S, molecular radical(s) in $^2\Sigma$) there will be an attractive singlet state, but a repulsive triplet. The situation is more complicated, but generally similar, when one or other of the isolated radicals possesses electronic angular momentum ($L>0$ or $\lambda>0$).

Usually, the kinetics and dynamics of radical-radical reactions will be dominated by motions on the attractive surface or surfaces which have no potential energy barrier preventing the formation of collision complexes. The fate of those collision complex determines the overall type of reaction that can occur. If the lowest energy surface is also the the entrance channel, then the only reaction possible at thermal energies is association. On the other hand, if there are one or other exits of lower energy than the entrance channel, then the collision complexes are likely to dissociate mainly via the lowest energy pathway yielding two product species. This second general class of reaction includes four-centre eliminations, like

$$CH_3 + CF_3 \rightarrow (CH_3CF_3)^{\dagger} \rightarrow CH_2 = CF_2 + HF$$

as well as the simplest cases where one new bond replaces one broken bond:

$$O + OH \rightarrow (O_2H)^{\dagger} \rightarrow O_2 + H.$$

Disproportionation, e.g.

$$C_2H_5 + C_2H_5 \rightarrow C_2H_4 + C_2H_6$$
$$H + HO_2 \rightarrow H_2 + O_2$$

may occur via collision complexes or directly, i.e. by H-atom abstraction from one radical by the other.

Association of radicals can occur only if the radical-radical collision complex loses internal energy before it redissociates. Under normal circumstances, this occurs by collisions with a third-body species and the reaction rate therefore depends on total pressure. Such a mechanism is impossible in the super-rarified environment of interstellar space. However, the kinetics of such reactions are of indirect interest to astrochemists on two counts. First, treatments of radiative association [22], which is implicated in the formation of molecular species in interstellar clouds, have much in common with those of three-body association [23]. Second, the rate constants for radical association in the limit of high pressure correspond to those for formation of the energised associated molecule, since all such species are collisionally stabilised in the limit of high pressure. Consequently, the values of k_{ass}^{∞} and how they vary with temperature provide an important test of theories of reactions occurring over attractive potential energy surfaces [6].

In practice, it is very difficult to study radical-association reactions over the full range from the low to the high pressure limit. However, for several systems, OH(OD) + NO, OH(OD) + NO_2, CN + NO, we have measured [13,15] rate constants for loss of the 'unstable' radical in both states v=0 and v=1. The rates for v=1 are found to be: (i) much faster than those for v=0, at least for total pressures below 1 atm; (ii) to be independent of total pressure, unlike those for v=0; and (iii) to show a mild negative temperature dependence. The rate constant for loss of CN(v=1) in collisions with NO is given by [15(c)]

$$k(v=1) = (4.95 \pm 0.2) \times 10^{-11} \exp(130 \pm 30/T) \text{ cm}^3 \text{ molecule}^{-1} \text{s}^{-1}$$

at temperature between 295 and 761 K. The explanation for these results is that v=1 loss is by vibrational relaxation which occurs virtually every time a radical-radical complex forms. The relationship between reaction and vibrational relaxation in radical-radical systems is discussed in detail elsewhere [6].

The results referred to in the previous paragraph were all obtained using pulsed photolysis to generate radicals and time-resolved LIF to observe the decays in their concentrations. The stability of NO and NO_2 meant that no change had to be made to the methods used to study radical-molecule reactions. In addition, we have just completed [15(c)] a study of the reaction

$$CN + O_2 \rightarrow NCO + O; \quad \Delta H^\circ_{298} = -27 \text{ kJ mol}^{-1}$$

Between 295 and 761 K,

$$k(v=0) = (1.25\pm0.16) \times 10^{-11}\exp(+200\pm50/T) \text{ cm}^3 \text{ molecule}^{-1}\text{s}^{-1}$$
$$k(v=1) = (1.86\pm0.08) \times 10^{-11}\exp(+140\pm20/T) \text{ cm}^3 \text{ molecule}^{-1}\text{s}^{-1}$$

The results are in quite good agreement with the theoretical predictions made by Clary [24], if one assumes that reaction occurs only across the lowest doublet surface of NCOO.

Three direct methods of studying reactions between dissimilar, unstable radicals have emerged based on the discharge-flow and pulsed photolysis methods. The first of these, used, for example, by Kaufman's group [11] to study

$$D + OH \rightarrow OD + H$$
$$O + HO_2 \rightarrow OH + O_2$$

simply extends the flow method by including means of generating radicals on both the flow lines through which gases enter the main flowtube. The main worry with such experiments is probably whether surface-catalysed reaction occurs. The fact that neither radical is rapidly removed when they are separately in the flowtube does not necessarily guarantee that the walls are inactive when both radicals are present.

The pulsed photolysis method can also be adapted to investigate the kinetics of reactions between unstable radicals. For example, Pilling and his coworkers have studied the association of H atoms with methyl radicals [25], making use of the fact that when acetone is photolysed at 193 nm, as well as producing methyl radicals there is a minor channel yielding H atoms:

$$CH_3COCH_3 + h\nu \rightarrow 2CH_3 + CO$$
$$\rightarrow H + CH_2COCH_3$$

The H atoms decay in an excess of CH_3 radicals and the absolute concentrations of CH_3 can be determined by observing their uv absorption. The method is complex, it requires that the reaction zone is illuminated evenly, producing the same concentrations of radicals through the volume which is sampled, and it can be difficult to establish the radical concentration which is required.

Probably the best method when it can be applied is that which combines pulsed photolysis with discharge-flow techniques. The latter method is used to generate the radicals which are to be in excess and to find their absolute concentration. A precursor to the second radical (which must resist attack by the first radical) is then added to the flowing gas and a conventional pulsed photolysis experiment is performed. Experiments of this type have been used to examine the reactions of OH, CN and CH with excess H, D, N or O atoms [26-30]. The rate constants for the O + OH reaction obtained by this method in the temperature range 250-515 K are compared in Fig. 2 with the results of calculations by Clary [31].

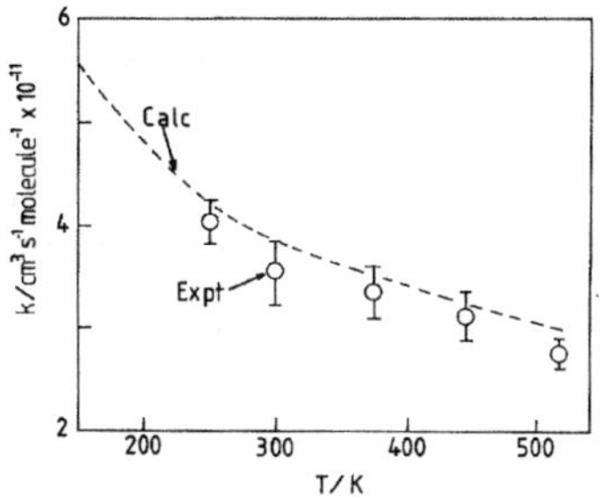

Fig. 2 Comparison of experimental [27] and theoretical [31] rate constants for the reaction $O(^3P) + OH \rightarrow O_2 + H$

Table 2. Rate Data for some Non-Associative Reactions between Small Free Radicals : $k = CT^n$

Reagents	10^{11} k (T=298 K) cm^3 $molecule^{-1}s^{-1}$	n	Range of (T/K)	$\Pi_{reag} g_{el}$ (T=298 K)
N + NO	3.4	0	200-400	12.5
D + OH	5.3	-0.63	250-515	6.0
N + OH	5.3	-0 .25	250-515	12.1
O + OH	3.1	-0.36	250-515	20.4
N + CH	2.1	-	-	25.3
O + CH	9.5	-	-	15.0
O + CN	1.8	-	-	13.5
CH + O_2	2.45	-0.50	294-761	6

Table 2 lists the available rate data for non-associative reactions between a number of small radicals. At 298 K, the rate constants are all within about a factor of ten of the simple collision theory estimate of the rate constant for all collisions. As was explained earlier, it is also necessary to remember that not all collisions will occur on attractive potential surfaces. The last column of Table 2 lists the product of the electronic partition functions of the two radical reagents. Taking this factor and the likely degeneracy of surfaces over which reaction can occur, it appears that reaction occurs at a rate close to the estimate of simple theory for collisions on the surfaces without barriers. In some cases, e.g., N + NO and N + CH, it is necessary to remember that conservation of total spin will not permit reaction to occur via the lowest electronic state of the system.

Studies of the temperature dependence of non-associative, radical-radical reactions are comparatively rare. From those which have been carried out, it is clear that reactions accelerate as the temperature is lowered. The observed variation of rate constant with temperature can be expressed as $k = CT^n$ but care should be taken in extrapolating data obtained at $T > 250$ K to the extremely low temperatures present in interstellar clouds.

In conclusion, it is worth re-emphasising the comparative scarcity of kinetic data for the type of radical-radical reactions which might play a role in astrochemistry. Even for the comparatively well-studied reactions, there may be only two or three independent measurements, and they do not always agree. For example, the two measurements at 298 K on the isotope exchange reaction:

$$D + OH \rightarrow OD + H$$

yield rate constants which differ by a factor of 2.5. There have been very few measurements below room temperature and only the H atom recombination reaction has been studied below ca 220 K [32]. The most promising way of extending rate measurements to very low temperatures would appear to involve the use of a fast flow reactor similar to that employed in flowing afterflow or SIFT studies of ion-molecule reactions.

Acknowledgements

I should like to thank those members of my research group who have contributed to the experiments from our own laboratory which are described in this paper: B.D. Cannon, M.J. Howard, J. de Juan, J.S. Robertshaw, I.R. Sims, B. Veyret and M.D. Williams. I should also like to acknowledge the financial support of our programme received from SERC and from Shell Research.

References

[1] E. Herbst and W. Klemperer, Astrophys. J., 185, 505 (1973).
[2] A. Lifshitz, M. Bidani and H.F. Carroll, J. Chem. Phys., 79, 2742 (1983); M.J. Rabinowitz and W.C. Gardiner, Jr., Chem. Phys. Lett., 124, 63 (1986).
[3] D.L. Baulch, R.A. Cox, R.F. Hampson, Jr., J.A. Kerr, J. Troe and R.T. Watson, J. Phys. Chem. Ref. Data, 13, 1259 (1984).
[4] A. Dalgarno, Phil. Trans. R. Soc. (London), A303, 513 (1981).
[5] I.W.M. Smith, Kinetics and Dynamics of Elementary Gas Reactions (Butterworths, London, 1980) chap.2.
[6] M.J. Howard and I.W.M. Smith, Prog. Reac. Kinetics, 12, 55 (1983).
[7] D.C. Clary, this volume, chapter
[8] D. Smith and N.G. Adams in Gas Phase Ion Chemistry, ed. M.T. Bowers (Academic Press, New York, 1979) vol.1, chap.1.
[9] ref. [5], chap.5.
[10] C.J. Howard, J. Phys. Chem., 83, 3 (1979).
[11] F. Kaufman, J. Phys. Chem., 88, 4909 (1984).
[12] (a) B.D. Cannon, J.S. Robertshaw, I.W.M. Smith and M.D. Williams, Chem. Phys. Lett., 105, 380 (1984); (b) I.W.M. Smith and M.D. Williams, J.C.S. Faraday Trans.2, 82, 1043 (1986).
[13] I.W.M. Smith and M.D. Williams, J.C.S. Faraday Trans.2, 81, 1849 (1985).
[14] J.S. Robertshaw and I.W.M. Smith, J. Phys. Chem., 86, 785 (1982).
[15] (a) J. de Juan, I.W.M. Smith and B. Veyret, Chem. Phys. Lett., 132, 108 (1986); (b) J. de Juan, I.W.M. Smith and B. Veyret, J. Phys. Chem., 91, 69 (1987); (c) I.R. Sims and I.W.M. Smith, J.C.S. Faraday Trans.2, in press (1988); (d) I.R. Sims and I.W.M. Smith , to be published.
[16] D. Gutman, N. Sanders and J.E. Butler, J. Phys. Chem., 86, 66 (1982).
[17] J. Hagele, K. Lorenz, D. Rhasa and R. Zellner, Ber. Bunsenges. Phys. Chem., 87, 1023 (1983).
[18] W.K. Bischel, B.E. Perry and D.R. Crosley, Chem. Phys. Lett., 82, 85 (1981) and Appl. Optics, 21, 1419 (1982).
[19] A.E. Croce de Cobos, H. Hippler and J. Troe, J. Phys. Chem., 88, 5083 (1984).
[20] I.W.M. Smith, Ber. Bunsenges. Phys. Chem., 86, 95 (1982).
[21] A.F. Wagner and R.A. Bair, Int. J. Chem. Kinetics, 18, 473 (1986).
[22] G. Winnewiser and E. Herbst, Topics in Current Chemistry, 139, 121 (1987).
[23] J. Troe, J. Chem. Phys., 66, 4745 and 4758 (1977).
[24] D.C. Clary, Mol. Phys., 53, 3 (1984).
[25] M. Brouard, M.T. Macpherson, M.J. Pilling, J.M. Tulloch and A.P. Williamson, Chem. Phys. Lett., 113, 413 (1985).
[26] M.J. Howard and I.W.M. Smith, J.C.S. Faraday Trans.2, 78, 1403 (1982).
[27] M.J. Howard and I.W.M. Smith, J.C.S. Faraday Trans.2, 77, 997 (1981).
[28] K.J. Schmatjko and J. Wolfrum, Ber. Bunsenges. Phys. Chem., 82, 419, (1978).

[29] I. Messing, T. Carrington, S.V. Filseth and C.M. Sadowski, Chem. Phys. Lett., 74, 56 (1980).

[30] I. Messing, S.V. Filseth, C.M. Sadawski and T. Carrington, J. Chem. Phys., 74, 3874 (1981).

[31] D.C. Clary and H.-J. Werner, Chem. Phys. Lett., 112, 346 (1984).

[32] D.O. Ham, D.W. Trainor and F. Kaufman, J. Chem. Phys., 53, 495 (1970).

POLYCARBON AND HYDROCARBON IONS AND MOLECULES IN SPACE

Diethard K. Bohme
Department of Chemistry and
Centre for Research in Experimental Space Science
York University
North York, Ontario
Canada M3J 1P3

ABSTRACT. The growth of polycarbon and hydrocarbon ions and molecules possible in interstellar clouds and circumstellar envelopes is viewed in terms of results of laboratory gas-phase measurements for selected ion/molecule reactions. Emphasis is given to the formation of these carbonaceous species as chains, rings, sheets and spheres of C atoms.

1. INTRODUCTION

A significant fraction of the carbon present in space is entrained in isolated polycarbon and hydrocarbon molecules and ions. Evidence for this comes, in part, from the detection by radio-astronomers of a large variety of carbonaceous molecules in dark interstellar clouds such as TMC-1 (1) and in circumstellar envelopes such as that of IRC+10216 (2). Many of these carbonaceous molecules are highly unsaturated in composition and chain-like in structure. Prominent among them are the homologous series of cyanopolyacetylene molecules, $HC_{2n}CN$ (n=1-5), the first member of which was discovered in 1971, the linear hydrocarbon radical molecules, C_nH (n=2-6), and the linear polycarbon monoxide and monosulphide molecules, C_3O and C_3S, discovered only very recently. Notwithstanding a possible observational bias (3), these detections, together with the failure to observe related but more hydrogenated molecules (4), imply an interstellar and circumstellar chemistry which is directed towards the production of highly unsaturated and chain-like hydrocarbons. In contrast, hydrocarbon-ring molecules have been much harder to find and only a few are known. The detections of cyclopropenylidene, c-C_3H_2, and the cyclic radical, c-C_3H, have been reported in only the past two years (5,6). The search for other ring molecules such as substituted benzenes (7), for example, has yielded negative results.

It has been suggested that free polycylic aromatic hydrocarbon (PAH) molecules can account for unidentified interstellar emission features in the infrared (8) and for the diffuse interstellar absorption bands in the visible (9). Also, the possibility of the interstellar and circumstellar presence of hollow, cage-like polycarbon molecules with spheroidal shells of hexagonal, graphite-like sheets of carbon atoms has

T. J. Millar and D. A. Williams (eds.), Rate Coefficients in Astrochemistry, 117–133.

been proposed recently (10). It may be supposed that these spheroidal molecules coexist with the long chain-like species and the PAH molecules, and even share with them an interlocking chemistry.

All of these observations and speculations have raised exciting questions about the chemical formation and reactions of polycarbon and hydrocarbon ions and molecules in space. Both homogeneous gas-phase ion/molecule reactions and heterogeneous catalytic surface reactions on grains have been proposed as chemical sources and sinks. Here we explore the feasibility of growing these polycarbon and hydrocarbon ions and molecules from atoms and small molecules in the gas phase by ion chemistry. The exploration is based primarily on results of laboratory measurements for selected ion/molecule reactions (11). No attempt is made to provide quantitative measures of the relative importance of these processes in interstellar synthesis. Such information should be provided by chemical models appropriate for the physical and chemical conditions of the specific interstellar or circumstellar environment of interest. A number of research groups are firmly entrenched in detailed studies of such models.

2. CHAIN-LIKE IONS AND MOLECULES

2.1. Polycarbon and Hydrocarbon Ions and Molecules

2.1.1. Polycarbon ions and molecules. Long chain molecules composed purely of carbon have not yet been identified in space, but their existence has been proposed. For example, Douglas has invoked long-chain carbon molecules to account for the origin of diffuse interstellar lines seen in the optical spectra of stars which lie behind low-density interstellar clouds (12). Douglas suggests that these chain molecules may contain from 5 to 15 carbon atoms and that they may be derived by the photo-dehydrogenation of hydrocarbon chain molecules formed in these environments, or released into them from dense clouds. Ionic mechanisms for the growth of hydrocarbon molecules are discussed in the next section. We consider here the possibility of growing chain molecules composed purely of carbon directly from carbon atoms. Such growth has been proposed by Suzuki for interstellar regions of low optical depth where carbon is mostly ionized (13).

The Suzuki scheme involves three steps: 1. the radiative association of C^+ with C_n molecules, 2. the hydrogenation of the resulting carbon cluster ions $C_{n+1}{}^+$, and 3. the neutralization of the hydrogenated carbon cluster ions $C_{n+1}H^+$:

$$C^+ + C_n \longrightarrow C_{n+1}{}^+ \xrightarrow{H_2} C_{n+1}H^+ \xRightarrow{e} C_{n+1}$$

No experimental evidence is available for the initial step of radiative association. Theory suggests that the rate constant for the radiative association step increases as the number of carbon atoms increases and becomes substantial for low temperatures at n = 4 (14). With regard to the second step, several research groups have shown that $C_n{}^+$ reacts with H_2 to form C_nH^+ for n = 2–5 (11). Also, laboratory measurements in which carbon cluster ions are generated by direct laser vapourization of graphite have shown that the larger cluster ions, n = 3-9, also may be hydrogenated by reactions with molecular hydrogen (15). The following bimolecular reactions were observed with deuterium (a second, minor channel leading to the elimination of C_3 was observed for n = 6,8):

$$C_n^+ + D_2 \longrightarrow C_nD^+ + D \quad (n = 3\text{-}9)$$

The structures of the reacting C_n^+ cluster ions are not known, but semi-empirical quantum-chemical calculations have shown that carbon-cluster ions with n up to 9 have linear electronic ground states (16). The reactivity of the cluster ions with deuterium showed an interesting trend with n: the relative reaction rate was observed to decrease as the cluster size increases with the even clusters appearing to be less reactive than the odd clusters (15). An abrupt drop in reactivity was observed between n = 9 and 10 which persists up to n = 19. This drop was explained by a change in structure of the carbon cluster ions from a linear chain, in which terminal carbon atoms are present which have carbene character, to a monocyclic structure, in which there are no reactive carbene sites (15). There is some uncertainty about the final neutralization step in the Suzuki scheme. It is not known whether the recombination of C_nH^+ with electrons results in H atom elimination or the rupture of a C-C bond, or whether the structure of the carbon skeleton of the C_nH^+ ion is preserved in the neutral C_n product even if the carbon skeleton remains intact. The structures and stabilities of C_n cluster molecules are currently a subject of much theoretical and experimental investigation. The most recent theoretical results for small carbon clusters (n = 2-10) suggest that odd-numbered clusters have linear structures and are more stable than the adjacent even-numbered clusters. In contrast, many of the even-numbered clusters are predicted to have cyclic structures, but C_4, C_6 and C_{10} also appear to have linear structures which are close in energy to the cyclic forms (17).

It should be noted that a sharply reduced efficiency for the hydrogenation of carbon cluster ions at n = 10 implies an upper limit at n=9 for growth of polycarbon chain molecules in the manner suggested by Suzuki. We suggest here that secondary and higher-order end-on association reactions of C_nH^+ chain ions with C_n chain molecules could significantly increase this limit, but there is no data available for such reactions. Some data is available for similar reactions of polycarbon ions with acetylene and polyacetylene chain molecules which generate C_nH^+:

$$C_n^+ + C_{2m}H_2 \longrightarrow C_{2m+n}H^+ + H$$

Reactions of this type have been observed to be rapid for n = 1-4, m = 1 and for n = 4, m = 2 at 300K (11). As yet there is no indication about the structures of the $C_{2m+n}H^+$ product ions in these reactions or about limits to growth in this fashion.

2.1.2. Hydrocarbon ions and molecules. In contrast to polycarbon chain molecules which cannot be detected by radioastronomy, the closely related molecules of the type C_nH, for which rotational transitions are allowed, have been observed for n = 2-6 (5). These observations point towards the extraterrestial existence of the homologous series of chain-like hydrocarbon molecules of the type C_nH. The closely related polyacetylenes also cannot be detected by radioastronomy, but it is important to note that methylacetylene and methyldiacetylene have been observed in TMC-1 (1). Acetylene and ethylene have been identified by IR in the molecular envelope of IRC+10216 (2). Except for methane, no observations have been reported for fully-saturated hydrocarbon molecules (1).

Gas-phase measurements of reactions of polycarbon and hydrocarbon ions with molecular hydrogen have shown that the hydrogenation of these ions one H atom at a time often encounters thermodynamic or kinetic bottlenecks (11). For example, C_2^+ efficiently adds hydrogen in two successive by H-atom transfer reactions with molecular hydrogen to form $C_2H_2^+$ (which may neutralize by charge transfer to form acetylene), but the next H-atom transfer reaction to form $C_2H_3^+$ (which may yield acetylene by recombination with electrons or proton transfer) is not efficient. One vibrational quantum of internal energy appears to be required for this reaction to overcome its thermodynamic barrier (18). Hydrogenation in this fashion of the larger C_n^+ cluster ions is even less favourable and appears to become inefficient already after the addition of only one H atom. This chemistry is therefore biased in favour of the formation of the unsaturated hydrocarbons C_n and C_nH (if neutralization of C_nH^+ proceeds by charge transfer). The occurrence of hydrogenation through the addition of H_2 by radiative association, as has recently been demonstrated for CH_3^+ (19), would have different implications for the likelihood of formation of saturated hydrocarbon molecules.

Formation of polyacetylene molecules by ion chemistry requires the formation of the acetylenic ions $C_{2n}H_3^+$ and $C_{2n}H_2^+$ which may neutralize by recombination with electrons, proton transfer or charge transfer. Two obvious routes for the growth of polyacetylene cations involve the sequential addition of two-carbon units either directly to the ion as follows (11):

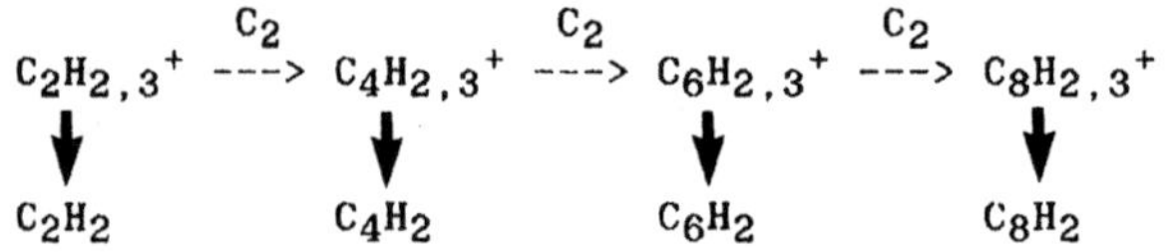

or indirectly by repeated reaction of a two-carbon cation with the polyacetylene as follows (11):

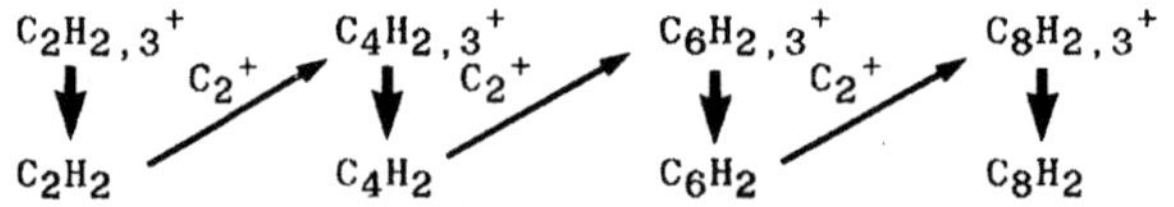

Both routes may also operate with polyatomic units containing an even number of carbon atoms, such as neutral or ionized C_2 and C_4 units. At the low pressures of the interstellar medium the addition of carbon units may occur either by radiative association or by condensation. Only a very few of these reactions actually have been characterized in the terrestrial laboratory. The available experimental results provide insight primarily into the kinetics of the initiation steps, but do suggest that polyacetylenic ions are readily generated by sequential reactions involving two- or four-carbon units (11).

It also appears that polyacetylenic ions provide a direct route to the formation of methyl-substituted polyacetylenes. For example, the following reactions are implied by measurements of the production of $C_3H_5^+$ from C_2H^+, $C_2H_3^+$ and methane (20):

$$C_{2n}H^+ + CH_4 \longrightarrow (CH_3\text{-}C_{2n}H)H^+ + h\nu$$

$$C_{2n}H_3^+ + CH_4 \longrightarrow (CH_3\text{-}C_{2n}H)H^+ + H_2$$

These reactions are likely to proceed by C-H bond insertion and so result in ions which may neutralize by recombination with electrons or proton transfer to form methylpolyacetylenes.

The following product options apply generally to reactions of molecular hydrocarbon ions with hydrocarbon molecules (11):

$$C_nH_x^+ + C_mH_y \longrightarrow C_{n+m}H_{x+y}^+$$

$$\longrightarrow C_{n+m}H_{x+y-1}^+ + H$$

$$\longrightarrow C_{n+m}H_{x+y-2}^+ + H_2$$

$$\longrightarrow C_{n+m-k}H_p^+ + C_kH_q$$

Association and condensation with elimination of H, H_2 or a small hydrocarbon fragment ($k < n$) will lead to growth of the ionic carbon skeleton, while elimination of small ionic fragments will lead to growth of the neutral carbon skeleton. Numerous reactions between hydrocarbon ions and hydrocarbon molecules have been investigated in the laboratory (11). Reactions of particular importance to interstellar chemistry are those of "terminal" hydrocarbon ions, e.g. C^+, CH_3^+ and $C_2H_2^+$, which do not hydrogenate readily in hydrogen and so are available for reactions with the less abundant interstellar hydrocarbon molecules.

For reactions of C^+ with hydrocarbon molecules experiments indicate the possibility of two competing bimolecular reaction channels (11):

$$C^+ + C_nH_m \longrightarrow C_{n+1}H_p^+ + H_q$$

$$\longrightarrow C_{n+1-x}H_p^+ + C_xH_q$$

Here growth of the carbon skeleton competes with its fragmentation. Systematic measurements with homologous series of hydrocarbons have shown that dissociative channels become increasingly preferred as the length of the carbon skeleton increases (21). So there seems to be a limit to molecular growth in this fashion.

Reactions of CH_3^+ and $C_2H_2^+$ have been measured with CH_4, C_2H_2, C_2H_4, C_2H_6 and C_4H_2. 100% growth of the carbon skeleton has been observed with CH_4 and C_2H_2. The product ions of these two reactions are possible sources for C_2H_4, C_3H_2, C_3H_3, C_3H_4, C_4H and C_4H_2. In the reactions of CH_3^+ and $C_2H_2^+$ with C_2H_4, C_2H_6 and C_4H_2 several non-productive channels compete with ionic growth, including hydride transfer, charge transfer and H_2 transfer (11). Little has been done to explore systematically, for homogeneous series of molecules, the dependence of the extent of the competition with ionic growth on the size of the hydrocarbon molecule, as has been done for C^+, for example. Such an experimental campaign could be quite instructive.

It is also known that radical cations of olefins and acetylenes in general, or of polyolefinic and polyacetylenic molecules, undergo condensation and/or association reactions with their parent molecules. Consecutive additions of the parent molecule are often initiated which lead to high molecular-weight polymeric hydrocarbon ions. For example, the following sequences have been identified in diacetylene (22):

$$C_4H_2^+ \dashrightarrow C_8H_4^+ \dashrightarrow C_{12}H_6^+ \dashrightarrow C_{16}H_8^+ \dashrightarrow C_{20}H_{10}^+$$

$$C_4H_2^+ \dashrightarrow C_6H_2^+ \dashrightarrow C_{10}H_4^+ \dashrightarrow C_{14}H_6^+ \dashrightarrow C_{18}H_8^+$$

$$C_4H_2^+ \dashrightarrow C_8H_2^+ \dashrightarrow C_{12}H_4^+ \dashrightarrow C_{16}H_6^+$$

$$C_4H_2^+ \dashrightarrow C_6H_2^+ \dashrightarrow C_8H_2^+ \dashrightarrow C_{12}H_4^+ \dashrightarrow C_{16}H_6^+$$

The association reactions in these sequences are likely to proceed by collisional stabilization under laboratory conditions and would need to proceed by radiative association to be effective in space. The ions produced in the sequence initiated by $C_8H_4^+$ all showed two populations, one reactive and one unreactive with C_4H_2. It has been suggested that these two populations correspond to straight-chain and cyclic isomers, respectively.

2.2. Polycarbon Monoxide and Monosulphide Ions and Molecules

The polycarbon monoxide C_3O and polycarbon monosulphides C_2S and C_3S are some of the latest molecules which have been detected in interstellar space (1, 23). They point toward the extraterrestrial existence of families of carbon-chain molecules of the type C_nO and C_nS. Laboratory measurements suggest that a number of ion/molecule pathways are available for their formation in the gas phase.

The preferred pathway in interstellar and circumstellar environments may well be the direct reaction of cluster ions of the type C_nH^+ with donor molecules containing oxygen or sulphur atoms (24). These cations are most suited to adding O or S atoms because of their carbene character which allows them to coordinate to non-bonded electron pairs. This is evident from the following reactions of the carbene cation $:C_3H^+$ which have been been observed for OX = H_2O, CH_3OH, CO_2, OCS, N_2O and O_2, and for SX = H_2S and OCS (25):

$:C_3H^+ \; + \; :OX \; \text{---->} \; HC_3O^+ \; + \; X$

$:C_3H^+ \; + \; :SX \; \text{---->} \; HC_3S^+ \; + \; X$

Neutralization of the ionic products of these reactions, by proton transfer or recombination with electrons, can establish tricarbon monoxide, :C=C=C=O, and tricarbon monosulphide :C=C=C=S (24).

Analogous bimolecular reactions with higher members of the homologous series of ions $:C_{2n+1}H^+$, which are potential sources of carbon-chain molecules of the type $:C_{2n+1}O$ and $:C_{2n+1}S$, remain to be investigated.

Coordination of unbonded electron pairs may also occur with carbon monoxide (or carbon monosulphide) in association reactions in which the adduct is stabilized by collision or by the emission of radiation (26):

$:C_nH^+ \; + \; :CO \; (:CS) \; \text{---->} \; HC_{n+1}O^+ \; (HC_{2n+2}S^+)$

$(:C_n:)^{+\cdot} \; + \; :CO \; (:CS) \; \text{---->} \; (:C_{n+1}O)^{+\cdot} \; ((:C_{n+1}S)^{+\cdot})$

Addition reactions of this type have been observed for CO reacting with C_nH^+ for n = 2-5 and C_n^+ for n = 2-6 in a helium bath gas at 296K (26). The neutralization of the products of these association reactions by recombination with electrons, proton transfer or charge transfer can produce polycarbon monoxides or polycarbon monosulphides.

Other radiative association reactions of CO which have been invoked to explain the formation of interstellar C_3O include those with the abundant interstellar ions $C_2H_2^+$ and $C_2H_3^+$ (27). In this case neutralization of the adduct ions by recombination with electrons requires complete dehydrogenation to yield C_3O.

Neutralization reactions, in general, may not always act to dehydrogenate the hydrocarbon ion. For example, charge transfer will preserve the degree of hydrogenation while hydride transfer will increase the hydrogen content. Thus, the neutralization of the ions HC_nO^+ and HC_nS^+ described above may also lead to carbon-chain radicals of the type $(HC_nO)\cdot$ and $(HC_nS)\cdot$ or molecules of the type H_2C_nO and H_2C_nS. Finally, it should also be noted that these latter molecules may be formed directly as neutral products in the ion/molecule reaction. For example, the following reaction channels have been observed (25):

$C_3H^+ \; + \; OCS \; \text{---->} \; CS^+ \; + \; (HC_3O)$

$C_3H^+ \; + \; CH_3OH \; \text{---->} \; CH_3^+ \; + \; (H_2C_3O)$

While the neutral products were not directly identified for these reactions, they are likely to correspond to the radical HC·-C=C=O and the ketene-like molecule, H_2C=C=C=O, respectively.

2.3. Cyanopolyacetylene Ions and Molecules

The presence of the homologous series of cyanopolyacetylene molecules up to $H(C{\equiv}C)_5$-CN, and the related C≡C-CN and CH_3-C≡C-CN molecules, is perhaps the most remarkable feature of the chemical compostion of interstellar and circumstellar environments. Four distictive mechanisms have been put forward for the formation of cyanopolyacetylenes by ion chemistry in the gas phase (28). In all of them the formation of the cyanopolyacetylene is achieved ultimately through the neutralization of the ion $H_2C_{2n+1}N^+$ by recombination with electrons or by proton transfer. The schemes differ in the manner of growth of the carbon backbone and in the manner of entrainment of nitrogen.

The coexistence of polyacetylenes and cyanopolyacetylenes in the molecular cloud TMC-1 is suggestive of an interlocking chemistry such as the entrainment of CN in polyacetylene ions or neutral polyacetylenes:

$$C_{2n}H_{2,3}^+ \quad + \quad HCN\ (CN) \quad \text{---->} \quad H_2C_{2n+1}N^+ \quad + \quad H,H_2\ (h\nu,H)$$

$$HCN^+ \quad + \quad C_{2n}H_2 \quad \text{---->} \quad H_2C_{2n+1}N^+ \quad + \quad H$$

Experimental indications are that this entrainment is successful only for the formation of cyanoacetylene, unless studies of the reactions with CN prove otherwise.

Alternatively, entrainment can proceed first, and be followed by growth of the carbon chain by reactions with acetylene ions:

$$C_2H_{2,3}^+ \quad + \quad HC_{2n+1}N \quad \text{---->} \quad H_2C_{2n+3}N^+ \quad + \quad H,H_2$$

Laboratory measurements suggest that this path is short-circuited by proton-transfer and HCN-elimination reactions.

Growth also could be achieved (in the absence of acetylene) one carbon atom at a time as follows:

$$C^+ \quad + \quad HC_{2n+1}N \quad \text{---->} \quad C_{2n+2}N^+ \quad + \quad H$$

$$C_{2n+2}N^+ \quad + \quad CH_{4,3,2} \quad \text{---->} \quad H_2C_{2n+3}N^+ \quad + \quad H_2,H,h\nu$$

This scheme has gained credibility from measurements of the reaction of C^+ with HC_3CN to form C_4N^+ and the reaction of C_4N^+ with CH_4 to form some $C_5H_2N^+$ (28). Also, it is attractive in view of the chemical feedback which is possible.

Finally, the build-up of the carbon chain may be followed by reaction with N atoms in a manner described by the following reactions:

$$C_{2n+1}H^+ \quad + \quad N \quad \text{---->} \quad C_{2n+1}N^+ \quad + \quad H$$

$$C_{2n+1}N^+ \quad \xrightarrow{H_2} \quad HC_{2n+1}N^+ \quad \xrightarrow{H_2} \quad H_2C_{2n+1}N^+$$

$$C_{2n+1}H_3^+ \quad + \quad N \quad \text{---->} \quad H_2C_{2n+1}N^+ \quad + \quad H$$

The hydrocarbon ions required for this build-up can be derived from the hydrogenation of (odd) polycarbon-chain ions, the protonation of neutral (odd) polycarbon-chain molecules, H-elimination reactions of C^+ with polyacetylenes or H_2 elimination reactions of CH_3^+ with polyacetylenes (28). Experimental evidence for nitrogen entrainment in this fashion is available only for the reaction of $C_3H_3^+$ with N atoms. Nevertheless, this mechanism remains an attractive option, unless hydrogenation reactions with H_2 are required to build up the hydrogen content before neutralization to form the cyanopolyacetylene (the efficiency of hydrogenation appears to be variable and unpredictable). Reactions with N atoms are attractive in the chemistry of dark interstellar clouds because of the relatively high abundance of these atoms.

CH_3-C C-CN has also been observed in TMC-1 where it may be derived from cyanoacetylene or diacetylene. Possible ion/molecule reactions which may provide a source for this molecule include the following (11):

$$CH_3^+ + HC_3N \longrightarrow C_4H_4N^+ + h\nu$$

$$HC_3N^+ + CH_4 \longrightarrow C_4H_4N^+ + H$$

$$C_4H_{2,3}^+ + N \longrightarrow C_4HN^+ \ (C_4H_2N^+) + H$$

The product ions of the latter reactions require further hydrogenation before recombination with electrons can yield methylcyanoacetylene. Derivitization of the higher acetylenes and cyanoacetylenes could proceed in a similar fashion so that we might expect the future observation of methylcyanopolyacetylenes in TMC-1.

3. CYCLIC HYDROCARBON IONS AND MOLECULES

The exciting discoveries of interstellar hydrocarbon ring molecules have been very recent. Cyclopropenylidene, c-C_3H_2, appears to be widely distributed in the interstellar medium; millimetre-wave transitions of c-C_3H_2 have recently been identified in emission in several cold molecular dust clouds and in absorption in the direction of the galactic center (4). The detection of the related cyclic C_3H radical molecule has been reported just recently (5). Toward TMC-1 the column density of c-C_3H is comparable to that of linear C_3H, l-C_3H, and is about one order of magnitude lower than that of c-C_3H_2 (5).

Both c-C_3H_2 and c-C_3H have been proposed to be formed in the gas-phase by the dissociative recombination reaction of $C_3H_3^+$ which is known to have both linear and cyclic isomers (5):

$$\text{l,c-}C_3H_3^+ + e \longrightarrow \text{c-}C_3H_2 + H$$

$$\longrightarrow \text{c-}C_3H + (H + H)$$

$$\longrightarrow \text{l-}C_3H + (H + H)$$

Unfortunately there is no information available about the branching ratios for this recombination or their dependence on the geometry of $C_3H_3^+$. Alternatively, $C_3H_3^+$ may be neutralized by proton transfer to yield C_3H_2, but here again the geometry of the neutral product is uncertain. Both isomers of $C_3H_3^+$ can be produced by several ion-molecule reactions in the interstellar gas. The radiative association of C_3H^+ with H_2 should be effective in hydrogen-rich molecular clouds (29):

$C_3H^+ \ + \ H_2 \ \ ----> \ \ C_3H_3^+ \ + \ h\nu$

This reaction has not been measured in the laboratory but the analogous collisional association reaction in helium at 80 K produces the two isomers with about equal probability (29). The following bimolecular reaction channels producing $C_3H_3^+$ exclusively in the cyclic form also have been observed in the laboratory (24,29):

$C_3H^+ \ + \ CH_4 \ \ ----> \ \ c\text{-}C_3H_3^+ \ + \ CH_2$

$CH_3^+ \ + \ C_2H_2 \ \ ----> \ \ c\text{-}C_3H_3^+ \ + \ H_2$

$C^+ \ + \ C_2H_4 \ \ ----> \ \ c\text{-}C_3H_3^+ \ + \ H$

Clearly all of these reactions (and perhaps others) may contribute to the formation of $c\text{-}C_3H_3^+$ in the interstellar medium, but their relative contributions will depend on the physical and chemical conditions which prevail.

$c\text{-}C_3H_2$ can also be envisaged to be produced from charge transfer reactions of $c\text{-}C_3H_2^+$ with species of small ionization energies such as metal atoms (29):

$c\text{-}C_3H_2^+ \ + \ \text{metal} \ \ ----> \ \ c\text{-}C_3H_2 \ + \ \text{metal}^+$

The same ion may produce $c\text{-}C_3H$ by dissociative recombination with electrons or proton transfer, and one should also allow for the possibility of formation of $c\text{-}C_3H$ from C_3H^+ by charge transfer. All of these reactions have yet to be characterized in the laboratory.

The observation of cyclopropenylidene and its plausible production through the radiative association reaction of the carbene cation C_3H^+ with H_2 suggest the possible formation of substituted cyclopropenylidenes from analogous reactions of C_3H^+ with other interstellar molecules. For example, laboratory measurements have shown that C_3H^+ reacts very efficiently at close to the collision rate with the interstellar molecules HCN and HC_3N exclusively by collisional association with probable C-H bond insertion (30). The corresponding radiative association, followed by neutralization through recombination with electrons or by proton transfer, may produce the acyclic disubstituted carbenes $:C(C_2H)CN$ and $:C(C_2H)C_2CN$ or the cyclic isomers indicated below with R=H and X=CN and C_2CN:

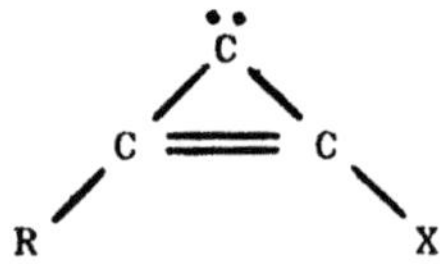

Furthermore, analogous association reactions observed in the laboratory with the carbene cation $:C_4N^+$ which is readily produced in space by the reaction of C^+ with HC_3N (11) imply the possible formation of cyclic carbenes of the type indicated above with R=CN and X=CN and C_2CN (or their acyclic isomers).

4. POLYCARBON AND HYDROCARBON SHEETS

Free PAH-related molecules are believed to be present in the interstellar medium as a mixture of partially hydrogenated ions and neutral molecules, including radical forms. Large polycyclic hydrocarbons are known from terrestrial chemistry to have properties which approach those of graphite, an allotrope of carbon of infinite planes or sheets of benzene rings. One may envisage growth of condensed benzenoid PAH molecules in the gas phase to proceed in at least two different ways: either directly by cross-bonding in the side-on approach of extended chain-like molecules such as cumulenes or polyacetylenes, or through the successive addition of hexagonal rings beginning with benzene as the embryo. Benzene may itself be formed by homogeneous gas-phase chemistry or, as has been suggested (6), may become available from reactions of acetylene on the surface of grains. Very little is known about the gas-phase chemistry of the growth of PAH molecules, but progress is beginning to be made in the characterization of the first few steps in such growth by ion chemistry.

4.1. Growth from Chain-Like Carbon or Hydrocarbon Ions and Molecules

Laboratory studies have now demonstrated the feasibility of the formation of benzene with gas-phase ion chemistry involving acyclic 3-carbon units. The final step in this formation would involve the neutralization of the benzenium ion (protonated benzene) by proton transfer or recombination with electrons. Formation of the benzenium ion in ionized allene and ionized propyne has been shown to be possible with the following bimolecular reactions (31):

$$H_2C{=}C{=}CH_2^+ \ + \ H_2C{=}C{=}CH_2 \ \text{---->} \ C_6H_7^+ \ + \ H$$

$$HC{\equiv}C{-}CH_3^+ \ + \ HC{\equiv}C{-}CH_3 \ \text{---->} \ C_6H_7^+ \ + \ H$$

Collisional activation and chemical reactivity studies have shown that 70% of the $C_6H_7^+$ ions produced in allene and 44% of the $C_6H_7^+$ ions produced in propyne have the benzenium structure. Interestingly, the structures of the remaining $C_6H_7^+$ ions appear to correspond to protonated fulvenes or protonated dimethylenecyclobutenes.

The feasibility of producing benzene with ion chemistry involving only two-carbon units is less certain. Both $C_2H_2^+$ and $C_2H_3^+$ are known to add two molecules of acetylene to from $C_6H_6^+$ and $C_6H_7^+$, respectively, but the structures of the $C_6H_n^+$ ions have not been elucidated, although it may be noted that the analogous trimerization of neutral acetylene is thought to lead to the formation of benzene via two Diels-Alder additions of acetylene (32). The dimeric intermediate $C_4H_4^+$ and $C_4H_5^+$ ions are also of interest as they may correspond to ionized and protonated cyclobutadiene, respectively, and in analogy with proposed neutral chemistry (32), could add methylacetylene, vinylacetylene and diacetylene to form ionized and protonated toluene, styrene and phenylacetylene. Similar processes may lead to the formation of the fused-ring ions of indene and naphthalene from a combination of acetylene, methylacetylene and vinylacetylene in the case of indene, and a combination of either acetylene and two vinylacetylene molecules, or acetylene, methylacetylene and methyldiacetylene in the case of naphthalene.

Other analogous combinations of larger chain-like carbon units can be envisaged to lead directly to larger fused-ring structures. For example, two intriguing possibilities for the growth of graphite-like molecules are shown below:

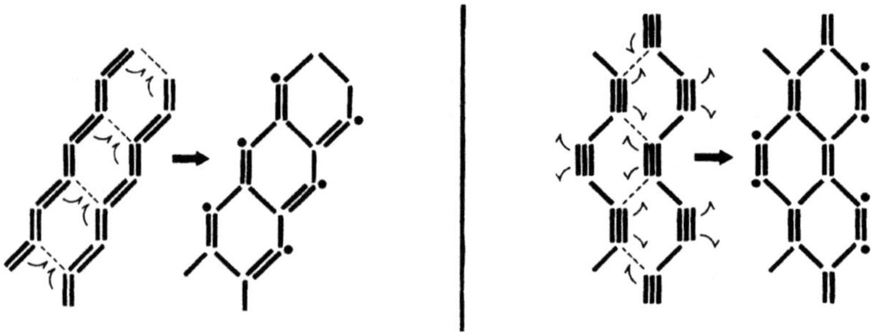

Side-on attack between cumulenes (left) or polyacetylenes (right), where one of the reactants may be ionic, can lead to carbon-ring formation through cross-bonding. This corresponds to the inverse of the mechanism proposed for the transformation of graphite at high temperatures (33) which is analogous to the transformation of benzene to acetylene. Indeed, the reactions in ionized allene and methylacetylene indicated above which lead to protonated benzene can be viewed as the lowest members in a series of such reactions. The next members in this series would be reactions in ionized $H_2C{=}C{=}C{=}CH_2$ and 1,3-pentadiyne which might lead to ionized naphthalene by the side-on combination mechanism.

4.2. Growth from Benzene

Growth of PAH molecules can be conceived to occur from benzene by the successive grafting of new hexagonal rings. A new hexagonal ring may be formed by the sequential addition of 1,2,3 or 4 carbon units through the development of a side chain which ultimately closes. Alternatively, the new hexagonal ring may be realized in one step through a condensation reaction with a polycarbon species as illustrated on the next page. This latter mechanism has been proposed for the high-temperature polymerization of PAH's (34) and is the basis for the perceived growth of the large aromatic ions which are believed to be precursors in the ionic formation of soot in hydrocarbon flames (35). The detailed features of the mechanism of such growth have not yet been demonstrated

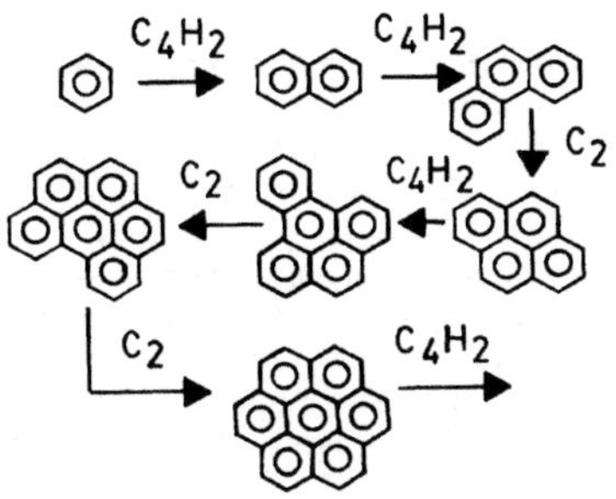

or characterized in the laboratory, but exploratory measurements have provided insight into the first step. The $C_6H_6^+$ cation derived from benzene has been found to add diacetylene at close to the collision rate in helium buffer gas at 296 K (36). The high efficiency suggests strong chemical bonding in the $C_{10}H_8^+$ adduct ion which may therefore be ionized naphthalene. Comparative charge-transfer reactivity studies with trimethylamine and styrene showed no difference between $C_{10}H_8^+$ produced from the addition reaction and $C_{10}H_8^+$ produced from the direct ionization of naphthalene (36).

Formation of PAH molecules through the growth of side chains with eventual cyclization may be more problematic. Rings may be formed which are not necessarily hexagonal and may contain 4,5,7, etc. C atoms. Also, the side chains are likely to be chemically fragile so that their growth may be impeded by competitive fragmentation. Successive addition of single-carbon units has been proposed for growth of side chains in interstellar gas clouds in which C or C^+ are relatively abundant (37). Other ionic building blocks in these environments might be CH_3^+, $C_2H_2^+$, $C_2H_3^+$ and $C_3H_3^+$. Alternatively, the growth might proceed through reactions of ionized aromatic molecules with neutral building-block molecules such as methane, acetylene and methylacetylene. Very little experimental information is available for reactions of aromatic molecules with C^+ ions or of aromatic ions with C atoms. A recent SIFT study has shown that 10% of the fast reaction of C^+ with benzene leads to condensation to form $C_7H_5^+$ with the remainder proceeding by charge transfer (67%) and fragmentation to form $C_5H_3^+$ and $C_3H_3^+$ (23%) (21). Other laboratory studies have demonstrated the occurrence of condensation in reactions of various hydrocarbon ions with benzene and substituted benzenes such as, for example, the reaction of $C_2H_3^+$ with benzene and the reactions of the linear isomer of $C_3H_3^+$ with benzene and ethylbenzene and the cyclic isomer of $C_3H_3^+$ with vinylbenzene (38).

5. POLYCARBON AND HYDROCARBON SPHERES.

Large PAH molecules resemble graphite-like sheets of hexagonal rings. Some, such as corannulene (39), have non-planar saucer shapes and contain pentagon rings. Also, it seems that when the edges of such a molecule lose their hydrogen and become reactive, the resulting polycyclic graphite-like molecule, if large enough, can spontaneously close into a spheroidal or cage-like geometry (40). Recent experimental studies of polycarbon clusters produced from laser vaporization of graphite and polycarbon ions produced in hydrocarbon flames suggest special stabilities for the fullerenes, C_n or C_n^+, with n = 24, 28, 32, 36, 50, 60, and 70 (41,42). The most stable and remarkable of these is C_{60} (or C_{60}^+), buckminsterfullerene, which has the symmetry of a European football. Kroto has suggested the ejection of such structures from stars into interstellar space, without being specific about the meachanism of their formation (43). In the context of the growth of ions and molecules discussed here, one can envisage formation of spheroidal ions or molecules from smaller units by the face-to-face approach of two sheets or saucers of carbon. Other mechanisms also are possible. For example, Smalley and co-workers have discussed a model for the step-wise growth of polycyclic aromatic molecules in sooting flames, in which C-C bonding is maximized by the incorporation of pentagons into the growing aromatic network. These pentagons generate curvature which brings the net back on itself to form a spheroidal shell. When some of the dangling bonds are occupied with hydrogen, the growing carbon net may close imperfectly and form a new concentric shell. The result of such growth would be a soot nucleus consisting of concentric, but slightly imperfect spheres (41). Dare we expect a similar morphology for grain-like species in space?

6. CONCLUSIONS.

Concomitant laboratory measurements, theoretical studies, and model calculations are beginning to provide some real insight into the homogeneous formation of hydrocarbon ions and molecules in interstellar clouds and circumstellar shells. The laboratory measurements and theoretical studies clearly imply that chain-like polycarbon and hydrocarbon ions and molecules, with and without atomic and molecular substituents such as O, S, CN and CH_3, can readily be grown from small carbon units and other small molecules in the presence of ionization. New experimental results continue to be incorporated in chemical models for dense interstellar clouds (44). However, there is a pressing need for a much more extensive experimental and theoretical data base for reactions involving large unsaturated hydrocarbon ions and molecules. Unfortunately, experiments involving such species become increasingly difficult to perform and define. The unsaturated hydrocarbon molecules become increasingly difficult to handle and to synthesize as they become larger. Also, reactant and product ions become increasingly difficult to identify since isomeric variations become much more numerous for large molecules. But we need not despair. One new technique which is most promising is that developed for studies of the condensation of gaseous carbon species derived from graphite by laser vaporization. Very recent experiments using technique both in the absence and presence of gases such as H_2, N_2, H_2O, NH_3 and CH_3CN, have provided valuable new insight into the formation of carbon-chain molecules such as polyacetylenes, cyanopolyacetylenes, dicyanopolyacetylenes and radical species such as $\cdot C_nH$ with up to 20 or more carbon atoms (45).

More measurements are needed on the kinetics for the hydrogenation of hydrocarbon ions in hydrogen, and for reactions of polycarbon and hydrocarbon ions with atoms such as N, C and O, with molecular radical species such as CN and with carbon chains such as C_n and C_nH. Still very little is known about the efficiency of the radiative association of ions with molecules. Also, very little is known about the products of the recombination of ions with electrons. Information of this kind has become critical for our understanding of the gas-phase chemistry of complex hydrocarbon ions and molecules.

Similar comments apply to the formation of cyclic, polycyclic and spherical aromatic hydrocarbon ions and molecules, but here the lack of experimental data is even more severe.

REFERENCES

1. W.M. Irvine, F.P. Schloerb, A. Hjalmarson and E. Herbst, in Protostars & Planets II, D.C. Black and M.S. Mathews (eds.), The Univ. of Arizona Press, Tucson, Arizona, 1985.

2. A. Omont, in Mass Loss from Red Giants, M. Morris and B. Zuckerman (eds.), D. Reidel Publ. Co., 1985, p. 269.

3. E. Herbst, in Origins of Life, 16, 3 (1985). D. Reidel Publ. Co.

4. H.W. Kroto, D. McNaughton, L.T. Little and N. Matthews, Mon. Not. R. astr. Soc. 213, 753 (1985).

5. H.E. Matthews and W.M. Irvine, Astrophys. J. 298, L61 (1985); P. Thaddeus, J.M. Vrtilek and C.A. Gottlieb, Astrophys. J. 299, L63 (1985).

6. S. Yamamoto, S. Saito, M. Ohishi, H. Suzuki, S.-I. Ishikawa and N. Kaifu, Astrophys. J. 322, L55 (1987).

7. J.H. Fertel and B.E. Turner, Astrophys. Letters 16, 61 (1975), and references therein.

8. A. Leger and J.L. Puget, Astron. Astrophys. 137, L5 (1984); L.J. Allamandola, A.G.G.M. Tielens and J.R. Barker, Astrophys. J. 290, L28 (1985).

9. G.P. van der Zwet and L.J. Allamandola, Astron. Astophys. 146, 76 (1985); A. Leger and L. d'Hendecourt, Astron. Astrophys. 146, 81 (1985).

10. H.W. Kroto, in Polycylic Aromatic Hydrocarbons and Astrophysics, A. Leger et al. (eds.), D. Reidel Publ. Co., 1987, p. 197.

11. D.K. Bohme, in Structure/Reactivity and Thermochemistry of Ions, P. Ausloos and S.G. Lias (eds.), D. Reidel Publ. Co., 1987, p. 219.

12. A.E. Douglas, Nature 269, 130 (1977).

13. H. Suzuki, Astrophys. J. 272, 579 (1983).

14. D.R. Bates, Astrophys. J. 267, L121 (1983).

15. S.W. McElvany, B.I. Dunlap and A. O'Keefe, J. Chem. Phys. 86, 715 (1987).

16. J. Bernhole and J.C. Phillips, Phys. Rev. B 33, 7395 (1986); J. Chem. Phys. 85, 3258 (1986).

17. K. Raghavachari and J.S. Binkley, J. Chem. Phys. 87, 2191 (1987).

18. S.E. Butrill, J.K. Kim, W.T. Huntress, P. LeBreton and A. Williamson, J. Chem. Phys. 61, 2122 (1974).

19. S.E. Barlow, G.H. Dunn and M. Schauer, Phys. Rev. Letters 52, 902 (1984).

20. N.G. Adams and D. Smith, Chem. Phys. Letters 47, 383 (1977).

21. D.K. Bohme, A.B. Rakshit and H.I. Schiff, Chem Phys. Letters 93, 592 (1982).

22. T.J. Buckley, L.W. Sieck, R. Metz, S.G. Lias and J.F. Liebman, Int. J. Mass Spectrom. Ion Processes 65, 181 (1985).

23. N. Kaifu, H. Suzuki, M. Ohishi, T. Miyaji, S. Ischikawa, T. Kasuga, M. Morimoto and S. Saito, Astrophys. J. 317, L111 (1987).

24. D.K. Bohme, Nature 319, 473 (1986).

25. A.B. Raksit and D.K. Bohme, Int. J. Mass Spectrom. Ion Processes 55, 69 (1983).

26. D.K. Bohme, S. Wlodek. L. Williams, L. Forte and A. Fox, J. Chem. Phys., 87, 6934 (1987).

27. E. Herbst, D. Smith and N.G. Adams, Astron. Astrophys. 138, L13 (1984).

28. D.K. Bohme, S. Wlodek and A.B. Raksit, Can. J. Chem. 65, 2057 (1987).

29. N.G. Adams and D. Smith, Astrophys. J. 317, L25 (1987).

30. D.K. Bohme, S. Dheandhanoo, S. Wlodek and A.B. Raksit, J. Phys. Chem. 91, 2569 (1987).

31. S.G. Lias and P. Ausloos, J. Chem. Phys. 82, 3613 (1985).

32. A.N. Hayhurst and H.R.N. Jones, J. Chem. Soc. Faraday Trans. 2 **83**, 1 (1987).

33. A.G. Whittaker, Science **200**, 763 (1978); R.B. Heimann, J. Kleiman and N.M. Salansky, Carbon **22**, 147 (1984).

34. S.E. Stein, J. Phys. Chem. **82**, 566 (1978).

35. H.F. Calcote, Combustion and Flame **42**, 215 (1981).

36. Unpublished results from the Ion Chemistry Laboratory at York University.

37. A. Omont, Astron. Astrophys. **164**, 159 (1986).

38. C. Lifshitz and B.G. Reuben, J. Chem. Phys. **82**, 3613 (1985); K.C. Smyth, S.G. Lias and P. Ausloos, Combustion Science and Technology **28**, 147 (1982).

39. W.E. Barth and R.G. Lawton, J. Am. Chem. Soc. **93**, 1730 (1971).

40. H.W. Kroto, Nature **329** 529 (1987).

41. Q.L. Zhang, S.C. O'Brien, J.R. Heath, Y. Liu, R.F. Curl, H.W. Kroto and R.E. Smalley, J. Phys. Chem. **90**, 525 (1987).

42. Ph. Gerhardt, S. Loffler and K.H. Homann, Chem. Phys. Letters **137**, 306 (1987).

43. H.W. Kroto, in Polycyclic Aromatic Hydrocarbons and Astrophysics, A. Leger et al. (eds.), D. Reidel Publ. Co., 1987, p. 197.

44. T.J. Millar, C.M. Leung and E. Herbst, Astron. Astrophys. **183**, 109 (1987).

45. H.W. Kroto, J.R. Heath, S.C. O'Brien, R.F. Curl and R.E. Smalley, Astrophys. J. **314**, 352 (1987).

STUDIES OF ION-MOLECULE REACTIONS AT T < 80K

B.R. Rowe
Laboratoire d'Aérothermique du C.N.R.S.
4ter route des gardes
92190 Meudon, France

ABSTRACT. Three new techniques are presented which allow rate coefficient measurements for ion molecule reactions in the temperature range of interstellar clouds. With an ion trap and a static drift tube cryogenic cooling by liquid helium is used. These techniques are therefore restricted to molecular hydrogen as neutral reactant. With the CRESU method the low temperature is obtained by supersonic expansion. This last technique has a high chemical versatility.

The results concerning radiative associations, fast and slow exothermic reactions and a reaction of very small endothermicity are presented. Some implications for interstellar chemistry are highlighted.

1. INTRODUCTION

The formidable development of radioastronomy over the last twenty years has led to the unambiguous identification of more than eighty molecules in interstellar space. Further discoveries will certainly come from new observation facilities, such as the IRAM radiotelescope, recently built at Veleta Peak in Spain. The more complex molecules have been detected in dense interstellar clouds, where the temperature is typically as low as 10K. A good knowledge of the physical and chemical processes leading to molecule formation in this temperature range is essential, not only for a detailed understanding of the molecule synthesis itself, but also in order to deduce important cloud parameters (such as isotopic ratios). These processes are certainly much more complicated than appeared previously in the first simple gas phase models, as recently pointed out by Williams (see the review paper in these proceedings). However, it is clear that the low temperature implies reactions without activation energy. From this point of view, the key role of ion-molecule reactions has been recognized for a long time now (Solomon and Klemperer, 1972 ; Herbst and Klemperer, 1973 ; Watson, 1974).

The rate coefficient of these reaction has been extensively studied for temperatures higher than 80K by various techniques. These include the Flowing Afterglow apparatus developed in the early sixties by Ferguson and coworkers at the NOAA laboratories in Boulder (E.E. Ferguson et al,1969), the Selected Ion Flow Tube

T. J. Millar and D. A. Williams (eds.), Rate Coefficients in Astrochemistry, 135–152.

(SIFT) primarily designed by Smith and Adams (1979) in Birmingham , ICR cells and beam experiments (Gentry and Giese, 1975). Numerous data at temperatures down to 80K are available today, many of them appearing in a very recent compilation (Anicich and Huntress, 1986).

The lack of experimental results at very low temperature necessitated the extrapolation of rate coefficients from higher temperatures. At the same time, important theoretical efforts were made to predict the temperature dependence of rate coefficients. Several calculations are restricted to this part of the potential surface corresponding to long range intermolecular forces (capture theories). They are generally performed from a classical or semiclassical viewpoint (Su and Bowers, 1979) and result in the so-called capture rate coefficient, which is in fact an upper limit (see the review by D.Clary in these proceedings). Other theoretical works have dealt with association reactions in order to link ternary and radiative associations and to predict the radiative rate coefficient at cloud temperatures (Bates, 1984, 1985 ; Herbst, 1982, 1987). On the other hand, very few calculations involving the complete reaction path have been performed (Spalburg et al, 1985 ; Barlow et al, 1986a).

In order to assess the validity of extrapolations and theoretical predictions, results in the range 10-80K are clearly required. Designing an experiment for these temperatures is a considerable challenge and only recently have three research groups succeeded in such measurements. G.Dunn and coworkers, at the JILA in Boulder, have used an ion trap (Barlow et al, 1986b) and H.Böhringer and F.Arnold (1983), at the Max Planck Institute of Heidelberg, a static drift tube. Both experiments are cooled by liquid helium and are therefore restricted to molecular hydrogen as neutral reactant, any other species being trapped by the apparatus walls. In our laboratory, an uniform supersonic jet is used as a flow reactor (Rowe et al, 1984a,b ; Dupeyrat et al, 1985). This technique, which was called CRESU standing for "Cinétique de Réactions en Ecoulement Supersonique Uniforme", does not require cryogenic cooling down to 27K and requires only liquid nitrogen cooling down to 8K. Measurements with condensable species are thus possible and the temperature dependences of more than thirty-five reactions have already been studied. In the present paper these new methods will be described and the results obtained at very low temperature reviewed. Some general implications for interstellar chemistry will be highlighted.

2. EXPERIMENTAL METHODS BELOW 80K

2.1. The ion trap

A sketch of the system used by Dunn and coworkers is shown on figure 1. A penning trap, centered in a superconducting solenoid is cooled to 11K by liquid helium. The electrostatic potential and magnetic field have a shape able to trap the ions for very long periods (several hours). The residual pressure is around 10^{-14} Torr and the working pressure 1-4 10^{-11} Torr. Ions are created by electron bombardment of a parent gas which is then pumped away. Afterward the reactant gas (i.e. hydrogen) is introduced. To be detected the reactant ions are excited (i.e. heated) on a short time by a small rf impulse at a frequency characteristic of the given m/e. This yields an increase of the ion signal which is proportional to their density, followed by a decrease

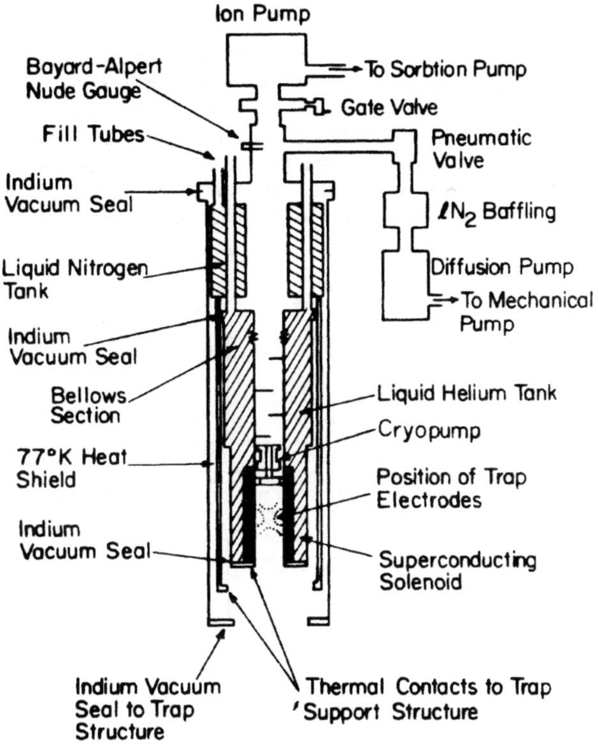

Figure 1 - Schematic view of the cooled ion trap (reprinted from Barlow et al - 1986)

due to the cooling by collisions with hydrogen. Repeating several heating and cooling cycles and monitoring the reactant ion density versus time yield the rate coefficient.

The ions being stored in the trap for very long periods, they are usually radiatively deexcited on their ground state (electronic and vibrational). This is a great advantage of the technique. In dihydrogen, due to the very slow ortho-para conversion, both rotational states J=1 and J=0 are present in this experiment. However it is unclear if they are in the normal 3:1 ortho-para ratio or if some conversion does occur on the apparatus walls.

Due to the ion detection technique, the trap method is not particularly well-suited for the determination of rate coefficient of very fast reactions (in this case, the characteristic time of a heating-cooling cycle is comparable to the reaction time) but is especially valuable for measurements of reaction rate coefficients between 10^{-11} and 10^{-15} cm^3s^{-1}.

2.2. The static drift tube

The selected-ion static drift tube built in the Max Planck Institute of Heidelberg is sketched in figure 2. Reactant ions are generated in an electron impact source, selected by a quadrupole mass spectrometer and then focused onto the entrance aperture of the drift tube. A homogeneous electric field is established inside the drift tube, which is filled with a buffer gas. Under the field influence, ions move through the tube and some of them pass through an orifice at the rear end plate. They are analysed by a second quadrupole mass filter and detected by a channeltron multiplier with a pulse counting device.

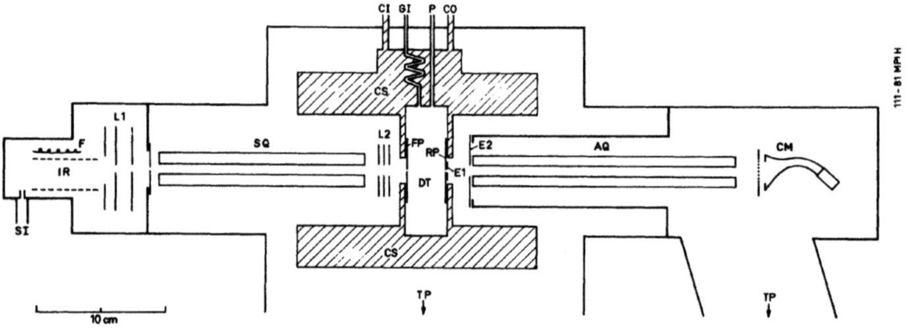

Figure 2 - Schematic view of the cooled static drift tube : primary ion source (F, filament ; IR, ionisation region ; SI, source gas inlet) ; SQ, primary ion selector quadrupole ; L1 and L2, focusing lenses ; DT, drift tube ; FP, front plate ; RP, rear plate ; E1 and E2 aperture plate, AQ, analyser quadrupole ; CM, channel multiplier ; CS, cryostat system ; GI, gas inlet ; P pressure-gauge port, CI and CO coolant inlet and outlet (reprinted from Böhringer and Arnold, 1983)

The rate coefficient is measured in the standard way for drift tube experiments by monitoring reactant and product ion densities versus the reactant gas density. The electric field within the tube yields an additional ion kinetic energy which is available for the reaction. In order to obtain the true thermal rate coefficient, measurements are performed for various values of the electric field/density ratio and the results extrapolated to zero drift energy. This apparatus can be operated at quite high pressure (up to eight Torr) therefore allowing the determination of rate coefficients for quite slow reactions with a limit around 10^{-13} cm^3s^{-1}.

2.3. The CRESU method

For the lowest temperatures, the ion trap and the static drift tube are limited to molecular hydrogen as neutral reactant. While, due to its abundance, hydrogen has a key role in interstellar chemistry, a good understanding of molecule synthesis clearly requires the knowledge of rate coefficient values for many other reactants. Also these experiments are slow and tedious and require a great deal of skill to operate.

With the CRESU method, developed at the Laboratoire d'Aérothermique in Meudon, a completely different approach to obtaining the very low temperature was used. A buffer gas in which reactions take place is cooled down by supersonic expansion. This means of cooling a gas appears quite obvious and has been used widely as a source of low temperature for spectroscopic and cluster studies (Hermann et al, 1982 ; Miller, 1984). But all these previous experiments were performed using the so-called free jet which exhibits very strong pressure and temperature gradients (Dupeyrat, 1979). Under such circumstances it appears quite impossible to obtain a rate coefficient at a given temperature. The most original and essential feature of the CRESU technique is its use of uniform supersonic jets as flow reactors. Such flows, generated by Laval nozzles, have been used for decades now by aerodynamicists for studies of rarefied gas flows around bodies (Maslach and Sherman, 1956). In their centre such flows are isentropic and temperature, density and velocity keep constant values. Outside the centre, within the so-called boundary layer, viscous effects dissipate energy and slow down the flow. If the viscous boundary layer is too large compared to the nozzle diameter, there is no longer an isentropic core (i.e. an uniform flow). This implies a minimum size of the nozzle for a given flow pressure. Low pressures (< 10^{-1}Torr) are also requested to avoid neutral dimer formation and strong ternary association of ions with the buffer gas. In order to satisfy these two conditions, exceptional pumping capacity is required. A convenient facility was in fact already available in our laboratory i.e. the rarefied gas wind tunnel SR3.

The calculation of a suitably contoured Laval nozzle is tedious and will normally require a solution of the Navier-Stokes equations, including heat transfer. In fact the calculation is performed under the boundary layer approximation (Dupeyrat et al, 1985). An isentropic core is calculated neglecting viscosity and thermal conductivity. Then, outside this core, the flow is calculated by using a perturbation method. Several nozzles have been calculated which use either helium or nitrogen/oxygen buffer gases. The flow conditions are summarized in table I.

A sketch of the apparatus is shown in figure 3. The buffer, parent, and reactant gases are driven to the nozzle reservoir, where they are mixed, through Tylan flow controllers. The reservoir can be either at room temperature or cooled down by liquid nitrogen. Close to the nozzle exit, the uniform supersonic flow is ionized by an electron beam in order to generate the reactant ions. These ions, as well as the product ions, are monitored by a quadrupole mass spectrometer with a channeltron and particle counting system, pumped by a turbomolecular pump. The sampling orifice is located at the tip of a skimmer in order to prevent shock wave problems. The flow being uniform, the equation which links the reactant ion density $[A^+]$ to the reactant neutral density [R] and the distance x between the electron beam and the sampling orifice is :

$$[A^+] = [A^+]_0 \exp(- k\,[R]\, x/v)$$

v being the flow velocity. Monitoring $[A^+]$ versus [R] or x yields the binary rate coefficient k (or the apparent binary rate for a termolecular process).

buffer gas	M	T (K)	P_∞ (torr)	velocity (cm s^{-1})	density (cm $^{-3}$)
He	3	20	$2.06\ 10^{-2}$	$7.87\ 10^{4}$	$9.64\ 10^{15}$
		67.5	$6.4\ 10^{-2}$	$1.51\ 10^{5}$	$9.2\ 10^{15}$
He	5	8	$8\ 10^{-3}$	$8.34\ 10^{4}$	$1.27\ 10^{16}$
		27	$3.9\ 10^{-2}$	$1.47\ 10^{5}$	$1.06\ 10^{16}$
N_2 - O_2	4	20	$5\ 10^{-3}$	$3.54\ 10^{4}$	$2.23\ 10^{15}$
		68	$3.3\ 10^{-2}$	$6.78\ 10^{4}$	$4.77\ 10^{15}$
N_2 - O_2	2	45	$1.8\ 10^{-2}$	$2.73\ 10^{4}$	$3.85\ 10^{15}$
		163	0.102	$5.18\ 10^{4}$	$6.0\ 10^{15}$

Table I - Different flow conditions used in the CRESU experiment

The buffer gas density is always sufficiently large to yield a relaxation time toward the Maxwell-Boltzmann velocity distribution always much smaller (typically 10^{-6} sec.) than the gas residence time (typically 10^{-3} sec). This ensures a real thermal equilibrium with a kinetic temperature T. Generally, the rotational relaxation time is nearly as small as the translational, and rotational states are in equilibrium at this temperature. A noticeable exception is molecular hydrogen, due to the large energy spacing in this case. There is usually no vibrational excitation of the reactant neutral since the highest temperature prevailing in the nozzle reservoir is room temperature. The reactant ions are formed by different processes: He^+, $O_2{}^+$ and $N_2{}^+$ are formed respectively in helium, oxygen and nitrogen buffer by electron impact ionization, presumably in excited states. However excited electronic states are quickly relaxed to the ground state by radiative cascade. On the other hand, collisions with the parent gas quench vibrational excitation very efficiently. In helium buffer, the electron beam produces mainly He^+ ions and metastable helium, He^m. N^+ and C^+ ions are formed by dissociative charge transfer of He^+ with a parent gas (respectively nitrogen and carbon monoxide. Only the ground $N^+(^3P)$ and $C^+(^2P^0)$ states are produced exothermically by these reactions. Penning ionization by He^m cannot yield these ions. The problem of fine structure states will be discussed later.

In the present state of the experiment, the CRESU method appears very similar to the old flowing afterglow, excepted for the nature of the flow. It has the same chemical versatility and is very simple to operate. It is especially suitable for measurements of reaction rate coefficients in the range $>10^{-12}$ cm^3s^{-1}. From this point of view, the ion trap and the drift tube complement the CRESU measurements.

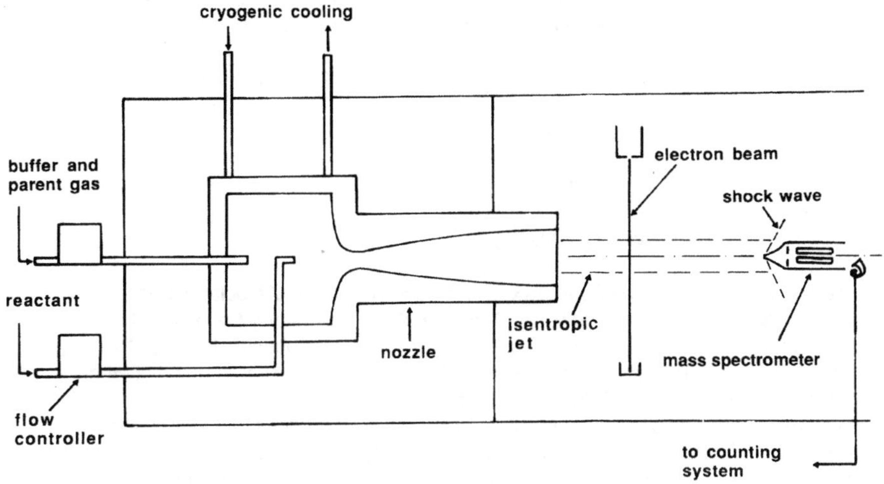

Figure 3 - Schematic view of the CRESU apparatus

3. RATE COEFFICIENT AT VERY LOW TEMPERATURES

3.1. Association reactions

Radiative association reactions are thought to be an important route for the synthesis of some interstellar molecules. This process is however slow, and, in laboratory experiments, is usually dominated by termolecular reactions. The extremely low pressure used in the cooled ion trap makes this technique especially valuable for direct study of radiative association. This method has provided the first unambiguous observations of radiative association concerning:

(1) $CH_3^+ + H_2 \rightarrow CH_5^+ + h\nu$

(2) $C^+ + H_2 \rightarrow CH_2^+ + h\nu$

(3) $HCO^+ + H_2 \rightarrow H_3CO^+ + h\nu$

with a rate coefficient k_r of 1.1 10^{-13} cm^3s^{-1} for reaction (1) (Barlow et al, 1984a,b) and upper limits of 1.5 10^{-15} cm^3s^{-1} and 2.5 10^{-15} cm^3s^{-1} for reactions (2) and (3) (Luine and Dunn, 1985). The value of k_r is nearly an order of magnitude larger than that available from theoretical calculations (Bates, 1983 ; Herbst, 1983). In these studies the effect of ortho-para hydrogen ratio on the rate coefficients was not evaluated and could be of some importance for interstellar chemistry.

The radiative and termolecular association processes can be modelized as:

(4) $A^+ + B \rightleftarrows (AB^+)^*$

(5) $(AB^+)^* \rightarrow AB^+ + h\nu$

(6) $(AB^+)^* + M \rightarrow AB^+ + M$

In the first step (4) an excited complex is formed which either dissociates back to the reactants, or is stabilized by radiation (5) or by collision with a third body (6). Then the radiative association can be linked to the termolecular process if the radiative lifetime of the complex and the third body stabilization efficiency are known (Bates, 1984, 1985 ; Herbst, 1982, 1987) . The termolecular rate coefficient has been extensively studied between 80 and 300K, generally displaying a T^{-n} temperature dependence. The question arises whether such a dependence is valid at the lowest temperatures. Böhringer et al (1983) have studied the association reaction:

(7) $He^+ + 2He \rightarrow He_2^+ + He$

from 30 to 350K. They found a $T^{-0.6}$ temperature dependence for the whole temperature range in good agreement with some theoretical calculations of Dickinson et al (1972). In previous measurements, they studied the reactions:

(8) $N_2^+ + 2N_2 \rightarrow N_4^+ + N_2$

(9) $O_2^+ + 2O_2 \rightarrow O_4^+ + O_2$

down to respectively 45 and 51K, finding a deviation from the power law below 80K. This behavior was not confirmed by further CRESU studies of these reactions down to 20K (Rowe et al, 1984b), while in good agreement with the cooled drift tube results at higher temperatures. For reaction (8) a $T^{-1.77}$ temperature dependence can be deduced in the 20-400K temperature range although only lower limits were deduced from the CRESU measurements on reaction (9).

The problem of intermediate complex is of some importance in ion-molecule chemistry and from this point of view, experimental study of association reactions can provide very useful information (Böhringer et Arnold, 1985 ; Böhringer, 1985). Much more low temperature work is clearly needed in this field.

3.2. Fast binary reactions

Ion-molecule rate coefficients k can be viewed as the product of a reaction probability P by a capture rate coefficient k_c. This corresponds to a division of the potential surface into two parts, one corresponding to the long range intermolecular forces driving the collision partners close together, the second to the short range forces, where the chemical bonds are effectively rearranged. If the distance r between the centre of mass of the ion and the neutral comes to zero during the collision driven by the long range forces, this collision is considered as leading to a capture. This happens when there is enough energy along the line of centres to overcome the so-called centrifugal barrier (Su and Bowers 1979). The reaction probability is very often close to unity for fast exothermic reactions, leading therefore to $k=k_c$. The value of k_c itself is dependent of the nature on the long range intermolecular potential. Three cases have been considered.

The ion-induced dipole potential

For a non-polar molecule, if the effects of the quadrupole moment and of anisotropic polarizabilities are neglected, the long-range potential has the form:

$$(10) \quad V(r)= -\alpha q^2/2r^4$$

where q, r and α are the ion charge, the distance between the centres of mass of the ion and the neutral, and the polarizability respectively. The rate coefficient calculations then yield the well known temperature independent Langevin value:

$$(11) \quad k = 2\pi q(\alpha/\mu)^{1/2}$$

Using the CRESU technique, the reactions of He^+ ions with N_2, O_2 and CO and of N^+ ions with O_2, CH_4, CO have been studied down to 8K (Rowe et al, 1985). The results, together with previous 300K and 88K values, show that over this entire temperature range the rate coefficient of these reactions is very close to the Langevin capture rate constant.

The quadrupole interaction potential

If, in addition to the ion induced dipole, the static quadrupole moment Q is considered, the long range potential takes the form:

$$(12) \quad V(r)= -\alpha q^2/2r^4 + Qq(3 \cos^2\Theta - 1)/2r^3$$

Θ being the angle that the electrical symmetry axis of the molecule makes with r. The 300K value can be calculated using the AQO formula (Su and Bowers, 1979) and various theories (Kosmas, 1985, Bates and Mendas, 1985) predict a small increase of the reactivity at low temperatures. Reactions of He^+, C^+ and N^+ ions with C_6F_6 and $c\text{-}C_6H_{12}$ have been examined at 27, 68 and 297K using both CRESU and SIFT techniques (Rebrion et al, 1987). These non-polar molecules present very large quadrupole moment. However, within the experimental uncertainties the rate coefficients were found temperature independent, as shown in table II. The discrepancy between theory and experiment could come from the failure of the adiabatic invariance hypothesis for the rotational motion of the neutral molecule.

reaction	27K	68K	300K
$He^+ + C_6F_6$	3.4	3.4	4.1
$C^+ + C_6F_6$	3.1	2.1	2.3
$N^+ + C_6F_6$	2.1	1.9	2.1
$He^+ + c\text{-}C_6H_{12}$	3.0	3.5	4.1
$C^+ + c\text{-}C_6H_{12}$	2.5	2.4	2.6
$N^+ + c\text{-}C_6H_{12}$	2.0	2.2	2.4

Table II - Rate coefficients (in $10^{-9}cm^3s^{-1}$) for reactions with molecules of high quadrupole moments. The 300K values are Birmingham SIFT measurements

The permanent dipole interaction potential

For the case of a polar molecule having a permanent dipole D, the long range potential takes the form:

$$(13) \quad V(r) = -\alpha q^2/2r^4 - qD\cos\Theta/r^2$$

where Θ is the angle the dipole makes with r. There are a very large number of theoretical calculations of the rate coefficient in this case. The very simple locked-dipole theory assumes $\Theta=0$ but greatly overestimates the rate coefficient. In the pioneering ADO (average dipole orientation) treatment (Su and Bowers, 1973, 1979) a mean value of Θ or cos Θ is calculated during the collision. This theory was improved in the AADO version which takes into account the conservation of total angular momentum (Su et al, 1978). However it has been suggested that they can underestimate the rate coefficient at the lowest temperatures (Marquette et al, 1985a; Clary et al, 1985). Several other models are linked either with the transition state theory (Chesnavitch et al, 1980), with an adiabatic invariant (Bates, 1981) or with trajectory calculations (Su and Chesnavich, 1982). The most recent theories are the ACCSA (adiabatic capture centrifugal sudden approximation) of Clary (1987) and the SACM (statistical adiabatic channel model) of J.Troë (1987).

All theories predict a large increase of the rate coefficient at very low temperature. Due to the condensable character of polar molecules, the available experimental data below 80K have been obtained using the CRESU technique and are summarized in table III. The 300K values have been obtained either from the litterature or from recent SIFT measurements by Smith and Adams. A large increase of the reactivity is effectively observed when the temperature decreases. These results have been systematically compared with the values deduced from the ACCSA and SACM theories and with the simple formula of Su and Chesnavich (1982) resulting from an empirical fit of several trajectory calculations (Marquette et al, 1985a ; Clary, 1987; Rebrion et al, 1988a and b). It appears that the theoretical results agree together for temperatures down to 20K. They fit the experimental data very closely for around 50% of the studied reactions as shown in figure (4). Otherwise the experimental values are lower than the theoretical ones as shown in figure (5) for the C^++HCl reaction. In these cases the reaction probability is lower than unity, indicating that short range intermolecular forces are involved in the reaction process. However, it can be noticed that the shape of the temperature dependence is quite well reproduced by the theory i.e. that the reaction probability is nearly temperature independent.

reaction	300K	163K	68K	27K
$He^+ + NH_3$	1.65		3.0	4.5
$He^+ + H_2O$	0.45		1.8	4.3
$C^+ + NH_3$	2.3		3.2	4.6
$C^+ + H_2O$	2.5		5.2	12
$N^+ + NH_3$	2.4	1.3	3.2	5.2
$N^+ + H_2O$	2.8	3.8	6.0	9.9
$He^+ + HCl$	3.3		4.6	11
$He^+ + SO_2$	4.3		6.5	8.2
$He^+ + H_2S$	2.8		4.6	5.5
$C^+ + HCl$	1.0		1.9	3.8
$C^+ + SO_2$	2.3		4.1	5.7
$C^+ + H_2S$	1.7		3.1	4.8
N^+ + trans $C_2H_2Cl_2$	1.8	2.3	1.7	1.9
N^+ + 1,1 $C_2H_2Cl_2$	2.2	2.7	3.0	3.4
N^+ + cis $C_2H_2Cl_2$	2.4	3.2	3.4	4.5

Table III - Reactions of He^+, C^+ and N^+ ions with polar molecules (excepted trans $C_2H_2Cl_2$). The listed rate coefficients are in units of $10^{-9}cm^3s^{-1}$

The case of N^+ reaction with various isomers of dichloroethylene is of special interest. With the trans isomer which has no permanent dipole moment the rate coefficient is very close to the Langevin value ($1.8\ 10^{-9}\ cm^3s^{-1}$) at all temperatures. For the polar isomers the rate coefficient is lower than the calculated capture rate but, at 300K, SIFT measurements shows that this is also true for reactions with H_3^+. Moreover the ratio of the experimental value of the rate coefficients for N^+ and H_3^+ is very close to the ratio of the square roots of the reactant reduced mass (Rebrion et al, 1988b). Then there are several indications that these reactions are effectively capture processes. The discrepancy between theory and experiment has been tentatively explained suggesting that the favoured angle of approach of the ion for reaction to occur is not co-linear with the dipole.

While in the present state of the CRESU experiment, data cannot be obtained below 27K for condensable species, extrapolation of the results down to 10K (a typical temperature for dense interstellar clouds) clearly shows an enhancement by an order of magnitude for the rate coefficients of very polar molecules. This has important implications for interstellar chemistry, as discussed by E.Herbst in these proceedings.

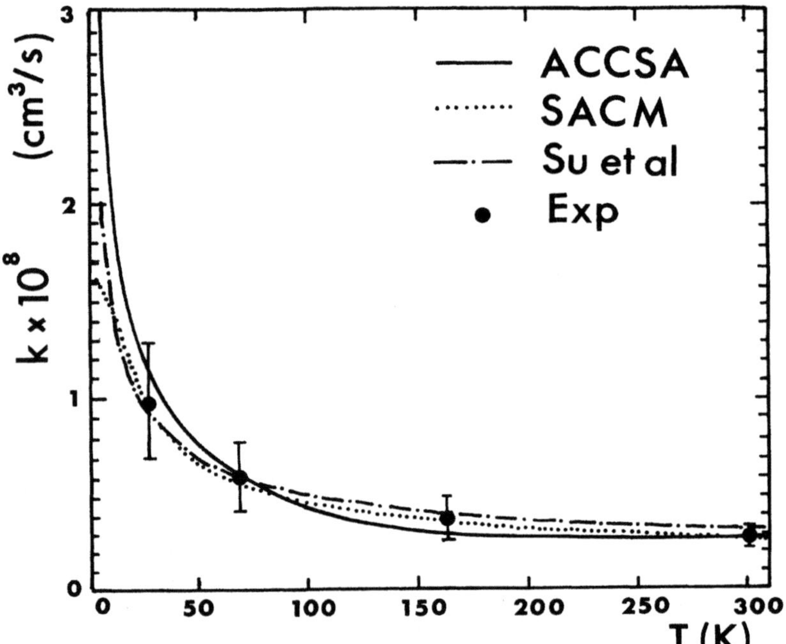

Figure 4 - The reaction of N^+ with H_2O

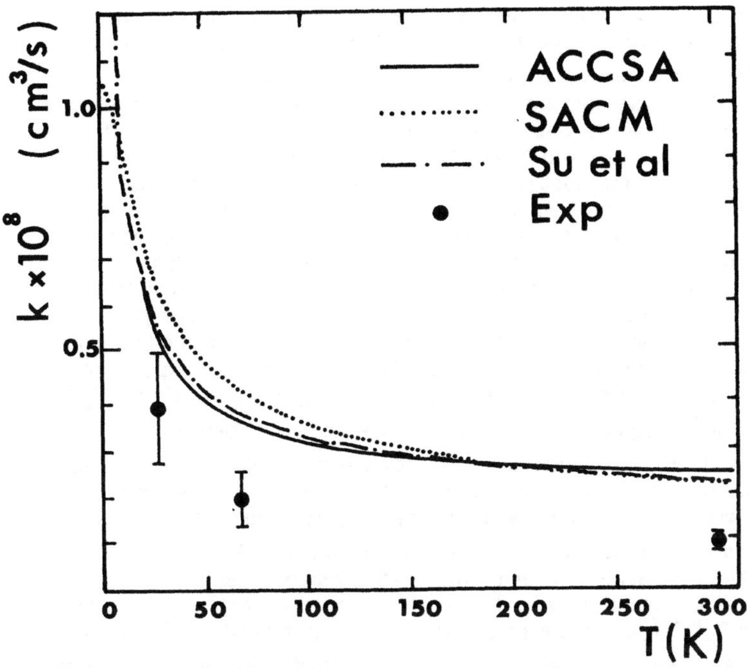

Figure 5 - The reaction of C^+ with HCl

3.3. Slow exothermic reactions

At room temperature exothermic ion-molecule reactions sometimes exhibit a rate coefficient order of magnitude smaller than the capture prediction. For these reactions a rapid increase of the reactivity with decreasing temperature is sometimes noticed. Low temperature measurements for such reactions are thus absolutely required for interstellar chemistry modeling. Moreover their kinetic behavior can reveal interesting details about the reaction dynamic since the whole potential energy surface of the reactant is clearly involved in this case.

The reaction of He^+ with H_2 is extremely slow at room temperature (k= 1.1 10^{-13} cm^3s^{-1}, (Johnsen, Chen and Biondi, 1980, Johnsen and Biondi, 1974). An increase at low temperature would have had very severe implications for interstellar chemistry. However measurements by the cooled drift tube (Böhringer and Arnold, 1986) have shown that the reaction rate coefficient keeps a very small value (lower than 2 10^{-13} cm^3s^{-1}) in the 18-408K temperature range. This is consistent with the value of 1.3 $10^{-14}cm^3s^{-1}$ measured at 13K with the ion trap (Barlow, 1984) and with various theoretical predictions (Preston et al, 1978, Hopper, 1980). These results clearly demonstrate that this reaction does not occur in interstellar conditions.

The reaction of NH_3^+ ion with hydrogen to form NH_4^+ could be involved in the formation of interstellar ammonia (Graedel et al 1982). A decrease of the rate coefficient with decreasing temperature was observed in the 80-600K temperature range by SIFT measurements (Smith and Adams, 1981). However the results of Böhringer (1985) and of Luine and Dunn (1985) down to respectively 20 and 11K show that below 80K the rate coefficient increases. Luine and Dunn make the hypothesis that the decrease at higher temperature is due to a small energy barrier. However at lower temperature the lifetime of the long-lived intermediate complex is sufficient to allow tunneling through the barrier. This influence of the complex lifetime increase was nicely demonstrated by pressure dependent results of Böhringer.

The existence of a transient reaction complex is often invoked to explain the increase of reactivity with decreasing temperature, which is found in numerous slow exothermic ion-molecule reactions (Ferguson, 1972). With the CRESU experiment such behavior has been studied for the reactions $O_2^+ + CH_4 \rightarrow CH_3O_2^+ + H$ and $N_2^+ + O_2 \rightarrow O_2^+ + N_2$ down to respectively 20 and 8K (Rowe et al, 1984a ; Barlow et al, 1986; Gaucherel et al, 1986). The extrapolation of these experimental data down to 0K yields respectively Langevin and half-Langevin value for the rate coefficient. Concerning the O_2^+ reaction with methane and deuterated methane a very complete description of the reaction mechanism has recently been proposed (Barlow et al , 1986) which allows modeling of the temperature dependence of the rate coefficient over the entire 20-500K temperature range. It implies the formation of two transient complexes: the weakly bound $O_2^+.CH_4$ complex is produced and then transformed into $CH_3.O_2H^+$ which, after the rapid loss of an hydrogen, yield the final methylene hydroperoxide ion.

Clearly there exist now too few results on this kind of reaction at very low temperature. Many more experimental and theoretical efforts are needed in order to obtain some general rules for slow reaction behavior.

3.4. The reaction of N^+ ions with hydrogen

The reaction of N^+ ions with molecular hydrogen to form NH^+ ions was thought to initiate ammonia formation in interstellar clouds. Recently several experimental studies, either at very low temperature (Marquette et al, 1985b ; Luine and Dunn, 1985) or at higher energies (Smith and Adams, 1985 ; Ervin and Armentrout, 1987), have shown that this reaction presents a very small activation energy which was unambiguously interpreted as an endothermicity. However the reported endothermicities lie between 11 meV (Smith and Adams) and 33 meV (Ervin and Armentrout). If in terms of thermochemistry these differences are extremely small (as an example it yields a bond energy of NH^+ between 3.51 and 3.531 eV, one of the better determinations for an ion), they yield extreme variations of the expected reaction rate coefficient in the thermodynamical conditions of interstellar clouds.

In order to derive such a small endothermicity from experimental results, the knowledge of the exact states of the reactants and products is absolutely crucial, as stated by Ervin and Armentrout. The energy values of the rotational levels of H_2 or of the fine structure levels of $N^+(^3P)$ are close to the expected endothermicity. Are these energies available to drive the reaction? It seems difficult to answer this kind of question with experiments where the energy of the reactant is well above the endothermicity. In order to obtain a deeper understanding of the reaction energetics, the reactions of N^+ with pure para-hydrogen and with HD have been studied at very low temperature using the CRESU technique (Marquette et al, 1988). The results can be described by the following exponential laws:

(14) $N^+ + n\text{-}H_2$: $k = 4.16\ 10^{-10}\ \exp(-41.9/T)\ cm^3s^{-1}$

(15) $N^+ + p\text{-}H_2$: $k = 8.35\ 10^{-10}\ \exp(-168.5/T)\ cm^3s^{-1}$

(16) $N^+ + HD$: $k = 3.17\ 10^{-10}\exp(-16.3/T)\ cm^3s^{-1}$

From these data various endothermicities were deduced for the reaction

(17) $N^+(^3P_0) + H_2(X^1\Sigma_g^+, J = 0) \longrightarrow NH^+(X^2\Pi, J = 0) + H(^2S)$

following various hypotheses and assuming a form of the reaction cross section similar to the model developed by Levine and Bernstein (1972) for endoergic reaction of an ion with a non-polar molecule. These calculations assume that the fine structure levels of N^+, $^3P_0$, $^3P_1$ and $^3P_2$, reach a Boltzman equilibrium at the temperature of the flow. It has been recognized that this is not the case for the rotational levels of H_2 (Marquette et al, 1988). However for normal and para-hydrogen there are conditions where only the lowest levels are populated (i.e. J=0 for para-H_2 and J=0,1 for normal H_2) allowing an unambiguous determination of the endothermicity. For HD rotational relaxation is much more efficient and thermal equilibrium is obtained. The endothermicity of reaction (17) is calculated taking into account the zero point energy of H_2, HD, NH^+ and ND^+ (Huber and Herzberg, 1979). All the results are summarized in table IV.

When it is assumed that rotational and fine structure energies are available to promote the reaction the CRESU results exhibit a very good internal consistency. Under these hypotheses the ion trap results obtained for normal hydrogen yield a slightly larger endothermicity of 21.2 meV. This could be due to a small ortho-para conversion in the ion trap.

In dense interstellar clouds the state of the reactants is thought to be the same as in reaction (17). Even if the excess of kinetic energy of N^+ ions in their formation process from He^+ ions is taken into account, the present results show that reaction (17) is bound to be a rather inefficient process for interstellar ammonia formation (Herbst et al, 1987).

Available energies	normal-H_2	para-H_2	HD
$E_k + E_r + E_f$	19.5	19.4	16.2
$E_k + E_r$	19.4	13.2	16.0
$E_k + E_f$	4.5	20.3	16.2
E_k	3.6	14.5	16.1

Table IV - Different endothermicities (in meV) assuming that different energies are available to promote the reaction : E_k = kinetic energy, E_r = H_2 or HD rotational energy, E_f = N^+ fine structure energy.

4. CONCLUSION

As shown by the present review experimental studies of ion-molecule reaction at extremely low temperatures have already provided several useful results for interstellar cloud chemistry. The rate coefficient values previously used in chemistry modeling have sometimes been confirmed as correct. This is the case for fast reactions with non-polar molecules which occur at near the Langevin rate at any temperatures. It was also confirmed that the reaction of He^+ ions with hydrogen is extremely slow down to 11K. But for fast exothermic reaction involving polar molecules it was pointed out that the reactivity was strongly underestimated in the models. The CRESU results show that the simple Su and Chesnavich formula (1982) often yields a good approximation of the rate coefficient in this case. However there exist more refined theories allowing calculation of the rate coefficient for each rotational state (Clary, 1987 ; Troë, 1987). Concerning the problem of interstellar ammonia formation, the low temperature measurements demonstrate that the reaction of atomic nitrogen ion with hydrogen is not an efficient route. Therefore it is still a problem to explain ammonia abundance in interstellar clouds.

Of the three experiments presented here the CRESU technique is certainly the most versatile and easy to use. Also it is not restricted to molecular hydrogen as neutral reactant. However the ion trap allows measurements of extremely slow reactions (down to 10^{-15} cm^3s^{-1}) and of radiative association. The static drift tube can be operated at the highest pressures, therefore sometimes reflecting the influence of this parameter. These new methods complement each other and this research field is still widely open. The study reported here on the reactivity of polar molecules needs to be extended to numerous systems. Much more data concerning radiative association, as well as data on isotope exchanges, tunnelling effects, slow reactions, is necessary for chemistry modeling.

The extension of very low temperature measurements toward neutral chemistry or relaxation problems would also be of extreme interest from an astrochemical point of view. The CRESU technique seems especially suitable for this purpose and some efforts are already being made in this direction.

Aknowledgements: I would like to thank my coworkers J.B. Marquette and C. Rebrion for their help in the preparation of this review and Elsevier Science Publishers, G. Dunn and H. Böhringer for the authorization of using Fig. 1 and Fig. 2.

REFERENCES

Adams N.G. and Smith D., 1985, Chem. Phys. Lett., **117**, 67.
Anicich V.G. and Huntress W.T.,Jr, 1986, Astrophys. J. Suppl., **62**, 553.
Barlow S.E., Dunn G.H. and Schauer K., 1984a, Phys. Rev. Lett., **52**, 902.
Barlow S.E., Dunn G.H. and Schauer K., 1984b, Phys. Rev. Lett., **53**, 1610.
Barlow S.E., 1984, Ph. D. Thesis, University of Colorado, Boulder CO 80309.
Barlow S.E., Van Doren J.M., De Puy C.H., Bierbaum V.M., Dotan I., Ferguson E.E., Adams N.G., Smith D., Rowe B.R., Marquette J.B., Dupeyrat G. and Durup-Ferguson M., 1986, J. Chem. Phys., **85**, 3851.
Barlow S.E., Luine J.A. and Dunn G.H., 1987, Int. J. Mass Spectrom. Ion Proc., **74**, 97.
Bates D.R., 1981, Chem. Phys. Lett., **82**, 396.
Bates D.R., 1983, Astrophys. J., **270**, 564.
Bates D.R. ,1984, Chem. Phys. Lett., **112**, 41.
Bates D.R., 1985, Astrophys. J., **298**, 382.
Bates D.R. and Mendas I., 1985, Proc. Roy. Soc. London, **A402**, 245.
Böhringer H. and Arnold F., 1982, J. Chem. Phys., **77**, 5534.
Böhringer H., Glebe W. and Arnold F., 1983, J. Phys. B. : At. Mol. Phys., **16**, 2619.
Böhringer H. and Arnold F., 1983, Int. J. Mass Spectrom. Ion Proc., **49**, 61.
Böhringer H. and Arnold F., 1985, J. Chem. Phys., **84**, 2097.
Böhringer H., 1985, Chem. Phys. Lett., **122**, 185.
Böhringer H. and Arnold F., 1986, J. Chem. Phys., **84**, 1459.
Chesnavich W.J., Su T. and Bowers M.T., 1980, J. Chem. Phys., **72**, 2641.
Clary D.C., Smith D. and Adams N.G., 1985, Chem. Phys. Lett., **119**, 320.
Clary D.C., 1987, J. Chem. Soc. Faraday Trans. 2, **83**, 139.
Dickinson A.S., Roberts R.E. and Bernstein R.B., 1972, J. Phys. B. : Atom. Mol. Phys., **5**, 355.
Dupeyrat G., 1979, Doctoral Thesis, Université de Paris 6.
Dupeyrat G., Marquette J.B. and Rowe B.R., 1985, Phys. Fluids, **28**, 1273.
Ervin K.M. and Armentrout P.A., 1987, J. Chem. Phys., **86**, 2659.
Ferguson E.E., Fehsenfeld. F.C. and Schmeltekopf A.L. 1969, in Advances in Atomic and Molecular Physics, vol. 5, D.R. Bates and I. Estermann Eds., Academic Press, New York, p. 1.
Ferguson E.E., 1972, in Ion-Molecule Reactions, Ed. J.L. Franklin, vol. 2, p. 363.
Gaucherel P., Marquette J.B., Rebrion C., Poissant G., Dupeyrat G. and Rowe B.R., 1986, Chem. Phys. Lett., **132**, 63.
Gentry W.R. and Giese G.F. 1975, J. Chem. Phys., **62**, 1364.
Graedel T.E., Langer W.D. and Frerking M.A., 1982, Astrophys. J. Suppl. Ser., **48**, 321.

Herbst E. and Klemperer W. 1973, Astrophys. J., **185**, 505.
Herbst E. 1982, Chem. Phys., **68**, 323.
Herbst E., 1983, Astrophys. J. Suppl. **53**, 41.
Herbst E., Defrees D.J. and Mc Lean A.D., 1987, Astrophys. J., **321**, 898.
Herbst E., 1987, Astrophys. J., **313**, 867.
Hermann V., Kay B.D. and Castelman A.W., J., 1982, Chem. Phys., **72**, 185.
Hopper D.G., 1980, J. Chem. Phys., **73**, 3289.
Huber K.P. and Herzberg G., 1979, Molecular Spectra and Molecular Structure, Constants of Diatomic Molecules, Van Nostrand Reinhold, p. 240.
Johnsen R. and Biondi M.A., 1974, J. Chem. Phys., **61**, 2112.
Johnsen R., Chen A. and Biondi M.A., 1980, J. Chem. Phys., **72**, 3085.
Kosmas A.M., 1985, J. Phys. (Paris), **46**, L799.
Levine R.D. and Bernstein R.B., 1972, J. Chem. Phys., **56**, 2281.
Luine J.A. and Dunn G.H., 1985, Astrophys. J. Lett., **299**, L67.
Marquette J.B., Rowe B.R., Dupeyrat G., Poissant G. and Rebrion C., 1985a, Chem. Phys. Lett., **122**, 431.
Marquette J.B., Rowe B.R., Dupeyrat G. and Roueff E., 1985b, Astron. Astrophys., **147**, 115.
Maslach G.J. and Sherman F.S., 1956, University of California, UTIAS Report n°WADC TR-56-341.
Miller T.A., 1984, Science, **223**, 545.
Preston R.K., Thompson D.L. and McLaughlin D.R., 1978, J. Chem. Phys., **68**, 13.
Rebrion C., Marquette J.B., Rowe B.R., Adams N.G. and Smith D., 1987, Chem. Phys. Lett., **136**, 495.
Rebrion C., Marquette J.B., Rowe B.R. and Clary D.C., 1988a, Chem. Phys. Lett., in press.
Rebrion C., Marquette J.B., Rowe B.R., Adams N.G., Smith D., Clary D.C. and Chakravarty C., 1988b, J. Phys. Chem., to be submitted.
Rowe B.R., Dupeyrat G., Marquette J.B., Smith D., Adams N.G. and Ferguson E.E., 1984a, J. Chem. Phys., **80**, 241.
Rowe B.R., Dupeyrat G., Marquette J.B. and Gaucherel P., 1984b, J. Chem. Phys., **80**, 4915.
Rowe B.R., Marquette J.B., Dupeyrat G. and Ferguson E.E., 1985, Chem. Phys. Lett., **113**, 403.
Smith D. and Adams N.G. 1979, in Gas Phase Ion Chemistry, vol. 1, Ed. M.T. Bowers , Academic Press, New York, p. 1.
Smith D. and Adams N.G., 1981, Mon. Not. Roy. Astr. Soc., **197**, 377.
Solomon P.M. and Klemperer W., 1972, Astrophys. J., **178**, 389.
Spalburg M.R., Los J. and Gislason E.A., 1985, Chem. Phys., **94**, 327.
Su T. and Bowers M.T., 1973, J. Chem. Phys., **58**, 3027.
Su T., Su E.C.F. and Bowers M.T., 1978, J. Chem. Phys., **69**, 2243.
Su T. and Bowers M.T., 1979, in Gas Phase Ion Chemistry, vol. 1, Ed. M.T. Bowers, Academic Press, New York, p. 83.
Su T. and Chesnavitch W.J., 1982, J. Chem. Phys., **76**, 5183.
Troë J., 1987, J. Chem. Phys., **87**, 2773.
Watson W.D., 1974, Astrophys. J., **188**, 35.

DRIFT TUBE STUDIES OF ION-NEUTRAL REACTIONS AND THEIR RELEVANCE TO INTERSTELLAR CHEMISTRY

David Smith and Nigel G. Adams
Department of Space Research
University of Birmingham
Birmingham, B15 2TT, UK

ABSTRACT. The principle of operation of drift tubes and their application to the determination of ion-neutral reaction rate coefficients, k, as a function of the ion/reactant molecule (E_r) and the ion/buffer gas (E_b) centre-of-mass energies are discussed. It is shown that drift tube data of k versus E_r, for atomic ion/neutral reactions can be used with confidence in modelling the ion chemistry of shocked interstellar gas. However, it is stressed that drift tube data relating to molecular ion reactions must be used with caution since internal excitation of the ions can occur in collisions with the buffer gas. Some consideration is given to the variation with E_b and E_r of the rate coefficients, k_3, for ternary association reactions and to the relevance of the data in estimating radiative association rate coefficients appropriate to shocked interstellar gas.

1. INTRODUCTION

Drift tubes are apparatuses with which it is possible to study the gas phase interactions between charged species (electrons and ions) and neutral species (atoms and molecules) at energies from thermal to a few electron volts (eV) in the centre-of-mass frame (see, for example, Lindinger and Smith 1983). In this paper, we will only be concerned with the interactions of positive ions with neutrals in drift tubes. The principle of operation of a drift tube is simple. Ions are either created in a buffer gas or injected into this gas from an external ion source and then drawn through the gas under the influence of a spatially uniform electrostatic field. The ions gain translational (kinetic) energy from the field, **E**, and lose translational energy in collisions with the buffer gas atoms or molecules (number density, N). When the gain and loss of kinetic energy are balanced then the ions have acquired a constant drift velocity, v_d, in the direction of **E** and this is dependent on the ratio **E**/N. It is a straightforward matter to measure v_d in most drift tube experiments. The v_d for numerous ionic species drifting in various buffer gases have been measured and these are reported in the literature as tables of v_d as a function of **E** /N (Ellis

T. J. Millar and D. A. Williams (eds.), Rate Coefficients in Astrochemistry, 153–171.

et al 1976, 1978). A swarm of ions in a drift tube thus possesses a random velocity distribution as well as the superimposed v_d. The mean kinetic energy of the drifting ions, E_{ion} , was calculated by Wannier (1953) to be

$$E_{ion} = {}^3/_2\ k_bT + \frac{m}{2} v_d^2 + \frac{M_{ion}}{2} v_d^2 \qquad (1)$$

where m and M_{ion} are the masses of the buffer gas particles and of the drifting ions respectively, T is the buffer gas absolute temperature and k_b is the Boltzmann constant. In reaction kinetics with which we are concerned here, the mean relative centre-of-mass energy of the reacting ion-neutral pair, E_r, is generally the important parameter and not E_{ion}. E_r is given by (McFarland et al 1973):

$$E_r = \frac{1}{2} Mm\ (v_i^2 + v_n^2)/(M + m) \qquad (2)$$

where M is the mass of the reactant neutral (which in practice is added in low concentration to the buffer gas), v_n^2 is the mean square velocity of the reactant neutral (obtained from ${}^3/_2k_bT = {}^1/_2\ Mv_n^2$) and v_i^2 is the ion mean square velocity (obtained from ${}^1/_2mv_i^2 = E_{ion}$ as given by Equation (1)). Substitution for v_n^2 and v_i^2 in Equation (2) gives:

$$E_r = [M/(m + M)]\ (E_{ion} - {}^3/_2k_bT) + {}^3/_2k_bT \qquad (3)$$

The validity of these expressions is discussed in detail elsewhere (Albritton et al 1977). The rate coefficients, k, for ion-neutral reactions studied using drift tubes are usually presented as functions of E_r. For atomic ions drifting in an atomic buffer gas, only elastic collisions occur between the ions and the buffer gas atoms (except, perhaps, for the internal excitation of fine structure levels in the ions) and it has been shown (Albritton et al 1977) that atomic ions drifting in helium possess a near-Maxwellian energy distribution described by the centre-of-mass energy, E_b, appropriate to the ion-buffer gas atom collisions (given by Equation (3), with M equal to m, the mass of the buffer gas atoms). So for atomic ions in helium, the energy of interaction of the ion with the buffer gas, E_b, and the energy of the interaction of the ion with the reactant neutral, E_r, can be correctly characterized. However, when the drifting ions are molecular, then internal excitation of the ion can occur during collisions with the buffer gas. Rotational excitation is facile in such collisions (and in some cases a rotational temperature can be ascribed to the ions in terms of E_b; see Section 2.2). At sufficiently high E/N, depending on the nature of the ion, and especially for heavy buffer gases (e.g. argon) then vibrational excitation of the ion is efficient (and for some ions can result in dissociation; see Section 2.2). Then, although the E_b can still be calculated, the internal energy of the drifting ions is often quite uncertain. These features of drift tubes have been discussed in detail previously, at least as far as they are currently understood (Viehland, Lin and Mason 1981, Viehland 1986). Thus, our objective here is not to review these features or the details of particular research programmes carried out using drift tubes, since such reviews are already available (Lindinger and Smith 1983, Ferguson 1984, Lindinger 1984). Rather, we are concerned with the question as to what extent, and under what astrophysical circumstances, may drift tube data be properly applied to elucidate interstellar chemistry, specifically diffuse and dense cloud chemistry and to the chemistry of the shocked regions of interstellar gas. Towards this end, we present some sample data relating to atomic ion reactions which are believed to be applicable to interstellar chemistry and also some other drift tube data relating to molecular ion reactions which illustrate the influence of internal excitation of the molecular ions on their reactivity.

2. SOME SAMPLE DATA OBTAINED FROM DRIFT TUBE EXPERIMENTS

2.1 Atomic Ion Reactions

As was previously stated, the study of the reactions of atomic ions using drift tubes is quite straightforward, provided that care is taken to ensure that the electronic state of the ions is known. In the interstellar medium, ions are generally in the ground electronic states since most excited electronic states of ions will be radiatively relaxed long before they undergo a collision. Some years ago, Elitzur and Watson (1980) suggested that the endothermic reaction

$$C^+(^2P) + H_2 \longrightarrow CH^+ + H - 0.4 \text{ eV} \quad (4)$$

could proceed in the shocked regions of interstellar gas where the E_r of the C^+/H_2 reactants is elevated sufficiently to overcome the endothermicity of the reaction thus making reaction (4) a source of CH^+ ions. It therefore became important to determine the rate coefficient for reaction (4), i.e. k(4), as a function of E_r. Initial studies in a selected ion flow drift tube (SIFDT) showed that k(4) increased rapidly with E_r reaching $4.5 \times 10^{-11} cm^3 s^{-1}$ at an E_r of 0.46 eV (Adams and Smith, see Draine and Katz 1986). On the basis of these measurements, it has been shown that reaction (4) can indeed be a significant source of CH^+ in the shocked regions of interstellar gas (Adams et al 1984, Draine and Katz 1986, Pineau des Forêts et al 1986). Recently, a very thorough study of reaction (4) has been carried out over an appreciable range of E_r using a SIFDT apparatus (Twiddy et al 1986). These data are reproduced in Figure 1 together with the k(4) derived from the guided beam data of Ervin and Armentrout (1986). Such data are very important in assessing the role that reaction (4) plays in interstellar shock chemistry (Draine and Katz 1986, Mitchell 1987). Treatment of the SIFDT data given in Figure 1 according to an Arrhenius law (i.e. plotting ln k(4) versus E_r^{-1})provides a value for the endothermicity of reaction (4) of 0.43 eV which is in acceptable agreement with the thermochemical value. This and similar agreements for other reactions are taken as justification for applying the Arrhenius law to non-thermal drift tube data. During the course of these SIFDT experiments, it was discovered

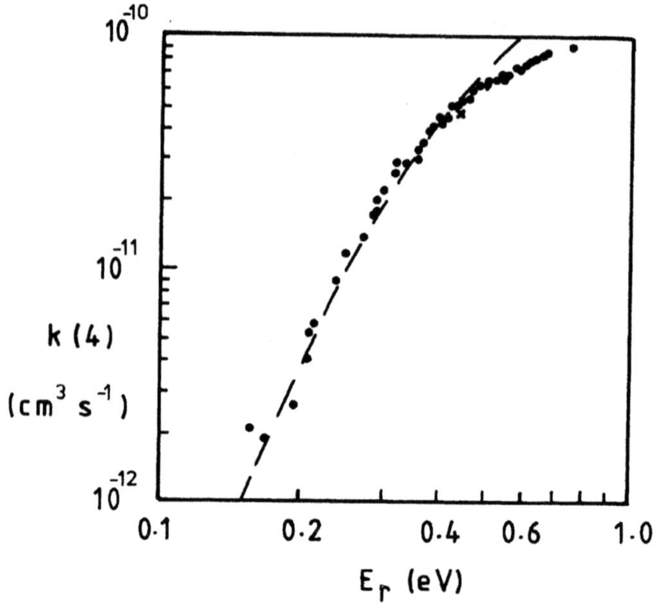

Figure 1. A plot of the rate coefficient for the reaction of C^+ with H_2, k(4), versus the C^+/H_2 centre-of-mass energy, E_r, obtained by Twiddy et al (1986) (black dots) using a SIFDT apparatus. The cross represents the value of k(4) at an energy of 0.46 eV obtained earlier by Adams and Smith also using a SIFDT apparatus (see Draine and Katz 1986). The carrier gas in both of these SIFDT studies was helium at a temperature of 300K. The dashed line is a representation of the k(4) derived from the cross section measurements of Ervin and Armentrout (1986).

that a fraction of the C^+ ions injected into the SIFDT were in their metastable ^4P state. The presence of these metastable ions confuses the determination of k(4) since the $C^+(^4P)$ fraction reacts rapidly with H_2 even when E_r is small. This problem was avoided by generating the C^+ ions in a high pressure electron impact ion source containing CO which reacts with $C^+(^4P)$ ions but not with the ground state $C^+(^2P)$ions. This simple example illustrates the pitfalls awaiting unwary experimenters even when (otherwise) relatively simple studies of atomic ion reactions are carried out. Note that the data in Figure 1 indicate that for E_r values of just a few tenths of an eV (and thus for modest interstellar MHD shocks; Draine 1980) k(4) can approach $\sim 10^{-10} cm^3 s^{-1}$ implying that reaction (4) must be considered as an important source of CH^+ in shocked regions. However, the reverse of reaction (4) may also occur in these regions; the rate coefficient for this reaction of CH^+ with H atoms, i.e. k(-4), has been studied by Federer et al (1984) over a limited range of

E_r. Using the kinetic data for these reactions which produce and destroy CH^+ ions together with the measured thermal rate coefficient for the reaction of CH^+ ions with H_2 (which also destroys CH^+ ions converting them to CH_2^+ ions) it is not possible to quantitatively account for the apparent overabundance of CH^+ in diffuse interstellar clouds. Thus Flower (in these Proceedings) has postulated that the CH^+/H_2 reaction may be inhibited under shocked conditions and that this reaction will not be so effective in destroying CH^+ ions. In response to this, we have studied the CH^+/H_2 reaction in a SIFDT apparatus and observed the k to be sensibly independent of the CH^+/H_2 centre-of-mass energy up to 0.35 eV. It therefore repudiates the postulate of Flower and so the CH^+ problem in diffuse clouds remains unsolved.

Once produced in interstellar gas, CH^+ ions can react rapidly with H_2 to produce CH_2^+ which also can react rapidly with H_2 to produce CH_3^+. Then the relatively slow radiative association reaction of CH_3^+ with H_2 (i.e. reaction (11) referred to later) followed by dissociative recombination of the CH_5^+ ions with electrons can produce CH_4. This is a probable source of CH_4 in dense interstellar clouds (Smith and Adams 1981b). The analogous reaction sequence involving S^+ ions and H_2 cannot proceed in cold interstellar clouds because each of the H-atom abstraction steps is quite endothermic:

$$S^+ + H_2 \longrightarrow SH^+ + H - 0.85\ \text{eV} \quad (5)$$

$$SH^+ + H_2 \longrightarrow SH_2^+ + H - 0.55\ \text{eV} \quad (6)$$

$$SH_2^+ + H_2 \longrightarrow SH_3^+ + H - 0.25\ \text{eV} \quad (7)$$

However, these reactions are potentially important in the synthesis of H_2S (following electron-ion recombination of SH_3^+) in the shocked regions of interstellar gas. Because of this, a drift tube study has been made of reactions (5), (6) and (7) and their reverse reactions (Millar et al 1986). The data obtained for k(5) as a function of E_r is reproduced in Figure 2 in the form of an Arrhenius plot. The slope of this plot indicates a value of ~0.7 eV for the endothermicity of the reaction which is in acceptable agreement with the value of 0.85 eV calculated using the available thermochemical data. This again illustrates the value of drift tubes for the study of atomic ion reactions. A rapid increase in k(5) is apparent with increasing E_r and, near the threshold energy for the reaction, k(5) has reached the large value of $\sim 10^{-10} cm^3 s^{-1}$. These studies also showed that k(6) and k(7) increase with increasing E_r, but in these cases the data are less readily interpreted because there was evidence that vibrational excitation of the reactant SH^+ and SH_2^+ ions occurred at the higher values of E_r. Nevertheless, the results are valuable in that they indicate the approximate values of k(6) and k(7) to be adopted for use in interstellar chemical models. The data for the rate coefficients of the reverse reactions, k(-5), k(-6) and k(-7) involving the molecular ions SH^+, SH_2^+ and SH_3^+ are also subject to some uncertainties because of vibrational excitation. Nevertheless, these data for reactions (5) and (6) have been used to approximately model the abundance of SH^+ in both magnetic (MHD) and non-magnetic interstellar shocks (Millar et al 1986).

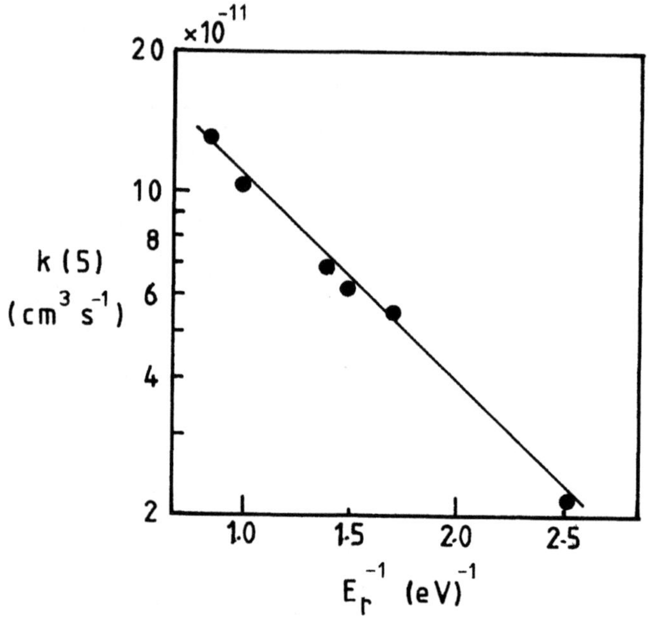

Figure 2. An Arrhenius-type plot of the rate coefficient for the reaction of S^+ and H_2, k(5), against E_r^{-1} , the reciprocal of the S^+/H_2 centre-of-mass energy. The data were obtained in helium buffer gas at a temperature of 300K using the Innsbruck selected ion drift tube, SIDT (Millar et al 1986).

For non-magnetic shocks, in which the forward (endothermic) reactions (4), (5) and (6) can only be driven thermally, the ion-chemical models predict that the column density of SH^+ should be must less than that observed for CH^+ , whereas in MHD shocks, in which streaming of the ions through the neutral gas can drive the $S^+ + H_2$ reaction (5) efficiently, the SH^+ column density should be comparable to that of CH^+. Thus, the presence or absence of SH^+ in regions in which CH^+ is observed could be an indicator of the type of shock which drives the ion chemistry (Millar et al 1986).

A final, most interesting example of drift tube studies of atomic ion reactions of relevance to astrochemistry relates to the slightly endothermic reaction :

$$N^+ + H_2 \longrightarrow NH^+ + H \qquad (8)$$

This reaction is usually considered to be the first stage in a sequence of H-atom extraction reactions of NH_n^+ ions with H_2 occurring in dense interstellar clouds (i.e. $N^+ \longrightarrow NH^+ \longrightarrow NH_2^+ \longrightarrow NH_3^+ \longrightarrow NH_4^+$). This sequence terminates with the NH_4^+ ion which then undergoes dissociative recombination producing NH_3. During ion trap experiments, primarily directed towards the study of the $NH_3^+ + H_2 \longrightarrow NH_4^+ + H$ reaction, Luine and Dunn (1985) noted that reaction (8) proceeded very slowly at a temperature of 11K ($k(8) = 3.5 \times 10^{-13} cm^3 s^{-1}$). Since k(8) had previously been measured to be much faster at 300K ($k(8) = 4.8 \times 10^{-10} cm^3 s^{-1}$; Adams, Smith and Paulson 1980), then it was correctly deduced that this reaction was slightly endothermic and the endothermicity, ΔE, was estimated to be about 8 meV. Subsequently, attempts were made to study this reaction in a SIFT experiment at temperatures less than 300K but these were unsuccessful since the reactant N^+ ions associated with the atoms of the helium carrier gas, (that HeN^+ is a stable species is itself an interesting observation). To avoid this problem, reaction (8) was studied in a SIFDT in helium carrier gas at 300K (Adams, Smith and Millar 1984, Adams and Smith 1985). The variation of k(8) with E_r thus obtained is shown in Figure 3 from which it can be seen that k(8) gradually increases as E_r increases. So, clearly, in this E_r regime (see Figure 3) an increase in the centre-of-mass (translational) energy alone promotes reaction (8). Note that these non-thermal SIFDT experiments were carried out at a fixed H_2 rotational "temperature" equivalent to the translational (carrier gas) temperature of 300K. Analysis of the data according to the Arrhenius law indicates an endothermicity, ΔE, of (11±3) meV which is somewhat greater than the previous estimate of Luine and Dunn given above. However, the closeness of these ΔE values is remarkable considering that the H_2 rotational temperature in the ion trap was uncertain and that the SIFDT experiment is non-thermal. SIFDT experiments were also carried out for the reactions of N^+ with D_2 and HD. The self-consistency of the ΔE values derived for the four reactions is also remarkable (Adams and Smith 1985) (note that the N^+ + HD reaction produces both NH^+ and ND^+ and is thus equivalent to two separate reactions). Recently, studies of reaction (8) and the $N^+ + D_2$ and HD reactions have been made at low temperatures using the CRESU technique. Careful consideration was given to the difference in reactivity of the ortho and para forms of H_2 and to the influence of the fine structure states of the N^+ ions on the derived ΔE for the reactions (see Marquette et al 1985 and the paper by Rowe in these Proceedings).

Stimulated by their drift tube data, Adams et al (1984) have considered the influence of kinetically-excited ions on the chemistry of interstellar clouds. They noted that N^+ ions are believed to be formed in dense clouds via the reaction :

$$He^+ + N_2 \longrightarrow He + N + N^+ + 0.28 \text{ eV} \qquad (9)$$

and that the 0.28 eV excess energy largely appears as kinetic energy in the N and N^+ products (0.14 eV in each of these species). For collisions between these nascent N^+ ions and H_2 molecules at 10K this is equivalent to an energy of about 40 meV in the centre-of-mass frame.

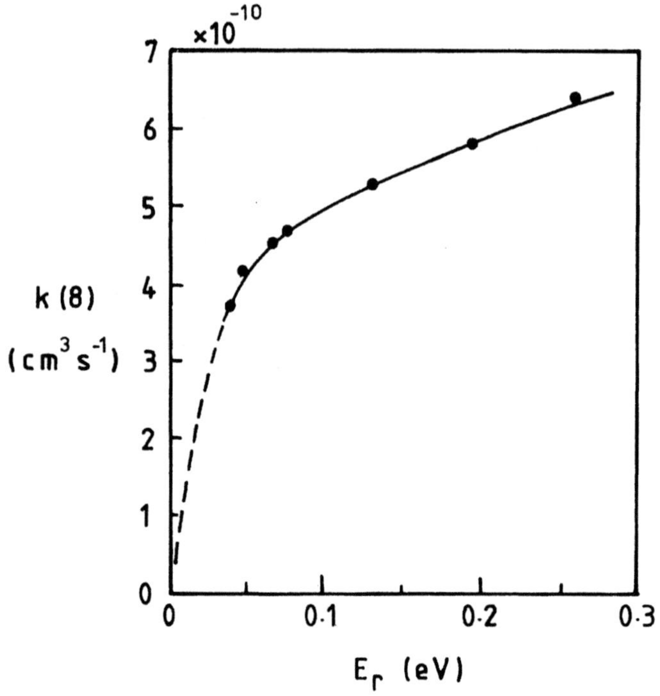

Figure 3. A plot of the rate coefficient for the reaction of N^+ with H_2, k(8), versus the N^+/H_2 centre-of-mass, E_r as obtained in helium carrier gas at 300K in a SIFDT apparatus (Adams et al 1984, Adams and Smith 1985). The SIFDT data are available down to an E_r of 0.039 eV which is equivalent to 3/2 kT at 300K. The dashed line indicates that k(8) reduces to very small values as E_r decreases towards zero (see text).

From this, it is to be expected, on the basis of the SIFDT data (Figure 3) that the appropriate value of k(8) will be much greater than that appropriate to thermalised N^+ and H_2 reactants at interstellar temperatures. Following this work, Yee, Lepp and Dalgarno (1987) have taken account of the kinetic relaxation of the N^+ ions formed in reaction (9) due to elastic collisions with H_2 molecules and calculated the values of k(8) appropriate to the temperatures of dense interstellar clouds. For 10K, they calculated that $k(8) = 1.3 \times 10^{-12} cm^3 s^{-1}$ and, for 30K, that $k(8) = 1.6 \times 10^{-10} cm^3 s^{-1}$, which are much greater than the thermal values appropriate to these temperatures. Since it is likely that the ionic products of other ion-molecule reactions occurring in cold interstellar clouds are suprathermal then this implies that they may undergo reactions which are otherwise endothermic. Some other examples are given in the paper by Adams et al (1984).

2.2 Molecular Ion Reactions

It has been known for many years, following the pioneering work by the NOAA group and the Innsbruck group, that molecular ions can be vibrationally-excited in drift tubes especially at high **E**/N (see, for example, the reviews by Albritton 1979, Lindinger and Smith 1983 and Lindinger 1984). Studies of vibrational excitation of ions in drift tubes has been - and remains - fundamentally interesting and has the potential to provide information on the collisional excitation of molecular vibrations in the low energy regime. An especially interesting study is that of the vibrational excitation of N_2^+ and O_2^+ ions in collisions with helium atoms in a drift tube (Federer et al 1985b, Kriegel et al 1987). Vibrational excitation of molecular ions is more efficient in collisions with buffer gas atoms or molecules heavier than helium (e.g. N_2, Ar, etc.). This is demonstrated below. However, the occurrence of vibrational excitation is unfortunate from the viewpoint of astrochemistry, since vibrational excitation of ions in drift tubes can obscure the influence of kinetic excitation on the rate coefficients of molecular ion-neutral reactions. Such information is needed in some astrophysical situations such as in interstellar MHD shocks. In these situations, the mean free times between collisions of the ions with the ambient gas are usually longer than the radiative lifetimes of the vibrational states of the ions and therefore the ions will generally be vibrationally relaxed. Thus data are required on the variation of the rate coefficients with E_r for the reactions of ions in their ground vibrational state. So, drift tube data on molecular ion reactions must be applied with caution to astrophysical situations except that is for data obtained at low **E**/N where vibrational excitation of the molecular ions is minimal.

A classic example of drift tube data is that relating to the charge transfer reaction of CO_2^+ with O_2 which is given in Figure 4. These data illustrate some important points which must be emphasised when studying molecular ion reactions in drift tubes. The rate coefficient, k, for the reaction has been determined in a flow drift tube (FDT) as a function of the CO_2^+/O_2 reactant gas centre-of-mass energy, E_r, in three different buffer gases, helium, nitrogen and argon at 300K (Lindinger et al 1975). As is clearly evident, the k at low E_r are not significantly different in the different buffer gases but at higher E_r there are obvious differences. These differences are attributed to an increase in k with the degree of vibrational excitation of the CO_2^+ ions, this excitation apparently being much more efficient in collisions of the CO_2^+ ions with the heavier buffer gases. Thus, the CO_2^+ ions are vibrationally excited prior to their interactions with the O_2 molecules at the higher **E**/N and this vibrational excitation promotes the charge transfer reaction. Therefore, to plot k versus E_r, as is a very common practice, is somewhat misleading since both the state of internal excitation of the ion, and E_r, change with **E**/N. How can the degree of vibrational excitation in the ion be quantified ? Presumably this will be dependent on the ion/buffer gas centre-of-mass energy, E_b, which can

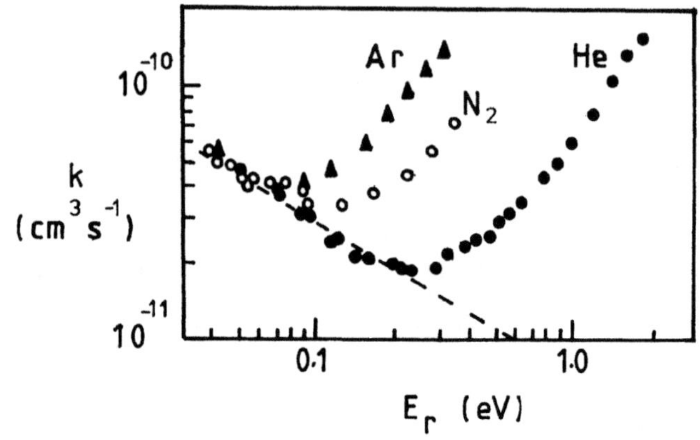

Figure 4. The rate coefficient, k, for the charge transfer reaction of CO_2^+ with O_2 plotted as a function of the CO_2^+/O_2 centre-of-mass energy, E_r. The data were obtained in helium, nitrogen and argon buffer gases (as indicated) at 300K in a FDT apparatus (Lindinger et al 1975). The dashed line is considered to describe the variation of k with E_r for CO_2^+ reactant ions that are not significantly vibrationally excited. The enhancement in the k above this line, k', is considered to be due to an increase in the reactivity of the CO_2^+ ions as they are vibrationally excited in collisions with the buffer gas atoms and molecules (see text).

readily be obtained by the procedure described in Section 1 (see equation (3)). But how can the separate influences of vibrational excitation in the reactant ion and of ion/reactant gas centre-of-mass (translational) energy E_r on the k for the reaction be distinguished ? Referring to Figure 4, the dashed line is considered to describe the variation of k with E_r for the reaction of CO_2^+ ions that are not vibrationally excited. (Note that the maximum value of E_r, for which data points fall on this dashed line, is greater for the helium (lighter) buffer gas than for the argon and nitrogen (heavier) buffer gases). This reduction of k with E_r is understandable in terms of the reduced interaction time of the CO_2^+ with the O_2 as E_r increases. The enhancement of k above the values described by the dashed line, which we designate as k', is then considered to be an indicator of the degree of vibrational excitation in the CO_2^+ ions which (as stated above) is dependent on E_b. The plot of ln k' versus E_b^{-1} (i.e. a pseudo-Arrhenius-type plot) is shown in Figure 5. Clearly, when plotted in terms of E_b, the k' data in the helium and argon buffer gases are not

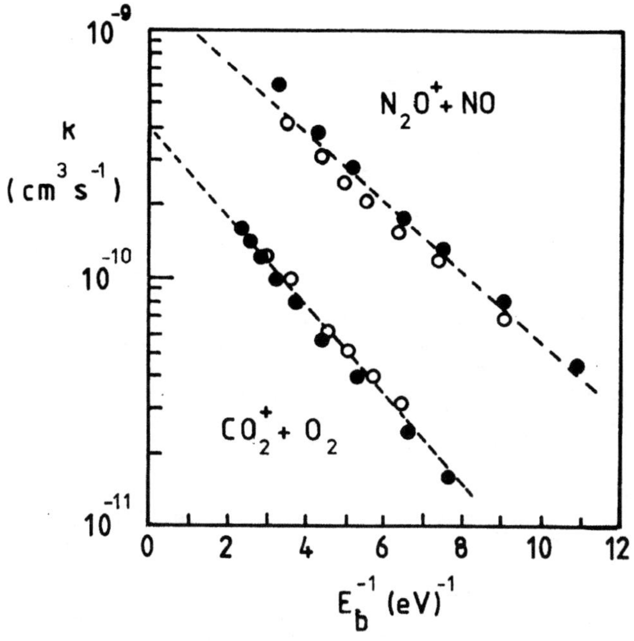

Figure 5. An Arrhenius-type plot of the enhancement in the k, i.e. k', as obtained from the data for the $CO_2^+ + O_2$ reaction given in Figure 4 for the helium (●) and argon (o) buffer gases, against E_b^{-1}, the reciprocal of the reactant ion/buffer gas atom centre-of-mass energy. Also shown are the corresponding data for the $N_2O^+ + NO$ charge transfer reaction. That the data points in the helium and argon buffer gases define the same line is taken as an indication that the equilibrium populations of excited ions are the same in both buffer gas for the same value of E_b (Lindinger 1987).

significantly different which indicates that, for a given E_b, the degree of vibrational excitation of the CO_2^+ ions is ultimately as great in collisions of these ions with helium atoms as it is with argon atoms. What may be different in the collisions of molecular ions with helium and argon atoms in drift tubes is the rate at which an equilibrium population of vibrational states is reached (determined, of course, by the relative rates of excitation and de-excitation in ion/buffer gas atom collisions) and this is also a function of E/N. This phenomenon is discussed by Kriegel et al (1987). Data for the $N_2O^+ + NO$ charge transfer reaction are also given in Figure 5 which also indicate that the equilibrium population of the vibrational states of N_2O^+ ions is the same in collisions with helium and argon atoms (for further discussion of this, see the recent paper by Lindinger 1987).

The role that vibrational energy in the reactant ion plays in determining the k of ion-neutral reactions has been investigated by several workers using flow tube techniques. It is known that vibrational excitation of ions can promote charge transfer reactions which are endothermic when the ions are in their ground vibrational states (Lindinger et al 1981, Smith and Adams 1981a). Thus, for example, $O_2^+(v=0)$ charge transfers only slowly with Xe, but this charge transfer becomes efficient for $O_2^+(v \geq 1)$ (Böhringer et al 1983). Also $NO^+(v=0)$ does not charge transfer with CH_3I but charge transfer occurs when NO^+ is excited to v=1 or greater (Dobler et al 1983, Federer et al 1985a). This can be readily shown by preparing the O_2^+ and NO^+ ions by electron impact onto O_2 and NO in a low-pressure SIFT ion source, and then studying their reactivity with Xe and CH_3I in the way conventional for SIFT experiments. Since the k for these reactions are very different for the ground and vibrational states of the reactant ions, then the reactions can be used as detectors (or monitors) of vibrationally-excited NO^+ and O_2^+, a principle which has been extended to other ionic species (Mauclaire et al 1987, Twiddy 1987). Using this 'monitor gas' technique, extensive measurements have been carried out of the quenching of the vibrational energy in $O_2^+(v)$ by several atomic and molecular gases (at Boulder, Böhringer et al 1983) and similarly for $NO^+(v)$ (at Innsbruck, Dobler et al 1983, Federer et al 1985a). A strong correlation was observed between the quenching rate coefficients and the polarizabilities of the quenching atomic/molecular neutrals. This phenomenon has been explained in terms of the strength of the binding in the ion/neutral intermediate complexes (Ferguson 1984, 1986). By performing such experiments in drift tubes, the variation of the quenching rate coefficients can be investigated in terms of E_r and E_b. More sophisticated studies have been made by the Innsbruck group who have designed a drift tube apparatus in which ions can be prepared in vibrationally-excited states by ion/neutral reactions prior to their injection into the drift tube. Thus, the proton transfer reactions of $N_2H^+(v=0)$, $N_2H^+(v=1)$ and $N_2H^+(v=2)$ with Ar and Kr atoms have been investigated (Villinger et al 1984). Using this approach (i.e. using argon and krypton as 'monitor gases' for vibrationally-excited N_2H^+ ions), they have shown that the astrophysically interesting reaction of H_3^+ with N_2 generates N_2H^+ ions predominantly in their ground vibrational state.

The degree of sophistication of flow tube, drift tube and flow drift tube techniques, especially following the development of selected ion injection sources, is such that now it is possible to study a very wide variety of ion-neutral reactions, including the reactions of state selected ions, doubly-charged ions, weakly-bound (cluster) ions, etc. The reviews by Adams and Smith (1983) and Lindinger and Smith (1983) survey the variety of reactions that have been studied. Not only can critical kinetic data be obtained, but also thermochemical data, and even the structures of ions can be determined by judicious analysis of the kinetic data. One example from many examples will suffice to illustrate these claims.

Recently, a SIFDT study (Tichy et al 1987) was made of the forward and reverse reactions:

$$N_4^+ + Ar \rightleftharpoons N_2Ar^+ + N_2 \qquad (10)$$

This required the generation of the weakly-bonded ions N_4^+ and N_2Ar^+ in a high pressure ion source and their injection into the helium carrier gas of the SIFDT without collisional break-up of the ions. This was successfully achieved and the rate coefficients k(10) and k(-10) were determined in the usual way over a restricted range of E_r, and thus of E_b, to ensure that significant vibrational excitation of the N_4^+ and N_2Ar^+ ions did not occur. For these restricted circumstances, it was judged that the kinetic data could be treated according to equilibrium thermodynamic laws. Thus a van't Hoff plot (i.e. [k(10)/k(-10)] versus E_r^{-1}) was constructed from which the enthalpy and entropy changes, ΔH and ΔS respectively, in the reaction were determined. The ΔH for the reaction was equated to the endothermicity of the reaction and hence to the difference in the bond energies of N_4^+ and N_2Ar^+ which was (0.09 ± 0.01)eV (the N_2^+...Ar bond being the weaker). The derived ΔS value was used to determine the structure of the N_2Ar^+ ions. This was done by expressing ΔS in terms of the total partition functions of the species involved in the reaction in which the only unknown was the moment of inertia of the N_2Ar^+ ion and hence its structure. In this way, N_2Ar^+ was shown to be linear. The details of this work are reported in the paper by Tichy et al (1987). It is interesting to note that at higher values of E_b the N_4^+ and the N_2Ar^+ ions could be readily dissociated and this is clear evidence that vibrational excitation of these ions occurred in collisions with helium atoms at the higher E/N (a phenomenon recognised in early SIFDT studies of N_4^+ ions; Smith, Adams and Alge 1984).

Much has been written concerning the role of ion/molecule radiative (binary) association in the synthesis of molecules in interstellar clouds (Herbst 1976, 1985b, Bates 1983, Smith and Adams 1981b). There seems to be little doubt that this is an important - perhaps crucial - process for synthesising large molecular ions (and therefore, following dissociative recombination, presumably polyatomic neutral molecules). Radiative association is also invoked as the most important process for the production of CH_2^+ in dense clouds (i.e. via $C^+ + H_2 \longrightarrow CH_2^+ + h\nu$). Yet binary ion-molecule association has rarely been observed in the laboratory. One reaction for which it has been observed (Barlow, Dunn and Schauer 1984) is :

$$CH_3^+ + H_2 \rightleftharpoons (CH_5^+)^* \longrightarrow CH_5^+ + h\nu \qquad (11)$$

Clearly, emission of radiation is needed to stabilize the excited $(CH_5^+)^*$ ion; this radiation has not yet been observed. The radiative association rate coefficient for reaction (11) i.e. $k_{rad}(11)$, has been determined to be $1.1 \times 10^{-13} cm^3 s^{-1}$ at 13K (Barlow et al 1984), although this is in conflict with a more recent experimental value obtained at a higher temperature by Gerlich (1987). Most of the k_{rad} values for radiative association reactions other than for reaction (11) are inferred from laboratory measurements of the analogous collisional (ternary) association rate coefficients, k_3. The collisionally stabilized analogue of reaction (11) is

$$CH_3^+ + H_2 \rightleftharpoons (CH_5^+)^* \; ; \; (CH_5^+)^* + M \longrightarrow CH_5^+ + M^* \qquad (12)$$

where the excited $(CH_5^+)^*$ ion is prevented from dissociating back to the reactants by a collision with a third body, M, which removes some internal energy from the $(CH_5^+)^*$. A simple kinetic model provides a value for the lifetime of $(CH_5^+)^*$ against unimolecular decomposition in terms of the k_3 for the reaction, and this lifetime, together with an estimate of the radiative lifetime of $(CH_5^+)^*$, provides an estimate of k_{rad}. The k_3 for many ternary association reactions have been measured including some over appreciable temperature ranges, notably using the VT-SIFT technique (Adams and Smith 1981, Herbst, Adams and Smith 1983). The k_{rad} deduced for reaction (11) using ternary association data and adopting a canonical value for the radiative lifetime of the $(CH_5^+)^*$of 1 ms is in acceptable agreement with the directly measured k_{rad}. The magnitudes of k_3 vary over orders-of-magnitude for the various reactants and these variations are reflected directly into the deduced k_{rad} values. Such studies indicate which radiative association reactions are important to interstellar molecular synthesis. The experimentally obtained variation of k_3 with temperature generally agrees with theoretical predictions, i.e. k_3 varies as $\sim T^{-r/2}$, where r is the total number of rotational degrees of freedom in the separated ion/neutral reactants (Bates 1979, Herbst 1979, also see Herbst in these Proceedings). Thus, k_3 for reaction (11) is predicted to vary as $\sim T^{-5/2}$ which is in close agreement with the experimental observations. Since the variation of k_{rad} with temperature is expected to mirror k_3 for the analogous reaction (except for a possible small adjustment to the power index to account for the temperature dependence of the collisional stabilization efficiency), then the value to interstellar chemistry of laboratory studies of temperature dependences of k_3 is clear.

It is worth asking whether radiative association occurs to a significant extent in shocked regions of the interstellar gas. The non-thermal situation of ions streaming through ambient gas in MHD interstellar shocks is in some sense rather similar to ions drifting through the buffer gas in a drift tube. However, there are also obvious differences between these situations, one being (as has been mentioned before) the time interval between collisions; the mean free time between collisions in the interstellar environment is sufficiently long that most ionic species radiate away the internal (vibrational and rotational) energy which they may have gained in collisions with ambient neutral species. This cannot occur to a significant extent at the much higher pressures in drift tubes and thus these ions possess the internal energy which they have gained in collisions with buffer gas particles. Nevertheless, studies of ternary association reactions in drift tubes can be used to study the variation of k_3 with E_r and E_b and to explore the separate influences of internal and translational energies. Thus, a preliminary study has been undertaken in a VT-SIFDT of several association reactions, principally involving CH_3^+ ions and including reaction (12) where M = helium atoms (Adams and Smith 1987). Measurements as a function of E_r can be made in a VT-SIFDT for different (fixed) temperatures of the helium carrier gas. It is advantageous to operate at low carrier gas temperatures (e.g. 80K) at which the k_3 at zero E/N is large and so the variation of k_3 over a wider range of E_r (and E_b) can be studied (before the k_3 becomes too small to be accurately measured) than can be achieved at higher gas temperatures (e.g. 300K). From quite general considerations, it is to be expected that k_3 will reduce with the increasing E_r since the energy content of the intermediate complex also increases. This is indeed observed to be the case as is indicated by the ln-ln plots shown in Figure 6 for the collisional association reactions of CH_3^+ ions with CO and H_2. The linearity of the plots for the data obtained at a helium temperature of 80K indicates a power law relationship between k_3 and energy. However, it must be noted that the CO data are plotted in terms of E_r, but the H_2 data are plotted in terms of E_b. This is done in accordance with the predictions of a simple theoretical analysis of these association reactions (which is based on the theoretical approaches due to Bates 1979 and Herbst 1979 for the truly thermal situation referred to above). This simple theory predicts that k_3 for the $CH_3^+ + CO + He$ association reaction should vary as $\sim E_b^{-3/2}$ and that k_3 for the $CH_3^+ + H_2 + He$ association reaction should vary as $\sim E_r^{-3/2}$ and these predictions are clearly borne out by the experiments. These results should be compared with the temperature dependences of these two reactions ($k_3 \sim T^{-5/2}$ for both reactions). The paper by Adams and Smith (1987) provides more details of these interesting results. Also plotted in Figure 6 are the data for the CO reaction obtained at a helium temperature of 300K. As can be seen, the plot of ln k_3 versus ln E_b departs from linearity at the larger values of E_b and this is attributed to the excitation of vibrations in the CH_3^+ ions during their collisions with the helium carrier buffer gas atoms, a phenomenon which is more evident at larger E/N and, equivalently, at larger E_b. This rapid reduction of k_3 is more evident when the reaction is studied in argon buffer gas and, again, this is explained as being due to the more effective excitation of vibrations in the CH_3^+ ions by collisions with argon atoms.

What are the implications of these observations to radiative association in shocked interstellar clouds ? Since the k_3 for all ternary association reactions decrease with increasing temperature and, presumably, with increasing E_b and E_r then so will k_{rad}. The magnitude of the reduction in k_{rad} will obviously depend on the type and strength of the shock and the particular reaction under consideration. For example, for a 5 km s^{-1} MHD shock, Draine and Katz (1986) indicate the interaction energy of C^+ with H_2 to be ~0.1 eV. As can be seen from the data in Figure 6, an increase of this order in E_r for the $CH_3^+ + H_2$ reaction results in a decrease in k_3 by about an order-of-magnitude and therefore a similar decrease in k_{rad} will be appropriate in such an MHD shock. However, it must be re-stated that the rotational excitation resulting from ion/neutral collisions in interstellar gas will, for most ionic species, be radiated away before the next collision. So the interactions occur between rotationally cold and translationally hot molecular ions. Thus the available limited laboratory drift tube data can at best only give an indication of the magnitude of the reduction in k_{rad} in shocked interstellar gas. Some discussion of this phenomenon has been given by Herbst (1985b).

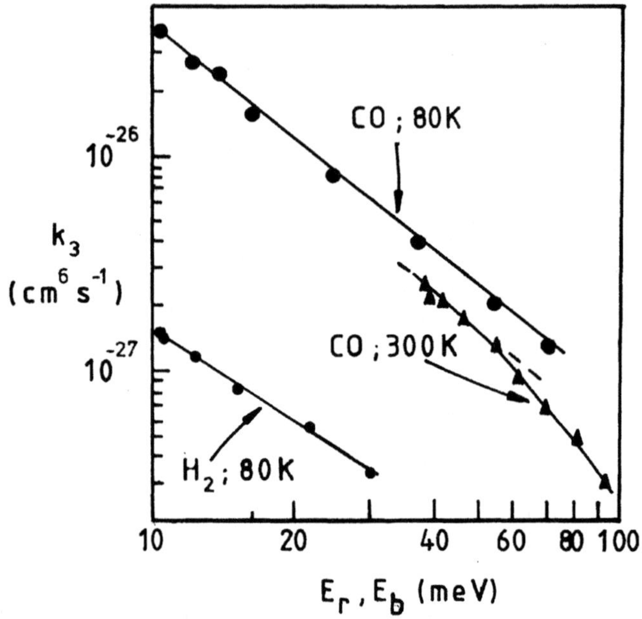

Figure 6. Plots in ln-ln form of the rate coefficients, k_3, for (i) the association reaction of CH_3^+ with CO in helium at both 80K and 300K plotted as a function of the CH_3^+/He centre-of-mass energy, E_b, and (ii) for the association reaction of CH_3^+ with H_2 in helium at 80K plotted as a function of the CH_3^+/H_2 centre-of-mass energy, E_r. The linearity of the plots indicates that for the CO reaction, $k_3 \sim E_b^{-1.5}$ and for the H_2 reaction, $k_3 \sim E_r^{-1.5}$ in accordance with simple theoretical ideas. The increase, at higher E_b, in the slope of the plot for the data obtained for a helium temperature of 300K is attributed to vibrational excitation of the CH_3^+ at the higher values of E_b which are accessible at this gas temperature. This vibrational excitation reduces the probability of the association of CH_3^+ with the CO (see the text and Adams and Smith 1987).

3. CONCLUDING REMARKS

The data relating to ion/neutral reactions, as obtained from drift tubes at centre-of-mass energies ranging from thermal (~10-30meV) to about

1eV, finds valuable applications to the estimation of the rates of the reactions occurring in shocked interstellar gas, particularly in MHD shocks where the ions are streaming through the ambient gas. Laboratory drift tube data relating to atomic ion reactions can be adopted with confidence in ion-chemical models of shocked regions since complications arising from internal excitation of the ions can not occur. However, drift tube data relating to molecular ion reactions must be used more judiciously, since at high E/N in the drift tube, the rate coefficients for the molecular-ion/neutral reactions can be greatly influenced by rotational and vibrational excitation of the ions in collisions with buffer gas atoms or molecules, an excitation which the ions do not generally possess when reacting in the interstellar cloud environment. Nevertheless, drift tube studies of molecular ion reactions can provide an insight into the relative influence of vibrational and rotational energy (in the reactant ions) and translational energy on reaction rate coefficients and products. Such insights will then allow the thoughtful astrophysicist and astrochemist to judge - rather than guess - the rate coefficients which are appropriate to the particular conditions which exist in the various regions of the interstellar medium.

ACKNOWLEDGEMENTS

Our thanks are due to the Science and Engineering Research council and the United States Air Force for providing financial support for this work.

REFERENCES

Adams, N.G. and Smith, D. 1981 *Chem. Phys. Letts.*, **79**, 563.
Adams, N.G., and Smith, D. 1983, in *Reactions of Small Transient Species*, Eds. A. Fontijn and M.A.A. Clyne, (London, Academic), p.311.
Adams, N.G., and Smith, D. 1985 *Chem.Phys. Letts.* **117**, 67.
Adams, N.G., and Smith, D. 1987, *Int. J. Mass Spectrom. Ion Processes*, **81**, 273.
Adams, N.G., Smith, D., and Paulson, J.F, 1980, *J. Chem. Phys.*, **72**, 288.
Adams, N.G., Smith, D., and Millar, T.J. 1984, *Mon Not. R. astr. Soc.* **211**, 857.
Albritton, D.L. 1979, in *Kinetics of Ion-Molecule Reactions*, Ed. P.Ausloos, (New York, Plenum), p.119.
Albritton, D.L., Dotan, I., Lindinger, W., McFarland, M., Tellinghuisen, J., and Fehsenfeld, F.C. 1977, *J. Chem. Phys.*, **66**, 410.
Barlow, S.E., Dunn, G.H., and Schauer, M. 1984, *Phys. Rev. Letts.* **52**, 902.
Bates, D.R. 1979, *J. Phys. B.*, **12**, 4135.
Bates D.R. 1983, *Ap. J.*, **270**, 564.
Böhringer, H., Durup-Ferguson, M., Fahey, D.W., Fehsenfeld, F.C., and Ferguson, E.E. 1983, *J. Chem. Phys.*, **79**, 4201.

Dobler, W., Federer, W., Howorka, F., Lindinger, W., Durup-Ferguson, M., and Ferguson, E.E. 1985, *J. Chem. Phys.,* **79**, 1543.
Draine, B.T. 1980, *Ap. J.,* **241**, 1021.
Draine, B.T., and Katz. N. 1986, *Ap.J.,* **306**, 655.
Elitzur, M., and Watson, W.D. 1980, *Ap. J.,* **236**, 172.
Ellis, H.W., Pai, R.Y, McDaniel, E.W., Mason, E.A., and Viehland, L.A. 1976, *Atom. Data Nucl. Data Tables,* **17**, 177.
Ellis, H.W., McDaniel, E.W., Albritton, D.L., Viehland, L.A., Lin, S.L., and Mason, E.A. 1978, *Atom. Data Nucl. Data Tables,* **22**, 179.
Ervin, K.M. and Armentrout, P.B. 1986, *J. Chem. Phys.,* **84**, 6738, 6750.
Federer, W., Villinger, H., Howorka, F., Lindinger, W., Tosi, P., Bassi, D., and Ferguson, E.E. 1984, *Phys. Rev. Letts.,* **52**, 2084.
Federer, W., Dobler, W., Howorka, F., Lindinger, W., Durup-Ferguson, M., and Ferguson, E.E. 1985a, *J. Chem. Phys.,* **83**, 1032.
Federer, W., Ramler, H., Villinger, H., and Lindinger, W. 1985b, *Phys. Rev. Letts.,* **54**, 540.
Ferguson, E.E. 1984, in *Swarms of Ions and Electrons in Gases,* Ed. W. Lindinger, T.D. Mark and F. Howorka, (Wien, Springer-Verlag), p.126.
Ferguson, E.E. 1986, *J. Phys. Chem.,* **90**, 731.
Gerlich, D. 1987, private communication.
Herbst, E. 1976, *Ap. J.,* **205**, 94.
Herbst, E. 1979, *J. Chem. Phys.,* **70**, 2201.
Herbst, E. 1985a, *Astron. Astrophys.,* **153**, 151.
Herbst, E. 1985b, *Ap. J.,* **291**, 226.
Herbst, E., Adams, N.G., and Smith, D. 1983, *Ap. J.,* **269**, 329.
Kriegel, M., Richter, R., Lindinger, W., Barbier, L., and Ferguson, E.E. 1987, *J. Chem. Phys.,* in press.
Lindinger, W. 1984, in *Swarms of Ions and Electrons in Gases,* Ed. W. Lindinger, T.D. Mark and F. Howorka (Wien, Springer-Verlag) p. 146.
Lindinger, W. 1987, *Int. J. Mass Spectrom. Ion Processes.,* **80**, 115.
Lindinger, W., and Smith, D. 1983, in *Reactions of Small Transient Species,* Ed. A. Fontijn and M.A.A. Clyne, (London, Academic), p.387.
Lindinger, W., McFarland, M., Fehsenfeld, F.C., Albritton, D.L., Schmeltekopf, A.L. and Ferguson, E.E. 1975, *J. Chem. Phys.,* **63**, 2175.
Lindinger, W., Howorka, F., Lukac, P., Kuhn, S., Villinger, H., Alge, E., and Ramler, H. 1981, *Phys. Rev.,* **A23**, 2319.
Luine, J.A., and Dunn, G.H. 1985, *Ap. J.*(Letters), **299**, L67.
Marquette, J.B., Rowe, B.R., Dupeyrat, G., and Roueff, E. 1985, *Astron. Astrophys.,* **147**, 115.
Mauclaire, G., Heninger, M, Fenistein, S., Wronka, J., and Marx, R. 1987, *Int. J. Mass Spectrom. Ion Processes,* **80**, 99.
McFarland, M., Albritton, D.L., Fehsenfeld, F.C., Ferguson, E.E., and Schmeltekopf, A.L. 1973, *J. Chem. Phys.,* **59**, 6620.
Millar, T.J., Adams, N.G., Smith, D., Lindinger, W., and Villinger, H. 1986, *Mon. Not. R. astr. Soc.,* **221**, 673.
Mitchell, G.F. 1987, in *Astrochemistry,* Ed. M.S. Vardya and S.P. Tarafdar, (Dordrecht, Holland, Reidel), p.275.
Pineau des Forêts, G., Flower, D.R., Hartquist, T.W., and Dalgarno, A. 1986, *Mon. Not. R. astr. Soc.,* **220**, 801.
Smith, D., and Adams, N.G. 1981a, *Phys. Rev.,* **A23**, 2327.
Smith, D., and Adams, N.G. 1981b, *Int. Rev. Phys. Chem.,* **1**, 271.

Smith, D., Adams, N.G., and Alge, E. 1984, *Chem. Phys. Letts.*, **105**, 317.
Tichy, M., Twiddy, N.D., Wareing, D.P., Adams, N.G., and Smith, D. 1987 *Int. J. Mass Spectrom. Ion Processes*, **81**, 235.
Twiddy, N.D. 1987, private communication.
Twiddy, N.D., Mohebati, A., and Tichy, M. 1986, *Int. J. Mass Spectrom. Ion Processes*, **74**, 251.
Viehland, L.A. 1986, *J. Chem. Phys.*, **84**, 244.
Viehland, L.A., Lin, S.L., and Mason, E.A. 1981, *Chem. Phys.* **54**, 341.
Villinger, H., Futrell, J.H., Saxer, A., Richter, R., and Lindinger, W. 1984, *J. Chem. Phys.*, **80**, 2543.
Wannier, G.H. 1953, *Bell Syst. Tech. J.*, **32**, 170.
Yee, J.H., Lepp, S., and Dalgarno, A. 1987, *Mon. Not. R. astr. Soc.* **227**, 461.

LABORATORY STUDIES OF DISSOCIATIVE RECOMBINATION AND MUTUAL NEUTRALIZATION AND THEIR RELEVANCE TO INTERSTELLAR CHEMISTRY

Nigel G. Adams and David Smith
Department of Space Research
University of Birmingham,
Birmingham, B15 2TT
United Kingdom.

ABSTRACT. The laboratory techniques, which have been applied to the determination of positive-ion/electron dissociation recombination coefficients, q_e, and positive-ion/negative-ion mutual neutralization coefficients, q_i, are briefly discussed. These include the stationary afterglow (SA), the variable-temperature flowing afterglow/Langmuir probe VT-FALP) and the merged beam (MB) techniques. It is suggested that the data of choice for interstellar ion-chemical models are those obtained using the truly thermal afterglow techniques rather than the q_e deduced from cross section data obtained using the MB technique. A detailed discussion is presented of the various determinations of q_e for H_3^+ ions which is now known to be very small. Some recent values of q_e determined using the VT-FALP technique are given for positive ions of interstellar interest and, from the limited amount of data available concerning the temperature dependence of q_e, appropriate values of q_e for use in interstellar ion chemical models are suggested. Also briefly discussed are the most reliable values available for q_i. These are included in view of the recent suggestion that significant concentrations of negatively-charged polycyclic aromatic hydrocarbons may exist in dense interstellar clouds and could undergo mutual neutralization reactions with ambient positive ions.

1. INTRODUCTION

It is now generally accepted that many of the neutral molecules observed in interstellar clouds are formed in the gas phase by positive ion-molecule reactions which produce a wide variety of molecular positive ions which then dissociatively recombine with electrons to form the observed neutral molecules. The final neutralization step in the conversion of a positive molecular ion to a neutral molecule may also be the transfer of a proton from the ion to another molecule or the transfer of an electron to the molecular ion from a species of low ionization energy (e.g. a metal atom). Very recently, it has been proposed that negatively-charged macromolecules, PAH^- (i.e. negatively- charged polycyclic aromatic hydrocarbons, PAH's), may be present in interstellar

T. J. Millar and D. A. Williams (eds.), Rate Coefficients in Astrochemistry, 173–192.

clouds and that these can neutralize positive ions by mutual neutralization. Several techniques have been developed to determine the rate coefficients for dissociative recombination, q_e, and the rate coefficients for mutual neutralization, q_i, including one technique by which both q_e and q_i can be determined for some ion species down to a temperature of 95K, which approaches the temperature of cool interstellar clouds. These techniques are briefly reviewed in this paper and comments are made relating to the magnitudes of q_e and q_i for various ions and their variation with temperature and values of q_e and q_i appropriate to the low temperatures of interstellar clouds are suggested. Some remarks are also made regarding the products of dissociative recombination reactions for which little experimental data are available and for which predictions from the available theories are in conflict. Mention is made of laboratory experiments which are expected to provide data on the products of dissociative recombination and mutual neutralization reactions of some simple molecular ions.

2. TECHNIQUES USED FOR THE DETERMINATION OF POSITIVE ION/ELECTRON DISSOCIATIVE RECOMBINATION COEFFICIENTS

It is not our purpose in this brief review to discuss the details of the techniques which have been used to measure positive ion/electron dissociative recombination coefficients, q_e, since much has already been written on this topic (see the reviews by Bardsley and Biondi 1970, Smith and Adams 1983b, 1984, Mitchell and McGowan 1983, Mitchell 1986, Johnsen 1987). However, a brief critique of these techniques is required to help to clarify which of the available data can be used with confidence in interstellar ion/chemical models.

The q_e have been determined for many reactions over the past twenty years or so, stimulated in the early part of this period by a need to know the q_e for reactions considered to be important in the terrestrial ionosphere (Reid 1976, Smith and Adams 1980) and lately by the need to know the q_e for reactions considered to be important in interstellar clouds (Dalgarno and Black 1976, Smith and Adams 1981, 1984a). Most of the data have been acquired by exploiting two general types of techniques, i.e. collision-dominated afterglow plasma and single-collision merged beam techniques, but a few important measurements have been made using an ion trap technique.

The stationary afterglow (SA) technique was the first to be extensively used to determine q_e. It was developed by M.A. Biondi and R. Johnsen and their colleagues (Mehr and Biondi 1969, also see Bardsley and Biondi 1970). An afterglow plasma is created in a microwave cavity (by discharging a pure gas or a gas mixture) consisting of electrons and the positive ions for which the q_e is to be measured. The time decay of the electron density is determined by observing the changing resonant frequency of the cavity; the ion species in the plasma is determined by sampling the diffusive ion current arriving at an orifice, located in the walls of the cavity, behind which is a mass spectrometer. Hence the q_e for various ionic species have been measured. The whole cavity, and therefore the afterglow medium, can be heated and cooled to study the temperature variation of q_e for ground state positive ions under thermal equilibrium conditions. Also the electrons in the plasma can be selectively heated and thus the variation of q_e with the electron temperature can be studied separately. This pioneering SA work provided the only body of data on dissociative recombination for several years. However, some apparent problems with the technique have been revealed by check measurements made recently by the variable- temperature flowing afterglow/Langmuir probe (VT-FALP) technique which is briefly described below (see Alge, Adams and Smith 1983, Adams, Smith and Alge 1984 and Section 4.1). The SA technique necessarily has to operate with relatively high gas pressures (~30 Torr) in the cavity to inhibit ambipolar diffusional losses of ions and this can promote clustering of the ions under study with ambient molecules (ligands) especially at low temperatures. Indeed, this phenomenon has been exploited in the SA to determine the q_e for cluster ions (e.g. $H_3O^+(H_2O)_n$ ions) and this elegant work shows that clustering of ligands to 'core' ions enhances the q_e (Leu, Biondi and Johnsen 1973a, Biondi 1973, Huang, Biondi and Johnsen 1976). Clearly, such clustering is undesirable when attempting to determine q_e for simple ions. Also, if the ligands are only loosely bound to the core ions then the clusters may not be detected by the wall-sampling mass spectrometer due to collisional dissociation (especially since the cluster ions recombine very rapidly). This comment is cautionary in nature and is not intended to cast doubt on the bulk of the measurements made in the SA. Some of the strengths and limitations of the SA technique and the other techniques used to detemine q_e are given in Table 1.

The most recent technique for determining q_e is again an afterglow plasma collision dominated system, the variable-temperature flowing afterglow/Langmuir probe (VT-FALP) technique, developed by D. Smith and N.G. Adams in Birmingham (Alge, Adams and Smith 1983, Smith and Adams 1983, 1984 a,b). In the VT-FALP, the flowing afterglow plasma is created by a microwave discharge in the upstream region of the flowing carrier gas (usually helium) and distributed along a flow tube (~100cm long and ~8cm diameter). The afterglow plasma 'decays' with distance along the flow tube due both to loss of ions and electrons via ambipolar diffusion to the walls and by recombination losses in the volume of the plasma if the q_e is sufficiently large and if the charged particle number densities are sufficient high. Diffusive loss is minimised by operating at a helium (carrier gas) pressure of only about 1 Torr, which is considerably lower than the pressure necessary in the SA experiments, and so clustering is much less likely in the VT-FALP experiments than it is in the SA experiments. The well-known chemical versatility of the flowing afterglow technique, demonstrated by the numerous studies of ion-molecule reactions that have been carried out with it and for which it is justly famous (Albritton 1978, 1979, Adams and Smith 1983, Lindinger and Smith 1983), is used to create plasmas in which a particular positive ion species is the dominant, or essentially the only, positive ion species. This is done by introducing appropriate gases at various axial positions into the helium (or helium/argon, etc.) afterglows and adjusting the flow rates of these ion source

TABLE I
Techniques used for Studying Dissociative Recombination and Mutual Neutralization

Technique	Temperature/ Energy Range	Pressure Range (Torr)	Strengths and Limitations
Dissociative recombination			
Stationary Afterglow (M.A. Biondi, R. Johnsen)	200-700K $T_e \lesssim 5000K$	10-40	Truly thermal q_e determined Range of ions restricted. High pressure, therefore clustering problems. Some difficulty in characterising and determining elevated T_e.
Flowing Afterglow/ Langmuir Probe (D. Smith, N.G. Adams)	95-600K	$\lesssim 1$	Truly thermal q_e determined. Wide range of ion types can be studied in their ground states. Low pressure, therefore clustering inhibited.
Ion Trap (G.H. Dunn)	0.04-4eV		$\bar{\sigma_e}$ determined. Long storage times, therefore internal states of ions relaxed.
Merged Beams (J.B.A. Mitchell, J.W. McGowan)	0.01-1eV		$\bar{\sigma_e}$ determined. Can be used to study many reactions which cannot be studied in plasmas. However, great care is required to avoid internal excitation of the ions.
Mutual Neutralization			
Stationary Afterglow (B.H. Mahan)	300K	>20	q_i is determined. High pressures, therefore cluster ions are a potential problem. Extrapolation to zero pressure is required.
Flowing Afterglow/ Langmuir Probe (D.Smith, N.G.Adams)	95-600K	$\lesssim 1$	Truly thermal q_i determined Many reactions can be studied.
Merged Beams (J.T. Moseley, J.R. Peterson)	0.1-3000eV		$\bar{\sigma_i}$ determined: ions are internally excited.

gases whilst monitoring the ion content of the afterglow using the downstream mass spectrometer. Thus, the VT-FALP can be used to determine the q_e for a much wider variety of positive ions than the SA, and, importantly, the ion source gas is not exposed to an electrical discharge as is inevitable in the SA and this can cause problems which are sometimes unforeseen. The VT-FALP experiments presently can be carried out over the wide temperature range from 95 to about 600K and so q_e can be determined for ground state molecular ions under truly thermal conditions for many ionic species over this temperature range. The axial gradient of the electron and/or the ion number density is determined by a small Langmuir probe which can be positioned at any point along the axis of the flow tube, and from this gradient together with a measured value of the plasma velocity the q_e can be determined. A mass spectrometer sampling orifice is located on the axis at the downstream end of the flow tube in order to identify the ion types present in the plasma. The VT-FALP technique has also been very successfully applied to the determination of electron attachment coefficients (Smith, Adams and Alge 1984, Alge, Adams and Smith 1984, Adams, Smith and Herd 1988) and to ion-ion mutual neutralization coefficients (see Section 3). This technique will also soon be used, in conjunction with spectroscopic techniques, to identify the neutral products of recombination reactions (see Section 4.5).

The other major technique of note for the study of dissociative recombination is the merged beam (MB) technique developed by J.W. McGowan and J.B.A. Mitchell (Auerbach et al 1977). Using this single-collision experiment, the cross sections for dissociative recombination reactions, σ_e, are determined as a function of the centre-of-mass interaction energy, ostensibly from a few tens of meV to many eV. Boltzmann averaging is then carried out to derive q_e values from the σ_e versus energy data. The greatest advantage of this MB technique is that data can be obtained for some positive ions which cannot be studied easily using afterglow plasma techniques because of the reactivity of these ions with the source gases present in the plasma (e.g. CH^+, although an extension to the VT-FALP technique is now allowing the q_e to be estimated for reactive ions. This is mentioned in Section 4.2). The greatest problem with the MB is the likelihood of internal (vibrational) excitation of the positive ions in the beam, although, with careful ion source design, this problem can be alleviated as the MB studies of the σ_e (and hence q_e) for H_3^+ recombination have shown (Section 4.2). A further potential problem is in accurately defining the interaction energies at the low interaction energies appropriate to temperatures at and below room temperature, which are the very energies relevant to interstellar cloud temperatures. Such uncertainties can result in large errors in the calculated q_e values. In some early published papers, q_e values appriopriate to temperatures as low as 10K were reported but these data are now considered to be suspect. Nevertheless, MB experiments are providing a valuable body of data on dissociative recombination reactions.

Finally, worthy of mention is the ion trap (IT) technique developed by G.H. Dunn and co-workers at Boulder, Colorado (Walls and Dunn 1974). In the IT technique, positive ions are trapped in a combination of electric and magnetic fields (a Penning-type quadrupole ion trap) in which they can be stored for periods of minutes, or hours if necessary, to allow any internal excitation in the stored ions to be dissipated by collisions or by radiation. A beam of electrons of variable energy is then passed through this ion space charge which neutralizes it at a rate dependent on σ_e and the electron current. The rate of reduction in the ion space charge is sensed by the system and thus σ_e is determined as a function of the interaction energy. This technique has been used to determine the σ_e for the ionospherically important ions O_2^+ and NO^+ and for the interstellar ion species NH_4^+, over interaction energy ranges from a few tens of meV to a few eV and hence the q_e have been derived (see Section 4.1).

3. TECHNIQUES FOR DETERMINING BINARY MUTUAL NEUTRALIZATION RATE COEFFICIENTS

It has been proposed from time-to-time, that negative ions are present in interstellar clouds in sufficient concentrations to allow significant fractions of the ambient positive ions to be neutralized by these negative ions (Dalgarno and McCray 1973, Herbst 1981). This is the process of mutual neutralization which is often termed binary ionic recombination. The important question has always been whether negative ions can form efficiently in interstellar clouds in binary collisions between electrons and attaching neutral species (i.e. by radiative attachment). This process has been considered for the formation of H^- ions in dense interstellar clouds and found to be very slow; otherwise it would have initiated a gas phase route to H_2 production when coupled with the process of associative detachment ($H^- + H \longrightarrow H_2 + e$). However, for large polyatomic molecules, spontaneous (radiative) attachment may occur and indeed has been shown to do so for some molecules (Woodin, Foster and Beauchamp, 1980). Thus, it has recently been proposed that radiative attachment to polycyclic aromatic hydrocarbons (PAH's) could occur rapidly in dense interstellar clouds producing PAH^- negative ions which could then undergo mutual neutralization with ambient positive ions (Omont 1986). If the abundance of PAH's (and PAH^-) is high then this could be the most important loss process for ions in these regions (Lepp and Dalgarno 1987).

Whilst, historically, the process of ternary (three-body) ionic recombination has received more attention, recently much more effort has been put into determining binary ionic recombination coefficients, q_i. Useful reviews of the experimental measurements and the theory of ionic recombination have been given by Bates (1979) and by Flannery (1982). The first comprehensive experimental study was carried out by B.H. Mahan and his colleagues in a system rather similar to a SA system (Mahan and Person 1964). In these experiments, ionization was created in a rare gas containing a small admixture of an electron attaching gas. The decay of the positive ion/negative ion plasma so formed was monitored and neutralization coefficients were derived. These experiments confirmed the magnitude of q_i to be $\sim 10^{-7} cm^3 s^{-1}$ at 300K (much earlier work by Sayers and colleagues (Sayers 1962) had also indicated this).

Unfortunately, these experiments of Mahan and Person could not be performed at inert gas pressures less than 20 Torr at which the ionic recombination was enhanced by the three-body process. Thus,the binary q_i were obtained by extrapolation of high pressure data to zero pressure. Also, no means of mass identification of the reactant ions was available and it seems likely that cluster ions were present in the plasma (the formation of clusters is enhanced at the high pressures of these experiments). Nevertheless, these experiments represented a major contribution to the study of ionic recombination. The work has been reviewed by Mahan (1973).

The merged beam (MB) technique was developed in the late 1960s by J.R. Peterson and his colleagues (Moseley, Olson and Peterson 1975) to determine the q_i for reactions considered to be important in the lower terrestrial atmosphere. The technique is similar to the MB technique described in Section 2 except that, in this case, beams of positive ions and negative ions which are derived from separate ion sources, are merged in a magnetic field and then the cross sections for mutual neutralization σ_i are determined as a function of the ion/ion centre-of-mass energy. Then, the q_i are obtained by Boltzmann averaging. These experiments suffer from the same problems as the MB experiments by which the q_e are determined, i.e. it is difficult to avoid internal excitation of the reactant ions and the determinations of the q_i are very sensitive to errors in determining the centre-of-mass interaction energy at the low energies for which the data were primarily intended. Clearly, at higher energies this is less of a problem. Using this MB technique, a good deal of useful data has been obtained and this is summarised in the review by Moseley et al (1975).

The VT-FALP technique has been used by D. Smith and N.G. Adams and their colleagues to carry out a comprehensive study of the q_i for a wide variety of reactions. Again, it is the extraordinary chemical versatility of the flowing afterglow, coupled with the simple but very effective Langmuir probe technique, that has allowed these studies to be successfully carried out. Positive ion/negative ion plasmas are created by the addition of a suitable gas (or commonly two different gases) either into the afterglow or upstream of the microwave discharge to create both the required positive and negative ion species. For example, the addition of NO_2 to the He/Ar afterglow creates an NO^+/NO_2^- plasma which, at sufficiently high ion number densities, decays by the process of mutual neutralization. The axial gradients of the positive ion and negative ion number densities are determined using the Langmuir probe and the q_i are determined in a way identical to that by which q_e is determined (see Section 2). Again, the mass spectrometer is used to confirm the identity of the recombining ions. Reactions between combinations of (both positive and negative) atomic ions, simple molecular ions and cluster ions have been studied, including many reactions of atmospheric importance, some over an appreciable temperature range. This work has been reviewed by Smith and Adams (1983, 1984b) and is summarised in Section 5.

4. EXPERIMENTAL DATA FOR DISSOCIATIVE RECOMBINATION

Here, the main objective is to present q_e data of relevance to interstellar chemistry. A fairly detailed discussion of the q_e for H_3^+ ions, i.e. $q_e(H_3^+)$, will be presented because of the importance of H_3^+ to interstellar physics and chemistry and because of the conflicting values obtained for $q_e(H_3^+)$ using the various experimental techniques. The q_e for several polyatomic ion species recently measured with the VT-FALP technique will also be presented and information on the temperature dependences of q_e, gleaned from the available data, will be discussed and used to indicate the q_e values appropriate to interstellar cloud temperatures. First, however, $q_e(O_2^+)$ and $q_e(NO^+)$ data will be briefly discussed in order to provide an indication of the reliability of the q_e data obtained by the various techniques described in the previous section (O_2^+ and NO^+ are, of course, not thought to be involved to any significant extent in interstellar chemistry).

4.1 Dissociative Recombination of O_2^+ and NO^+

$q_e(O_2^+)$ has been determined over a range of temperatures using all the techniques discussed in Section 2 (Stein et al 1964, Kasner and Biondi 1968, Mehr and Biondi 1969, Plumb, Smith and Adams 1972, Cunningham and Hobson 1972, Walls and Dunn 1974, Mul and McGowan 1980). There is unanimity between the SA, VT-FALP and IT data which all indicate that $q_e(O_2^+) \sim 2 \times 10^{-7}(300/T)^{0.7} cm^3 s^{-1}$. The $q_e(O_2^+)$ derived from MB data on σ_e is somewhat out-of-step with this, a $T^{-0.5}$ variation being indicated (Mul and McGowan 1980). Nevertheless, $q_e(O_2^+)$ is considered to be the must accurately known dissociative recombination coefficient. Dissociative recombination of O_2^+ occurs in the terrestrial ionosphere and is considered to be the source of the auroral red and green lines of atomic oxygen. Because of this, a good deal of experimental and theoretical effort has been given to determining the electronic states of the product oxygen atoms (Zipf 1980a,b, Bates and Zipf 1980, Guberman 1983, Vallee et al 1987b).

Consistency is also good between the values for the magnitude and temperature dependence of $q_e(NO^+)$ obtained using the SA, FALP and the IT techniques and even the value inferred from ionospheric observations (Weller and Biondi 1968, Walls and Dunn 1974, Torr, St. Maurice and Torr 1977, Alge, Adams and Smith 1983). However, for several years a perplexing discrepancy existed between all these data and the data obtained in the SA at elevated electron temperatures (Huang, Biondi and Johnsen 1975). Very recently, it has been recognised that this discrepancy is due to errors in deriving the elevated electron temperatures in the earlier SA experiment (Dulaney, Biondi and Johnsen 1987). Thus $q_e(NO^+)$ and its temperature dependence ($\sim T^{-0.85}$) is now well established. Again, the MB data differs both in the magnitude and in the indicated temperature dependence (Mul and McGowan 1980) which again is $T^{-0.5}$ (a temperature dependence which MB experiments indicate for the derived q_e for most ionic species; see Mitchell and McGowan 1983). It must be said, therefore, that the q_e and the temperature dependences derived from MB experiments must been treated with caution. This will be referred to again in the next section which concerns $q_e(H_3^+)$.

4.2 Recombination of H_3^+

H_3^+ ions are important in the chemistry of interstellar clouds. In dense clouds, they are formed in the reaction of H_2^+ (produced by cosmic ray ionization of H_2) with H_2 (e.g. see Dalgarno and Black 1976, Smith and Adams 1981). It is known that the excess energy in the reaction appears largely as vibrational excitation of the H_3^+ ion (Kim, Theard and Huntress 1974) but, in the interstellar environment, this internal energy is dissipated radiatively and in collisions with the ambient H_2 molecules. Thus, it is ground vibronic state H_3^+ ions which react with the ambient molecules (proton transfer reactions being the most likely with, for example, CO, N_2, CS, HCN, etc. producing the observed interstellar ions HCO^+, N_2H^+, HCS^+, H_2CN^+, etc.) or with electrons. The relative importance of these loss processes for H_3^+ ions clearly depends on their rate coefficients and the relative abundances of the reactive molecules and electrons. It has been known for many years from numerous experiments that exothermic proton transfer reactions proceed with unit efficiency, i.e. the rate coefficients are equal to the collisional limiting values (Bohme 1975). It is worthy of note here, however, that only relatively recently has it been shown that the collisional rate coefficients for the reactions of ions with polar molecules increase rapidly with decreasing temperature, a fact which must of course be accounted for in interstellar ion chemical calculations (Adams, Smith and Clary 1985). Concerning the H_3^+ interaction with electrons, it had been confidently assumed from the time at which interest in interstellar ion chemistry began that this resulted in rapid dissociative recombination following the early SA measurements of $q_e(H_3^+)$ in 1973. These measurements indicated that $q_e(H_3^+)$ at 300K is $2.3 \times 10^{-7} cm^3 s^{-1}$ (Leu, Biondi and Johnsen 1973b) and, from inferences made from the temperature variation of the q_e for other species, that q_e is significantly greater at the low temperatures of interstellar clouds. Consequently, for several years, $q_e(H_3^+)$ values of $\sim 10^{-6} cm^3 s^{-1}$ were included in the ion chemical models predicting molecular abundances in diffuse and dense interstellar clouds and were included in the model which was used to provide estimates of the electron number densities in dense clouds (Watson 1978).

It was perhaps not too surprising therefore, in view of the importance of $q_e(H_3^+)$, that when the first VT-FALP study of dissociative recombination in 1984 indicated that $q_e(H_3^+)$ and $q_e(D_3^+)$ for ground state ions were very small ($< 2 \times 10^{-8} cm^3 s^{-1}$ over the temperature range 95-550K; Smith and Adams 1984, Adams, Smith and Alge 1984), it was received with scepticism not to say disbelief (even though clear indications that $q_e(H_3^+)$ was indeed small were reported in a much earlier experiment (Persson and Brown 1955), a fact which we were made aware of after the publication of our VT-FALP paper). Fortunately, however, Michels and Hobbs (1984) had been carrying out theoretical work on the potential energy curves for H_3^+ in parallel with the VT-FALP work and were predicting that $q_e(H_3^+)$ would be small if not negligible for ground state H_3^+ ions. They also predicted that $q_e(H_3^+)$ would be larger for vibrationally-excited H_3^+ ions, a feature which was discernible in the VT- FALP data (Adams et al 1984). This support from theory, and the perceived importance of $q_e(H_3^+)$ to interstellar and planetary ion physics and chemistry, meant that the VT-FALP data were then considered seriously, even though the first MB experiments on H_3^+ also indicated that $q_e(H_3^+)$ was large (Mitchell et al 1984). Thus the SA group (Macdonald, Biondi and Johnsen 1984) were prompted to remeasure $q_e(H_3^+)$ and, whilst their new value at 300K was marginally smaller ($1.7 \times 10^{-7}cm^3s^{-1}$) than their earlier value, it was in no way possible to reconcile it with the VT-FALP upper limit value. Subsequently, we have used the VT-FALP in a different mode (Adams and Smith 1987), involving the comparison of the recombination loss rates of H_3^+, HeH^+ and He^+ ions in the afterglow plasma, to lower the upper limit of $q_e(H_3^+)$ to $10^{-11}cm^3s^{-1}$ (and incidentally to show that $q_e(HeH^+)$ was similarly small, as predicted much earlier by Flower and Roueff (1979)). Very recently, further MB experiments have indicated, in accordance with the theoretical predictions, that $\sigma_e(H_3^+)$ reduces dramatically as the vibrationally-excited H_3^+ components in the ion beam are reduced and a value of $q_e(H_3^+)$ $\sim 1 \times 10^{-8}cm^3s^{-1}$ for $H_3^+(v \lesssim 1)$ was indicated at a H_3^+/electron centre-of-mass energy of ~0.01 eV (Hus et al 1987). Also, the SA group (Johnsen 1987) have reported very recently that the large $q_e(H_3^+)$ reported previously by them can be accounted for by the presence of impurity ions in the afterglow (they indicate these ions to be CH_5^+ although in our view H_5^+ ions are more likely to be involved under the high pressure conditions of their experiments; Adams et al 1984). Finally, following the stimulus we received at the meeting at UMIST which is reported by these Proceedings, we have made efforts to further analyse our VT-FALP data and to place closer limits on $q_e(H_3^+)$. This is a difficult problem in view of the very small degree of recombination of H_3^+ (if any) that occurs in the afterglow. Again, only an upper limit can be obtained but we now consider a more judicious limit to be $\lesssim 10^{-10}cm^3s^{-1}$. The chronology of the various determinations of $q_e(H_3^+)$ is given in Table II.

Clearly, further work is needed to more closely determine the actual values of $q_e(H_3^+)$ over a range of temperatures, in view of the importance of even small $q_e(H_3^+)$ values to diffuse interstellar chemistry (E.F. van Dishoeck in these Proceedings) and to Jovian ionospheric chemistry (A. Dalgarno in these Proceedings). However, experimental techniques have yet to be developed to do this.

4.3 Dissociative Recombination of Triatomic and Polyatomic Ions

Until quite recently, little q_e data were available for most astrophysically-important ions. The $q_e(HCO^+)$, $q_e(H_3O^+)$ and $q_e(NH_4^+)$ had been measured using the SA technique (Leu, Biondi and Johnsen 1973a, c, Huang, Biondi and Johnsen 1976, Ganguli et al 1987) and the σ_e for several interstellar ion species had been obtained using the MB and IT techniques (see Heppner et al 1976, Dubois, Jeffries and Dunn 1978 and the review by Mitchell and McGowan 1983). With the arrival of the VT-FALP technique, the opportunity became available to determine the q_e for many more interstellar ion species under truly thermalized conditions

TABLE II
Studies of the Dissociative Recombination of H_3^+ Ions

Recombination Coefficient, q_e (cm^3s^{-1})	Technique	Reference	Comment
$<3 \times 10^{-8}$ 300K	Stationary Afterglow	Persson and Brown 1955	
2.3×10^{-7} 300K	Stationary Afterglow	Leu et al 1973b	Impurity ions
$<2 \times 10^{-8}$ 90-550K	Flowing Afterglow/ Langmuir Probe	Adams et al 1984	Diffusive loss restricted upper-limit
$H_3^+(v=0)$	Theory	Michels and Hobbs 1984	Predicts q_e finite for $H_3^+(v>3)$
1.7×10^{-7} 300K	Stationary Afterglow	Macdonald et al 1984	Impurity ions present
$\sim 3.5 \times 10^{-7*}$ 300K	Merged Beam	Mitchell et al 1984	Ions in excited states
$\sim 1 \times 10^{-8*}$ 300K	Merged Beam	Hus et al 1987	$H_3^+(v\leq 1)$
$\lesssim 1 \times 10^{-10}$ 90-500K	Flowing Afterglow/ Langmuir Probe	Adams and Smith 1988	From comparison with $q_e(He^+)$ and $q_e(HeH^+)$

q_e deduced from σ_e data; see text

and over appreciable temperature ranges in some cases. That H_3^+ ions do not recombine at a significant rate in the VT-FALP plasma provides a simple method for determing the q_e for ions XH^+ which can be formed via proton transfer from H_3^+ to the parent molecules, X, thus :

$$H_3^+ + X \longrightarrow XH^+ + H_2 \qquad (1)$$

Such reactions are rapid when the proton affinity (PA) of the molecule X, i.e. PA(X), exceeds that of H_2. Since PA(H_2) is quite small, then PA(X)> PA(H_2) for many species X, and thus the addition of X to a non-recombining H_3^+/electron plasma in the VT-FALP readily creates a recombining XH^+/electron plasma in which $q_e(XH^+)$ can be determined. If X is not too condensible, the $q_e(XH^+)$ can be determined at temperatures down to 95K. Thus $q_e(HCO^+)$, $q_e(N_2H^+)$ and $q_e(CH_5^+)$ have been determined at 95 and 300K (Adams et al 1984; the 300K values are given in Table III). Some information on the temperature variation of the q_e for various ionic species is given in Section 4.4.

TABLE III

Some positive ion/electron recombination coefficients, q_e, of relevance to interstellar chemistry determined at 300K using the VT-FALP technique; also included are the corresponding values (when available) determined in stationary afterglow (SA) experiments at 300K and those deduced for a temperature of 300K from ion trap (IT) and merged beam (MB) cross section data

		Recombination Coefficient (cm^3s^{-1})		
Ion	VT-FALP	SA	IT	MB
HCO^+	1.1(-7)[a]	2.0(-7)[b] 2.4(-7)[c]		
N_2H^+	1.7(-7)[a]			7.5(-7)[d]
HCO_2^+	3.4(-7)[e]			
H_2CN^+	3.5(-7)[e]			
$C_3H_5^+/C_6H_5^+$*	3.5(-7)[e]			
H_3S^+	3.7(-7)[e]			
HN_2O^+	4.2(-7)[e]			
$C_2H_5^+$	7.4(-7)[e]			
CH_3SH^+	7.7(-7)[e]			
$CH_3OH_2^+$	~9(-7)[e]			
HCS_2^+	9.1(-7)[e]			
H_3O^+	1.0(-6)[f]	1.0(-6)†	2(-6)[g]	6.3(-7)[h]
CH_5^+	1.1(-6)[a]			7.0(-7)[i]
$C_2H_5OH_2^+$	~1.1(-6)[e]			
NH_4^+	1.4(-6)[j]	1.5(-6)[k]	0.6-1.1(-6)[l]	
$CH_3NH_3^+$	1.4(-6)[e]			

* $C_3H_5^+/C_6H_5^+$ ratio was ~1:1
† obtained at 540K (Leu, Biondi and Johnsen 1973a)

a Adams et al 1984, b Leu et al 1973c, c Ganguli et al 1987, Mul and McGowan 1979, e Adams and Smith 1988, f Smith and Adams 1984a, g Heppner et al 1976, h Mul and McGowan 1980, i Mul et al 1981, j Alge et al 1983, k Huang et al 1976, l Dubois, Jeffries and Dunn 1978.

The values for $q_e(H_3^+)$ are given in Table II. An upper-limit value for $q_e(HeH^+)$ of $10^{-10}cm^3s^{-1}$ has been obtained using the VT-FALP technique.

In a very recent VT-FALP study (Adams and Smith 1988), the q_e for several ion species have been determined at 300K, some of which are considered to be involved in dense interstellar cloud chemistry and include the ions H_2CN^+ and HCO_2^+ which have been detected in dense interstellar clouds (these q_e are included in Table III). The details of the experimental technique are given in the paper by Adams and Smith (1988). As can be see from Table III, q_e generally increases with the complexity of the recombining ion, although this is by no means a 'golden rule'. At 300K, the q_e vary from about $10^{-7}cm^3s^{-1}$(e.g. $q_e(HCO^+) = 1.1 \times 10^{-7}cm^3s^{-1}$) for the simple ions to about $10^{-6}cm^3s^{-1}$ for the complex ions (e.g. $q_e(CH_3NH_3^+) = 1.4 \times 10^{-6}cm^3s^{-1}$), although $q_e(CH_5^+)$is also large at $1.1 \times 10^{-6}cm^3s^{-1}$.

4.4 Temperature Variation of the Dissociative Recombination Coefficients

Obviously from the viewpoint of interstellar chemistry, the q_e for many ionic species are required at appropriately low temperatures, T, which in some circumstances are as low as 10K. There is, as yet, no technique, to determine q_e at such low temperatures and so the best that can be done is to determine the variation of q_e with temperature at higher temperatures and then to rely on extrapolations to lower temperatures.

The limited available data indicate that q_e increases with decreasing temperature in accordance with theoretical expectations. The MB data is useful in that it indicates that σ_e and q_e increase with decreasing ion/electron centre-of-mass energy (Mitchell and McGowan 1983) albeit for ions of uncertain degrees of vibrational/rotational excitation. That the MB experiments almost always indicate that q_e varies as $T^{-0.5}$ (after appropriate treatment of the σ_e data - see Section 2) is presumably because the variation of q_e with centre-of-mass translational energy only, and not the ion internal temperature, is being indicated. In any event, where comparisons are available with the truly thermal SA or VT-FALP data then differences are very apparent. Then the data obtained under truly thermal conditions, where the degree of internal excitation is small and known, must be favoured. Theory predicts that q_e will vary with T according to the recombination process involved and a distinction is made between the 'direct process' for which $q_e \sim T^{-0.5}$ and the 'indirect process'for which $q_e \sim T^{-1.5}$ (Bardsley and Biondi 1970). The temperature power index as determined from truly thermal measurements of q_e usually falls between the values of -0.5 and - 1.5 indicated by theory! For example, $q_e(O_2^+) \sim T^{-0.7}$, $q_e(NO^+) \sim T^{-0.9}$ and, for the astrophysically important ions HCO^+ and N_2H^+, q_e varies as $\sim T^{-1}$ (albeit according to VT-FALP data obtained at only two temperatures). Some of the available data are represented in Fig. 1. It is to be noted that the increase in q_e with reducing temperature is more gradual when the q_e at 300K is relatively large. Thus, the difference between the q_e for ions of varying complexity will tend to be smaller at the low temperatures of dense interstellar clouds and a 'canonical value' for q_e at ~20K is likely to be $\sim 2 \times 10^{-6}cm^3s^{-1}$ (excluding, of course, the known 'abnormal ions' H_3^+ and HeH^+).

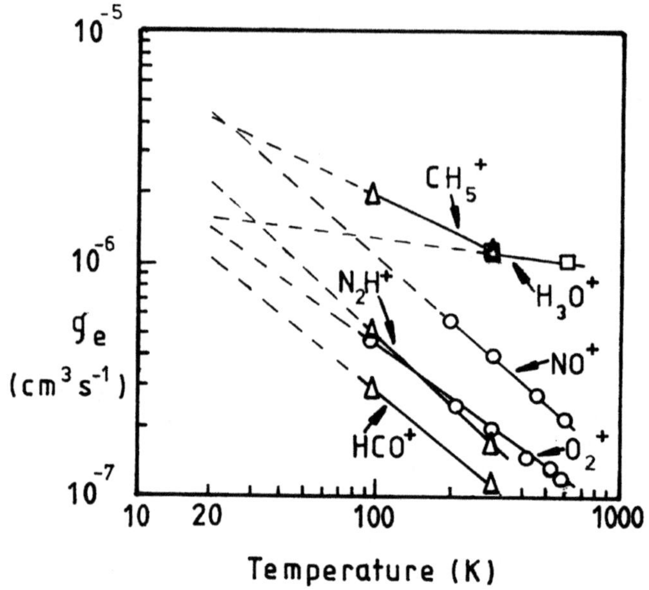

Figure 1. The variation with temperature of the ion-electron dissociative recombination coefficient, q_e, for some diatomic, triatomic and polyatomic ions. The dashed extrapolation to low temperatures of the power law behaviour at higher temperatures is intended to indicate the q_e appropriate to these recombination reactions at dense interstellar cloud temperatures of 20K. The data were obtained using the VT-FALP technique; ○ Alge et al 1983; △ Adams et al 1984; □ Smith and Adams, 1984b

Clearly, new techniques need to be developed to enable q_e to be determined at suitably low temperatures, especially for ions for which the precursor neutral molecules are condensible. Perhaps a variant of the CRESU technique could be developed, a technique which is so successfully being used to study ion-molecule reactions at temperatures as low as 8K (See B.R. Rowe in these Proceedings). However, it is fortunate that the q_e for most ionic species of interstellar importance appear likely to be similar at the low temperatures of dense interstellar clouds so that the use of a proper 'canonical value' for q_e will not result in major inaccuracies.

4.5 Products of Dissociative Recombination Reactions

The understanding of dissociative recombination will of course be incomplete, and the situation from an interstellar chemistry viewpoint unsatisfactory, until the neutral products of the important recombination reactions are known. Except for a few dissociative recombination reactions for which the energetics determine the products, e.g. $HCO^+ + e \longrightarrow CO + H$ (and similarly the $N_2H^+ + e \longrightarrow N_2 + H$), nothing is known about the neutral products of recombination of ions considered to be important in interstellar chemistry. Thus, it is necessary to use educated guesses or intuition when deducing the products for these reactions. Generally, for polyatomic hydrocarbon ions, it is assumed that recombination results only in the release of an H atom leaving the remaining neutral fragment intact. Some theoretical efforts have been made in this difficult area, notably by Herbst (1978) and by Bates (1986 and in these Proceedings), but these theories are in conflict as to the predictions for even 'simple' reactions. For example, for the important reaction:

$$H_3O^+ + e \longrightarrow OH + H_2 \quad (2a)$$
$$\longrightarrow H_2O + H \quad (2b)$$

the Herbst approach favours channel (2a) and OH production whereas the Batès approach favours channel (2b) and H_2O production. Clearly, it is critically important in interstellar molecular synthesis to know the products of reaction (2) and, inevitably, laboratory experiments must provide this information for reaction (2) as well as for many other recombination reactions.

The only laboratory data available on the product of dissociative recombination relate to a few internally-excited diatomic and triatomic ions (specifically for O_2^+, Zipf 1980a,b, Vallee et al 1987b; $N_2^+(v = 1)$, Queffelec et al 1985; NO^+, Kley, Lawrence and Stone 1977; CO_2^+, Gutcheck and Zipf 1973, Vallee et al 1986, and H_2O^+, Vallee et al 1987a, Rowe et al 1987) and in only one case (i.e. for $N_2^+(v = 1)$) was the state of internal excitation considered to be known (Queffelec et al 1985). However, none of these ions play a significant role in interstellar chemistry and what is required are the products of the dissociative recombination of the so-called terminating interstellar ions (i.e. ions which are unreactive with H_2) such as H_3O^+, H_2CN^+, NH_4^+, CH_5^+, other hydrocarbon ions, etc. to mention just a few (see Smith 1987). It is to this end, that a collaborative experiment has been initiated between Rennes University and Birmingham University which unites a VT-FALP apparatus (used to generate ground state ions of the kind mentioned above) with spectroscopic techniques. Vacuum ultraviolet (VUV) spectroscopy is to be used to detect any atomic radical products (e.g. H, N, O, etc.) of such dissociative recombination reactions and, in parallel with this, laser induced fluorescence (LIF) is to be used to detect the molecular radicals which may be released (e.g. OH from H_3O^+, NH_2 from NH_4^+, CN from H_2CN^+, etc.). It is hoped that these experiments will soon be providing data and will thus help to minimise the guesswork in this area.

5. EXPERIMENTAL DATA FOR MUTUAL NEUTRALIZATION REACTIONS

As is mentioned in Section 1, the proposal that polycyclic aromatic hydrocarbons (PAH's) with 20-100 carbon atoms may be responsible for the IR emission features of many objects, prompted Omont (1986) to suggest that negatively-charged PAH's i.e. PAH^-, may exist in dense clouds in number densities exceeding those of free electrons. Thus, in addition to positive ion/electron dissociative recombination, positive ion/negative ion (PAH^-) mutual neutralization could be an important process for ionization loss and for neutral molecule formation in dense interstellar clouds. In his paper, Omont (1986) considers the rate of PAH^- formation and gives a theoretical expression for the mutual neutralization rate coefficient, q_i. As was mentioned in Section 3, the most comprehensive laboratory studies of mutual neutralization (ionic recombination) have been carried out using the MB and VT-FALP techniques, the latter technique providing the most reliable data. Much of the VT-FALP work on ionic recombination was directed towards the study of reactions important in the terrestrial atmosphere. However, an important feature of the accumulated data is the relatively small variation in the q_i for reactant molecular ions of widely differing complexities which only span the limited range of $(4 - 10) \times 10^{-8} cm^3 s^{-1}$ at 300K (for example see the review by Smith and Adams 1983). Thus, for example, the q_i for the reaction

$$NO^+ + NO_2^- \longrightarrow NO + NO_2 \qquad (3)$$

is $6.4 \times 10^{-8} cm^3 s^{-1}$ at 300K whilst that for the reaction

$$H_3O^+(H_2O)_3 + NO_3^-(HNO_3) \longrightarrow \text{products} \qquad (4)$$

is little different at $5.7 \times 10^{-8} cm^3 s^{-1}$ at 300K. (It must be noted however that for reactions involving only atomic ions the q_i were much smaller, in accordance with theoretical expectations; Church and Smith 1978). The amount of data available is limited concerning the temperature variation of q_i (Smith and Church 1977, Smith and Adams 1983). For reaction (3) and for the neutralization reaction of NH_4^+ with Cl^-, the VT-FALP data indicate that q_i varies as $T^{-0.5}$ in accordance with the theoretical predictions of Olsen (1972). It is also important to note that the magnitude of q_i for reaction (3), determined using the VT-FALP, is smaller only by a factor of two than the upper-limit value predicted by the 'absorbing sphere' model of Olsen (1972), whereas the MB value of q_i for this reaction at an equivalent temperature is unreasonably large being some ten times greater than the theoretical prediction ! (and hence some twenty times greater than the VT-FALP value). On the basis of the $T^{-0.5}$ variations and the laboratory 300K data for large molecular ions then at 20K in a dense molecular cloud the q_i is predicted to be $\sim(2 - 4) \times 10^{-7} cm^3 s^{-1}$, i.e. a few times smaller than the q_e at the same temperature (see Section 4.4). Clearly, the relative importance of positive ion/electron recombination and positive ion/negative ion mutual neutralization will be dependent on the number densities of electrons relative to the number densities of PAH^- as well as on the q_e and q_i.

Very little laboratory work has been carried out to determine the neutral products of mutual neutralization reactions. It has been shown by optical emissions from the NO^+/NO_2^- plasma in the VT- FALP experiment that the products of reaction (3) are largely NO_2 and electronically-excited NO (Smith, Adams and Church 1978). From this study, it was shown that the electron transfer from the NO_2^- to the NO^+ occurred at an internuclear spacing of ~10Å. Since mutual neutralization reactions are generally less exoergic than positive ion/electron recombination reactions (by virtue of the finite detachment energy of the negative ion) then less fragmentation is to be expected following mutual neutralization and thus large neutral species can be formed via this process. Much more laboratory work is necessary in this area; such is planned in the Birmingham laboratory using the additional spectroscopic techniques mentioned at the end of Section 4.5 .

6. CONCLUDING REMARKS

There is little doubt that positive-ion/electron dissociative recombination is an important step in the production of neutral interstellar molecules. Hence to properly model the ion chemistry of interstellar clouds, it is necessary to know the values of the dissociative recombination coefficients, q_e, for a variety of positive ions at the low temperatures pertaining to interstellar clouds and also to know the identity of the products of such reactions. Recent VT-FALP work has indicated the magnitudes of q_e for several interstellar positive ion species and also has shown that when the q_e at 300K is relatively small then it increases relatively rapidly with decreasing temperature but when q_e at 300K is relatively large then it increases more gradually with decreasing temperature. Thus, at interstellar cloud temperatures, the q_e for the many varied ionic species are not likely to be very different; a 'canonical' value of $2 \times 10^{-6}cm^3s^{-1}$ seems to be appropriate to a temperature of 20K. Little is known about the products of dissociative recombination reactions and clearly more work is required before a thorough understanding of interstellar ion chemistry can be obtained. To this end spectroscopic diagnostics are about to be added to the VT-FALP apparatus and in the first instance the products of dissociative recombination of the relatively simple molecular ions H_3O^+, H_2CN^+ and NH_4^+ will be determined.

If, as has been proposed, significant concentrations of negative ions (in the form of PAH^- ions) exist in dense interstellar clouds then the additional process of mutual neutralization must be considered as a loss process for positive ions. The magnitudes of the mutual neutralization rate coefficients, q_i have been determined by VT-FALP experiments for a variety of reactant positive and negative ions, and the variation of q_i with temperature has been determined for two reactions and are seen to be in accordance with theoretical predictions. Thus, at interstellar temperatures, a q_i of about $(2 - 4) \times 10^{-7}cm^3s^{-1}$ appears to be appropriate. Little is known about the neutral products of such reactions but this will hopefully be rectified soon, again using spectroscopic techniques to probe ion-ion plasmas in the VT-FALP apparatus.

ACKNOWLEDGEMENTS

Our thanks are due to the United States Air Force and to the Science and Engineering Research Council for providing financial support for this work.

REFERENCES

Adams, N.G., and Smith, D. 1983 in *Reactions of Small Transient Species* Ed. A. Fontijn and M.A.A. Clyne (London, Academic), p.311.
Adams, N.G. and Smith, D. 1987 in *Astrochemistry* Ed. M.S. Vardya and S.P. Tarafdar, (Dordrecht, Holland, Reidel), p.1.
Adams, N.G., and Smith, D. 1988 *Chem. Phys. Letts.*, in press.
Adams, N.G., Smith, D., and Alge, E. 1984 *J. Chem. Phys.*, **81**, 1778.
Adams, N.G., Smith, D., and Clary, D.C., 1985, *Ap. J. (Letters)*, **296**, L31.
Adams, N.G., Smith, D., and Herd, C.R. 1988, *Int. J. Mass Spectrom. Ion Processes*, in press.
Albritton, D.L. 1978, *Atom. Data Nucl. Data Tables*, **22**, 1.
Albritton, D.L. 1979, in *Kinetics of Ion-Molecule Reactions*, Ed. P. Ausloos, (New York, Plenum) p. 119.
Alge, E., Adams, N.G., and Smith, D. 1983, *J.Phys. B.*, **16**, 1433.
Alge, E., Adams, N.G., and Smith, D. 1984, *J. Phys. B.*, **17**, 3827.
Auerbach, D., Cacak, R., Caudano, R., Gaily, T.D., Keyser, C.J., McGowan J.W., Mitchell, J.B.A. , and Wilk, S.F.J. 1977, *J. Phys. B.*, **10**, 3797.
Bardsley, J.N., and Biondi, M.A. 1970, *Adv. Atom. Mol. Phys.*, **6**, 1.
Bates, D.R. 1979, *Adv. Atom. Mol. Phys.* **15**, 235.
Bates, D.R. 1986, *Ap. J. (Letters)*, **306**, L45.
Bates, D.R., and Zipf, E.C. 1980, *Planet. Space Sci.*, **28**, 1081.
Biondi, M.A. 1973, *Comm. Atom. Mol. Phys.*, **4**, 85.
Bohme, D.K. 1975, in *Interactions between Ions and Molecules*. Ed. P. Ausloos, (New York, Plenum), p.489.
Church, M.J., and Smith, D. 1978, *J. Phys. D*, **11**, 2199.
Cunningham, A.J., and Hobson, R.M. 1972, *J. Phys. B.*, **5**, 1773, 2320, 2328.
Dalgarno, A., and Black, J.H. 1976, *Rept. Prog. Phys.*, **39**, 573.
Dalgarno, A., and McCray, R.A. 1973, *Ap. J.*, **181**, 95.
Dubois, R.D., Jeffries, J.B., and Dunn, G.H. 1978, *Phys. Rev.*, **A17**, 1314.
Dulaney, J.D., Biondi, M.A., and Johnsen, R. 1987, *Phys. Rev.* **A**, in press.
Flannery, M.R. 1982, in *Applied Atomic Collision Physics*, Vol. 3 Ed. H.S.W. Massey, B. Bederson and E.W. McDaniel, (New York, Academic) p. 141.
Flower, D.R., and Roueff, E. 1979, *Astron. Astrophys.*, **72**, 361.
Ganguli, B., Biondi, M.A., Johnsen, R., and Dulaney, J.L. 1987, to be

published.
Guberman, S.L. 1983, in *Physics of Ion-Ion and Electron-Ion Collisions*, Ed. F. Brouillard and J.W. McGowan, (New York, Plenum) p. 167.
Gutcheck, R.A., and Zipf, E.C. 1973, *J. Geophys. Res.*, **78**, 5429.
Heppner, R.A., Walls, F.L., Armstrong, W.T., and Dunn, G.H. 1976, *Phys. Rev.* **A13**, 1000.
Herbst, E. 1978, *Ap. J.* **222**, 508.
Herbst, E. 1981, *Nature*, **289**, 656.
Huang, C.M., Biondi, M.A., and Johnsen, R. 1975, *Phys. Rev.*, **A11**, 901.
Huang, C.M., Biondi, M.A., and Johnsen, R. 1976, *Phys. Rev.*, **A14**, 984.
Hus, H., Youssif, F., Noren, C., Sen, A., and Mitchell, J.B.A. 1987, *Phys. Rev. Letts.*, in press.
Johnsen, R. 1987, *Int. J. Mass Spectrom. Ion Processes*, **81**, 67.
Kasner, W.H. and Biondi, M.A. 1968, *Phys. Rev.*, **174**, 139.
Kim, J.K., Theard, L.P., and Huntress, W.T. Jr. 1974, *Int. J. Mass Spectrom. Ion Phys.*, **15**, 223.
Kley, D., Lawrence, G.M., and Stone, E.J. 1977, *J. Chem. Phys.*, **66**, 4157.
Lepp.˙ S., and Dalgarno, A., 1987, reprint.
Leu, M.T., Biondi, M.A., and Johnsen, R. 1973a, *Phys. Rev.*, **A7**, 292.
Leu, M.T., Biondi, M.A., and Johnsen, R. 1973b, *Phys. Rev.*, **A8**, 413.
Leu, M.T., Biondi, M.A., and Johnsen, R. 1973c, *Phys. Rev.*, **A8**, 420.
Lindinger, W., and Smith, D., 1983 in *Reactions of Small Transient Species*. Ed. A. Fontijn and M.A.A. Clyne, (London, Academic), p.387.
Macdonald, J.A., Biondi, M.A., and Johnsen, R. 1984, *Planet. Space Sci.*, **32**, 651.
Mahan, B.H. 1973, *Adv. Chem. Phys.*, **23**,1.
Mahan, B.H. and Person, J.C. 1964, *J. Chem. Phys.*, **40**, 392.
Mehr, F. J. and Biondi, M.A. 1969, *Phys. Rev.*, **181**, 264.
Michels, H.H., and Hobbs, R.H. 1984, *Ap. J. (Letters)*, **286**, L27.
Mitchell, J.B.A., 1986 in *Atomic Processes in Electron-Ion and Ion-Ion Collisions*, Ed. F. Brouillard (New York, Plenum), p.185.
Mitchell, J.B.A. and McGowan, J.W. 1983, in *Physics of Ion-Ion and Electron-Ion Collisions*, Ed. F. Brouillard and J.W. McGowan, (New York, Plenum), p.279.
Mitchell, J.B.A., Ng, C.T., Forard, L., Janssen, R., and McGowan, J.W. 1984, *J. Phys. B.*, **17**, L909.
Moseley, J.T., Olsen, R.E., and Peterson, J.R. 1975, in *Case Studies in Atomic Physics, Vol. 5*, Ed. M.R.C. McDowell and E.W. McDaniel, (Amsterdam, North Holland), p.1
Mul, P.M., and McGowan, J.W. 1980, *Ap. J.*, **237**, 749.
Mul, P.M., Mitchell, J.B.A., D'Angelo, V.S., Defrance, P., McGowan, J.W., and Froelich, H.R. 1981, *J. Phys. B.*,**14**, 1353.
Olson, R.E. 1972, *J. Chem., Phys.*, **56**, 2979.
Omont, A. 1986, *Astron. Astrophys.*, **164**, 159
Persson, K.B., and Brown, S.C. 1955, *Phys. Rev.*, **100**, 729.
Plumb, I.C., Smith, D., and Adams, N.G. 1972, *J. Phys. B.*, **5**, 1762.
Queffelec, J.L., Rowe, B.R., Morlais, M., Gomet, J.C., and Vallee, F. 1985 *Planet. Space Sci.*, **33**, 263.
Reid, G.C. 1976, *Adv. Atom. Mol. Phys.*,**63**, 375.
Rowe, B.R., Vallee, F., Queffelec, J.L., Gomet, J.C. and Morlais, M.

1987, *J. Chem. Phys.*, in press.
Sayers, J. 1962, in *Atom. Mol. Proc.* Ed. D.R. Bates, (New York, Academic), p. 272.
Smith, D. 1987, *Phil. Trans. R. Soc. Lond.,* in press.
Smith, D., and Adams, N.G. 1980, in *Topics in Current Chemistry.Vol.89*, Ed. S. Veprek and M. Venugopalan, (Berlin, Springer-Verlag), p.1.
Smith, D., and Adams, N.G. 1981, *Int. Rev. Phys. Chem.,* **1**, 271,.
Smith, D., and Adams, N.G. 1983, in *Physics of Ion-Ion and Electron -Ion Collisions,* Ed. F. Brouillard and J.W. McGowan, (New York, Plenum), p.501.
Smith, D., and Adams, N.G. 1984a, *Ap. J. (Letters),* **284**, L13.
Smith, D., and Adams, N.G. 1984b, in *Swarms of Ions and Electrons in Gases,* Ed. W. Lindinger, T.D. Mark and F. Howorka (Wien, Springer-Verlag), p. 284.
Smith, D., and Church, M.J. 1977, *Planet. Space Sci.*, **25**, 433.
Smith, D., Adams, N.G., and Church, M.J. 1978,. *J. Phys. B.,* **11**, 4041.
Smith, D., Adams, N.G., and Alge, E. 1984, *J. Phys. B.,* **17**, 461.
Stein, R.P., Schiebe, M. Syverson, M.W., Shaw, T.M., and Gunton, R.C. 1964, *Phys. Fluids,* **7**, 1641.
Torr, M.R., St. Maurice, J.P., and Torr, D.G. 1977., *J. Geophys. Res.*, **82**,3287.
Vallee F., Rowe, B.R., Gomet, J.C., Queffelec, J.L., and Morlais, M. 1986, *Chem. Phys. Lett.*, **124**, 317.
Vallee, F., Gomet, J.C., Rowe, B.R., Queffelec, J.L. and Morlais, M. 1987a in *Astrochemistry*, Ed. M.S. Vardya and S.P. Tarafdar, (Dordrecht, Holland, Reidel), p.29.
Vallee, F., Rowe, B.R., Queffelec, J.L., Gomet, J.C., and Morlais, M. 1987b, *J. Chem. Phys.*, preprint.
Walls, F.L., and Dunn, G.H., 1974, *J. Geophys. Res.*, **79**, 1911.
Watson, W.D. 1978, *Ann. Rev. Astron. Astrophys.*, **16**, 585.
Weller, C.S., and Biondi, M.A. 1968, *Phys. Rev.*, **172**, 198.
Woodin, R.L., Foster, M.S., and Beauchamp, J.L. 1980, *J. Chem. Phys.*, **72**, 23.
Zipf, E.C. 1980a, *J. Geophys. Res.*, **85**, 4232.
Zipf, E.C. 1980b, *Geophys. Res. Lett.*, **7**,645.

CHEMICAL PATHWAYS FROM ATOMIC SILICON IONS TO SILICON CARBIDES AND OXIDES.

Diethard K. Bohme, Stanislaw Wlodek, and Arnold Fox
Department of Chemistry and
Centre for Research in Experimental Space Science
York University
North York, Ontario
Canada M3J 1P3

ABSTRACT. Ion/molecule reactions initiated by ground-state atomic silicon ions are described which lead to Si-C and Si-O bond formation. They are viewed as the first steps in the chemical pathways leading from atomic silicon to the formation of silicon carbide and oxide molecules of the type which may be precursors for the formation of silicon carbide and silicate grain particles in carbon-rich and oxygen-rich astrophysical environments.

1. INTRODUCTION

A fundamental problem in astrophysics is to understand the evolution of isolated atoms to molecules and grain particles. The growth of carbon-rich and oxygen-rich molecules and grain particles in interstellar and circumstellar environments containing silicon appears to be of special interest and importance (1-3). For example, laboratory simulation experiments with a plasma jet apparatus have recently been used to explore the transition from atomized reactant gases consisting of Si, O, C, and H to solid particles containing Si-O and Si-C bonds (4). We report here the results of experimental studies directed towards gaining insight into the kinetics and mechanism of the first steps of such growth as it occurs in the presence of selective ionization. A Selected-Ion Flow Tube (SIFT) apparatus is used to generate ground-state Si^+ (^2P) ions and to react these ions in helium buffer gas (at ca. 0.35 Torr and 296 K) with molecules which may lead directly to the chemical bonding of silicon with carbon and oxygen. Secondary and higher order reactions which lead to further ionic growth are also monitored, including the sequential formation of ions in which the silicon becomes increasingly co-ordinated. A variety of possible interstellar and circumstellar molecules containing carbon and/or oxygen have been chosen as neutral reactants including methane, acetylene, diacetylene, amines, cyanides, carbon monoxide, water, dimethyl ether, alcohols, carboxylic acids, and acetone.

T. J. Millar and D. A. Williams (eds.), Rate Coefficients in Astrochemistry, 193–200.

2. EXPERIMENTAL

Measurements were made in the Ion Chemistry Laboratory of York University with the Selected-Ion Flow Tube (SIFT) technique (5,6). With this technique ground-state atomic silicon ions are selected from a suitable source, and injected into flowing helium gas to which is added the desired neutral reactant (7). Moderate pressures (ca. 0.35 Torr) of helium buffer gas were used. The reactant and product ions are monitored with a mass spectrometer as a function of the addition of the neutral reagent. The observations provide a measurement of the specific rate and products of the chosen reactions and information about the secondary and higher order ion chemistry at 296 +/- 2 K.

3. RESULTS

3.1. Formation of Si-C Bonds

3.1.1. Methane and Acetylene. Methane was observed to react only slowly with Si^+ (^2P), $k = 5 \times 10^{-13}$ cm^3 $molecule^{-1}$ s^{-1}), to form an adduct with an unknown structure (8). A weakly (electrostatically) bound adduct ion of the type $Si^+.CH_4$ is possible, or $H\text{-}Si^+\text{-}CH_3$ may be produced by C-H bond insertion. This resuly is in sharp contrast to that for the reaction of C^+ with methane which proceeds rapidly to form C-C bonds with the elimination of H and H_2 (9).

The reaction of $Si^+(^2P)$ with acetylene is much faster, $k = 3.5 \times 10^{-10}$ cm^3 $molecule^{-1}$ s^{-1}, and proceeds primarily by associaton with H atom elimination as has been observed with C^+ (8,9):

$$Si^+ + C_2H_2 \xrightarrow{0.7} SiC_2H^+ + H$$
$$\xrightarrow{0.3} SiC_2H_2^+$$

Further carbon accretion was observed to occur in the sequential reaction of SiC_2H^+ with acetylene with a rate constant of 2×10^{-10} cm^3 $molecule^{-1}$ s^{-1}:

$$SiC_2H^+ + C_2H_2 \xrightarrow{0.1} SiC_4H^+ + H_2$$
$$\xrightarrow{0.9} SiC_4H_3^+$$

Diacetylene, on the other hand, did not entrain silicon, yielding only a hydride ion to the silicon cation with a rate constant of 1.6×10^{-9} cm^3 $molecule^{-1}$ s^{-1} (8):

$$Si^+ + C_4H_2 \longrightarrow SiH + C_4H^+$$

3.1.2. Amines. The reactions of methyl-, dimethyl-, and trimethyl amine with atomic silicon ions were all observed to be fast, $k > 1 \times 10^{-9}$ cm^3 $molecule^{-1}$ s^{-1} (10). Two reaction channels were predominant. They can be understood in terms of Si^+ insertion into N-H and N-C bonds:

$$Si^+ + CH_3\text{-}NR_1R_2 \longrightarrow CH_3 + Si^+NR_1R_2$$

and hydride transfer to form immonium ions and the neutral SiH molecule:

$$Si^+ + H\text{-}CH_2\text{-}NR_1R_2 \longrightarrow SiH + CH_2NR_1R_2^+$$

The contribution of the hydride transfer was observed to increase with increasing stability of the immonium ions. Neither channel leads to Si-C bonding in the products. Si-N and Si-H bonds are formed instead. Si-C bond formation was apparent only for the minor reaction channels (less than 5%) which were observed. They lead directly to SiCH and $SiCH_3$ as follows:

$$Si^+ + CH_3NH_2/(CH_3)_2NH \longrightarrow SiCH + NH_4^+/CH_3NH_3^+$$

$$Si^+ + (CH_3)_3N \longrightarrow SiCH_3 + CH_2NHCH_3^+$$

$$\longrightarrow SiCH_2^+ + (CH_3)_2NH$$

3.1.3. <u>Cyanides.</u> A range in reactivity was observed for the reaction of Si^+ with the cyanides HCN, C_2N_2, CH_3CN and HC_3N (8). Hydrogen cyanide reacted only slowly, $k = 7 \times 10^{-12}$ cm^3 molecule^{-1} s^{-1}:

$$Si^+ + HCN \xrightarrow{0.8} (CHNSi)^+$$

$$\xrightarrow{0.2} (CNSi)^+ + H$$

It is uncertain whether the product ions are Si-C or Si-N bonded. Quantum chemical calculations indicate that Si-N bond formation may be preferred because of the lower energy of $SiNC^+$ compared to $SiCN^+$.

The reaction with methylcyanide was observed to be much more rapid, $k = 2.4 \times 10^{-9}$ cm^3 molecule^{-1} s^{-1}, and to proceed with both an addition and a condensation channel:

$$Si^+ + CH_3CN \xrightarrow{0.5} (CH_3CNSi)^+$$

$$\xrightarrow{0.5} (CH_2Si)^+ + HCN$$

In this case the condensation channel must lead to Si-C bond formation.

Cyanogen also showed both an addition and a condensation channel:

$$Si^+ + C_2N_2 \xrightarrow{0.55} (C_2N_2Si)^+$$

$$\xrightarrow{0.45} (CNSi)^+ + CN$$

Again, both Si-N and Si-C bond formation are possible in the product ions. The reaction is moderately fast, $k = 1.5 \times 10^{-10}$ cm^3 molecule^{-1} s^{-1}.

The reaction of Si^+ with cyanoacetylene mimicked the reaction with acetylene in that SiC_2H^+ was the predominant product:

$$Si^+ + HC_3N \xrightarrow{0.7} SiC_2H^+ + CN$$
$$\xrightarrow{0.3} (SiHC_3N)^+$$

This reaction is rapid, $k = 1.4 \times 10^{-9}$ cm^3 $molecule^{-1}$ s^{-1}, and represents another source for Si-C bonding.

3.2. Formation of Si-O Bonds

3.2.1. Carbon Monoxide. Atomic silicon ions were found to be unreactive towards CO, $k < 2 \times 10^{-14}$ cm^3 $molecule^{-1}$ s^{-1}.

3.2.2. Water, Dimethyl Ether, Alcohols and Carboxylic Acids. We have reported that Si^+ reacts rapidly with molecules containing hydroxyl groups to abstract the hydroxyl group (7):

$$Si^+ + R\text{-}OH \longrightarrow SiOH^+ + R$$

for R = H, CH_3, C_2H_5, HCO and CH_3CO. A second, less predominant, channel with methanol yielded the $SiOCH_3^+$ ion which was also the predominant product in the fast reaction of Si^+ with dimethyl ether (8):

$$Si^+ + CH_3OCH_3 \xrightarrow{0.9} SiOCH_3^+ + CH_3$$
$$\xrightarrow{0.1} CH_2OCH_3^+ + SiH$$

The silene cation $SiOH^+$ also was observed to be reactive. Five molecules of water were observed to add to $SiOH^+$. Predominantly bimolecular products were observed for the fast reactions of $SiOH^+$ with methanol, ethanol, formic acid, and acetic acid. Formation of $SiOCH_3^+$ predominated with methanol as in the reaction with Si^+:

$$SiOH^+ + CH_3OH \longrightarrow SiOCH_3^+ + H_2O$$

$SiH_3O_2^+$ was the main product ion observed for the reactions with ethanol and formic acid:

$$SiOH^+ + C_2H_5OH/HCOOH \longrightarrow SiH_3O_2^+ + C_2H_4/CO$$

Neutralization of this ion may yield silanoic acid, HSiOOH, or silicon dihydroxide, $Si(OH)_2$. These molecules may be formed directly in the reaction observed with acetic acid:

$$SiOH^+ + CH_3COOH \longrightarrow CH_3CO^+ + (SiH_2O_2)$$

Still further, higher order, chemistry was observed with methanol, ethanol and formic acid. The chemistry was clearly directed toward increasing the silicon/oxygen co-ordination. One sequence of reactions initiated by Si^+ in methanol produced the following ions:

Si^+ ----> $SiOCH_3^+$ ----> $HSi(OCH_3)_2^+$ ----> $HSi(OCH_3)_3H^+$

An analogous sequence was observed with ethanol:

Si^+ ---> $SiOC_2H_5^+$ ---> $HSi(OC_2H_5)_2^+$ ---> $HSi(OC_2H_5)_3H^+$

Still other reaction sequences initiated by Si^+ were identified which appeared to saturate the silicon centre. With ethanol the experimental data was consistent with the following additional bimolecular reaction sequences:

Si^+ ----> $SiOH^+$ ----> $HSi(OH)_2^+$ ----> $HSi(OH)_3H^+$

Si^+ ----> $SiOH^+$ ----> $HSi(OH)_2^+$ ----> $HSi(OC_2H_5)OH^+$

With formic acid complete saturation appears to be achieved in three bimolecular steps as with ethanol as follows:

Si^+ ----> $SiOH^+$ ----> $HSi(OH)_2^+$ ----> $HSi(OH)_3H^+$

or in four steps as follows:

Si^+ --> $SiOH^+$ --> $HSi(OH)_2^+$ --> $HSi(OCHO)OH^+$ --> $HSi(OCHO)(OH)_2H^+$

3.2.3. Acetone. Acetone was observed to react rapidly with Si^+, $k = 2.8 \times 10^{-10}$ cm^3 $molecule^{-1}$ s^{-1}, to produce both Si-C and Si-O bonds (8):

$$Si^+ + (CH_3)_2CO \xrightarrow{0.45} SiOH^+ + C_3H_5$$
$$\xrightarrow{0.30} CH_3CO^+ + SiCH_3$$
$$\xrightarrow{0.25} C_3H_6^+ + SiO$$

The secondary reaction of $SiOH^+$ with acetone proceeded exclusively by proton transfer and so did not lead to a larger silicon-containing ion.

4. SUMMARY AND CONCLUSIONS

The experimental studies reported here have led to the identification of a large number of gas-phase ion/molecule reactions and reaction sequences which act to build ions containing Si-C and Si-O bonds. Si-C bonds were observed to be established in the reactions of Si^+ with acetylene, cyanoacetylene, methylcyanide and acetone, while diacetylene and possibly methane were found to be resistant to silicon implantation. The results obtained with methanol, ethanol, and formic acid are striking in that they suggest that the sequential chemistry with these molecules initiated by Si^+ is inexorably directed toward increasing silicon-oxygen co-ordination.

Ionic growth proceeding in partially ionized environments such as interstellar clouds and circumstellar envelopes will compete with neutralization reactions such as proton transfer, charge transfer and recombination with electrons. For example, neutralization of $SiOH^+$ and SiC_4H^+ may proceed as follows:

$$SiOH^+ + M, e \dashrightarrow SiO + MH^+, H$$

$$SiC_4H^+ + M, e \dashrightarrow SiC_4 + MH^+, H$$

In these neutralization reactions (as written) the newly formed Si-C or Si-O bond is preserved, but this will not always be the case. Proton transfer may become dissociative when it is highly exothermic. Also, electron/ion recombination is often quite exothermic and conceivably may involve the rupture of heavy-atom bonds in the ion.

Neutralization of the ions containing Si and C generated by the chemistry identified in this study, viz. $SiCH_2^+$, SiC_2H^+, $SiCH_4^+$ and SiC_4H^+, provide possible sources for SiCH/HSiC, SiC_2, $SiCH_3$/$HSiCH_2$ and SiC_4, respectively. Two reactions form neutral $SiCH_3$/$HSiCH_2$ directly. Scheme 1 summarizes the <u>bimolecular</u> reaction chemistry identified in this study which leads to the growth of silicon carbide molecules.

Scheme 1. Limited reaction scheme for the synthesis of silicon carbide molecules initiated by atomic silicon ions. The double arrows represent neutralization reactions such as proton transfer and electron/ion recombination.

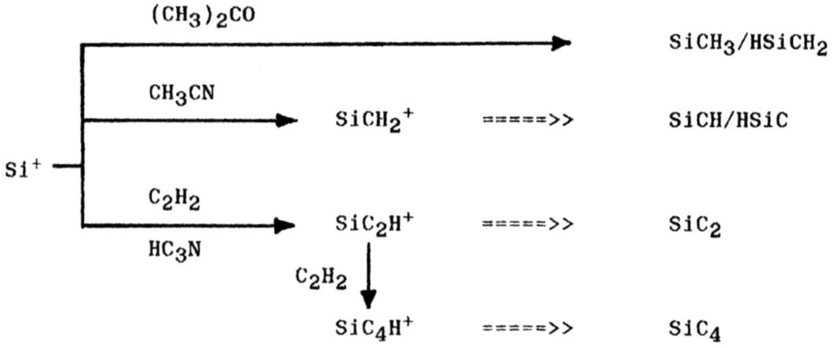

Similarly, the ions containing Si and O generated by the chemistry identified in this study provide possible sources for the following silicon oxides: SiO, $SiOCH_2$, $SiOC_2H_4$, SiO_2H_2, $HSi(OH)_3$, HSi(O)(OCHO), $HSi(O)(OC_2H_5)$, $HSi(OCHO)(OH)_2$, $Si(OCH_3)_2$, $Si(OCH_3)_3$, $Si(OC_2H_5)_2$, and $HSi(OC_2H_5)_3$. This is illustrated and summarized in schemes 2 and 3.

Several isomers are possible for many of the carbon and oxygen-containing molecular silicon ions which have been identified in these studies and the molecules which are formed by their neutralization. In some cases cyclic structures may be preferred as, for example, with SiC_2, $SiOCH_2$, and $SiOC(H)CH_3$, or even a three-demensional square pyramid structure in the case of SiC_4.

Scheme 2. Limited reaction scheme for the synthesis of silicon oxide molecules initiated by $SiOH^+$ in methanol and ethanol.

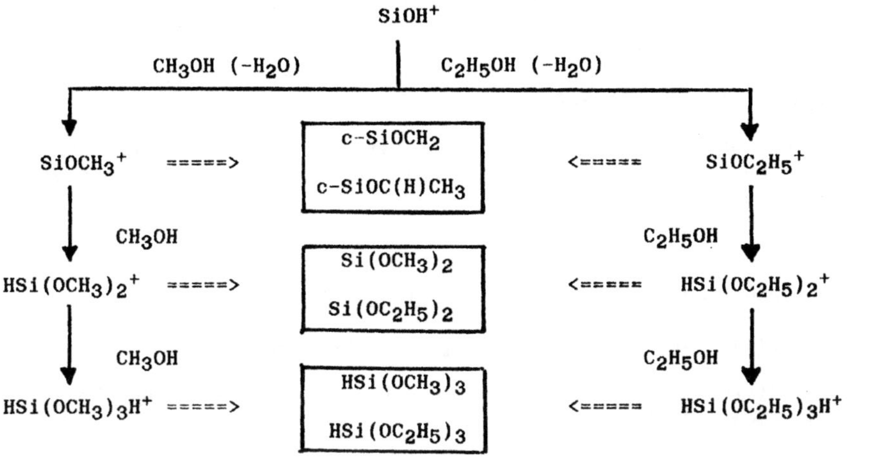

Scheme 3. Limited scheme for the synthesis of silicon oxide molecules ith ion chemistry initiated by Si^+ in ethanol and formic acid.

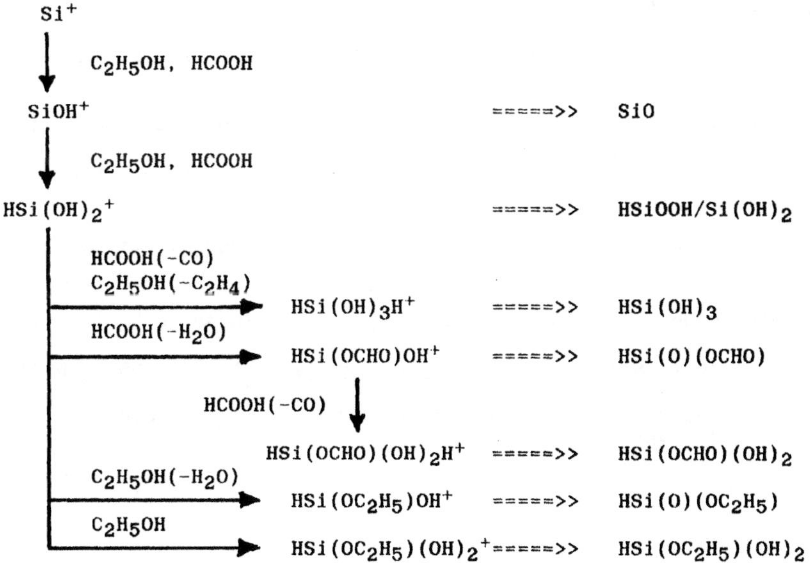

Many of the larger ions and molecules are likely precursors important in silicon carbide and silicate particle formation. For example, re-ionization of trihydroxysilane, followed by a dehydrative reaction with a carboxylic acid, may provide a source for tetrahydroxysilane which has been shown to be a building block for the condensational synthesis of hydrated silicate networks (11,12). Perhaps tetrahydroxysilane may even be formed in the gas phase by the reaction of trihydroxysilane with water vapour.

Acknowledgement: The authors thank the Natural Sciences and Engineering Research Council of Canada for financial support during the course of this research.

REFERENCES

1. J.L. Turner and A. Dalgarno, Astrophys. J. 213, 386 (1977).

2. T. J. Millar, Astrophys. Space Sci. 72, 509 (1980).

3. R.E.S. Clegg, L.J. van IJzendoorn and L.J. Allamandola, Mon. Not. R. astr. Soc. 203, 125 (1983).

4. T. Tanabe, T. Onaka, F. Kamijo, A. Sakata and S. Wada, Jap. J. Appl. Phys. 25, 1914 (1986).

5. G.I. Mackay, G.D. Vlachos, D.K. Bohme and H.I. Schiff, Int. J. Mass Spectrom. Ion Phys. 36, 259 (1980).

6. A.B. Raksit and D.K. Bohme, Int. J. Mass Spectrom. Ion Processes 55, 69 (1984).

7. S. Wlodek, A. Fox and D.K. Bohme, J. Am. Chem. Soc. 109, 6663 (1987).

8. Unpublished results from the Ion Chemistry Laboratory at York University.

9. D.K. Bohme, A.B. Raksit and H.I. Schiff, Chem. Phys. Lett. 93, 592 (1982).

10. S. Wlodek and D.K. Bohme, J. Am. Chem. Soc., in press.

11. Y. Abe and T. Misono, J. Polym. Sci. Polym. Lett. 22, 565 (1984).

12. R.H. Meinhold, H.P. Rothbaum and R.H. Newman, J. Colloid Interface Sci. 108, 234 (1985).

CHEMICAL PATHWAYS FOR DEUTERIUM FRACTIONATION IN INTERSTELLAR MOLECULES

Michael Henchman
Chemistry Department
Brandeis University
Waltham MA 02254
USA

John F. Paulson
Air Force Geophysics Laboratory
Hanscom Air Force Base
Bedford MA 01731
USA

David Smith and Nigel Adams
Department of Space Research
University of Birmingham
Birmingham B15 2TT
UK

W. Lindinger
Institute for Experimental Physics
University of Innsbruck
Karl Schönherrstrasse 3
A 6020 Innsbruck Austria

ABSTRACT. The formation of deuterated interstellar molecules is considered. This contribution (which, in the summarizing remarks of the conference, was admonished for being itself admonitory) has a simple purpose — to apply gas-phase ion chemistry to distinguish deuterium exchange reactions that *can* occur from those that *cannot*. Thus the exoergic reaction $HCO^+ + D \rightarrow DCO^+ + H$ is facile whereas the exoergic reaction $HCO^+ + HD \text{ —x→ } DCO^+ + H_2$ is not. Forbidden pathways involve energy barriers that can result from both reactants showing filled valence electron shells. A procedure is outlined for identifying such.

1. FRACTIONATION OF DEUTERIUM IN INTERSTELLAR MOLECULES

The fractionation of deuterium in interstellar molecules continues to excite considerable interest.[1-3] Cosmologists identify the cosmic D/H ratio as a parameter critical to the assessment of cosmological models. Astrophysicists can use the isotopic ratio of species found in interstellar clouds as a probe of the conditions in those clouds. Isotopic abundances can help ion chemists to map synthetic pathways for forming interstellar molecules. Finally to chemical kineticists, interested in the formation of interstellar molecules at temperatures approaching absolute zero, isotope effects offer a unique challenge — what is a minor perturbation at 300 K must exercise a profound influence at 10 K. Thus[4] the equilibrium constant for the reaction

$$HCO^+ + D = DCO^+ + H \qquad -\Delta H^\circ/R \cong 800\ \text{K} \qquad (1)$$

is ~ 6 at 300 K but at 10 K becomes $\sim 10^{33}$!

Eleven deuterated species have been identified [5,6] in interstellar space:

Molecules: HD, HDO, NH_2D, CCD, DCN, DNC, HDCO, CH_3OD, C_3HD, DC_3N, DC_5N
Ions: H_2D^+, N_2D^+, DCO^+

T. J. Millar and D. A. Williams (eds.), Rate Coefficients in Astrochemistry, 201–207.

The deuterium fractionation observed for these species must be measured against the benchmark of the cosmic D/H ratio and this is illustrated[7] for three examples below:

Cosmic D/H ratio		~2×10^{-5}	
DCO^+/HCO^+	2×10^{-4}	↔	1.4×10^{-2}
DCN/HCN	1.4×10^{-3}	↔	1.2×10^{-2}
DNC/HNC	0.05	↔	2.5

In general the extent of fractionation depends upon the location of the interstellar cloud. In many cases, the data confirm the thermodynamic expectation that the colder the cloud, the greater will be the fractionation. Yet in some notable cases, they do not.[8]

2. CHEMICAL PATHWAYS FOR DEUTERATING INTERSTELLAR MOLECULES

From the reaction sequence proposed for the synthesis of interstellar molecules in dense clouds,[3] it is straightforward to predict how deuterated species of increasing complexity could be formed. The starting material must be atomic deuterium, subsequently ionized by photoionization or charge exchange. This can react to form HD, which in turn can form H_2D^+ and CH_2D^+. The H_2D^+ can then form DCO^+ and so on. In this way a set of deuterated reactants is established — D, HD, H_2D^+, DCO^+, CH_2D^+ etc — that themselves can act as agents for deuterating interstellar molecules.

Species	Formation	Reaction Type
D	*Big Bang*	
HD	$D^+ + H_2 = HD + H^+$	Isotope Exchange
H_2D^+	$H_3^+ + HD = H_2D^+ + H_2$	Isotope Exchange
DCO^+	$H_2D^+ + CO = DCO^+ + H_2$	Proton Transfer
CH_2D^+	$CH_3^+ + HD = CH_2D^+ + H_2$	Isotope Exchange

These species — D, HD, H_2D^+, DCO^+, CH_2D^+ etc — can incorporate deuterium by different chemical reactions, such as isotope exchange, proton transfer, association, dissociative recombination, etc. We can illustrate this by writing general equations for two of these reaction types below — isotope exchange and proton transfer — using two of the deuterating species for each reaction — D and HD for isotope exchange and H_2D^+ and DCO^+ for proton transfer.

Isotope Exchange :
$$XH^+ + \begin{matrix} D \\ HD \end{matrix} \rightleftharpoons XD^+ + \begin{matrix} H \\ H_2 \end{matrix} \qquad (2)$$

Proton Transfer :
$$M + \begin{matrix} H_2D^+ \\ DCO^+ \end{matrix} \rightarrow MD^+ + \begin{matrix} H_2 \\ CO \end{matrix} \qquad (3)$$

Here XH^+ is a cation containing at least one hydrogen site (which is exchanged for deuterium) while M is a neutral molecule (which incorporates deuterium by accepting a deuteron).

Isotope exchange would seem to be a particularly simple route for incorporating deuterium into interstellar molecules, for example by reaction (2) wherever D or HD are present.

3. DEUTERIUM EXCHANGE REACTIONS

Reaction (2)

$$XH^+ + \begin{matrix} D \\ HD \end{matrix} \rightleftharpoons XD^+ + \begin{matrix} H \\ H_2 \end{matrix} \qquad (2)$$

is written to be exoergic according to the empirical rule that, for small species, the exoergic route places the heavier isotope in the larger product species. (Established from the study of many isotope exchange reactions,[9] this rule can fail for larger species[10]). This is a statement that the fractionation of deuterium in interstellar molecules via Reaction (2) is thermodynamically allowed. Since exoergic ion-molecule reactions proceed generally with high reaction efficiency, reaction (2) should incorporate deuterium efficiently into interstellar molecules.

The data tell a different story. Atomic deuterium does exchange efficiently with the few ions that have been studied[11,12]

$$XH^+ + D \rightarrow XD^+ + H \qquad (4)$$

whereas molecular deuterium[13] sometimes does exchange

$$\begin{matrix} H_3^+ \\ CH_3^+ \end{matrix} + \begin{matrix} HD \\ D_2 \end{matrix} \rightarrow \textit{exchange} \qquad (5)$$

but sometimes does not

$$\begin{matrix} HCO^+ \\ N_2H^+ \\ NH_4^+ \\ H_3CO^+ \end{matrix} + \begin{matrix} HD \\ D_2 \end{matrix} \text{—x→} \textit{no exchange} \qquad (6)$$

These exoergic reactions (6) fail because of energy barriers in the potential energy hypersurface. In order to be able to distinguish those reactions which can occur from those which cannot, it is necessary to infer from the chemistry the qualitative features of the reaction hypersurface.

4. CHARACTERIZING THE REACTION HYPERSURFACE

4.1. Reactants with Filled and Unfilled Valence Shells

It is useful to classify the reactants as either having filled or unfilled valence electronic shells. Reactants with filled valence electron shells, possess the electronic configuration characteristic of an inert gas (s^2, s^2p^6 or $s^2p^6d^{10}$) : examples are H_2, H_3^+, N_2, N_2H^+, CO, and HCO^+. Examples of species with unfilled valence shells are H, CH_3^+ etc. According to this classification, there are three possible combinations of reactants, as shown below with illustrative examples:

Filled / filled: $HCO^+ + HD = DCO^+ + H_2$ (7)

Filled / unfilled: $HCO^+ + D = DCO^+ + H$ (8)

Unfilled / unfilled: $CH_3^+ + D = CH_2D^+ + H$ (9)

Exoergic isotope exchange reactions that do not occur [Reaction (6)] are distinguished by both reactants having filled shells [Reaction (7)]. (This is a necessary but not sufficient condition since reactants with filled shells can still undergo facile exchange, e.g. H_3^+ + HD [Reaction (5)].[13])

4.2. Basins in the Reaction Hypersurface

Reactants with unfilled valence shells are *electrophilic*, reacting to fill the shells by sharing electrons, through forming chemical bonds. Thus for the filled / unfilled example (8), the intermediate is the H_2CO^+ -type ion with a binding energy of 1.0 eV. For the unfilled/unfilled case (9), the intermediate is the CH_4^+ -type ion with a binding energy of 1.75 eV.[14] The strongly bound intermediate is represented on the hypersurface as a *single, deep basin*, as shown in Figure 1.

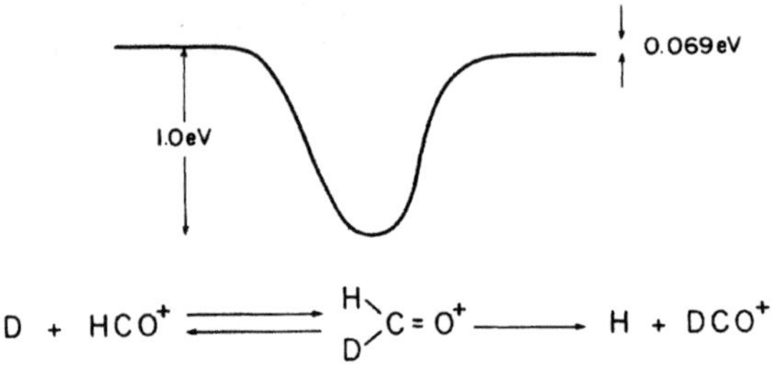

Figure 1. Representative slice through the reaction hypersurface for deuterium exchange for a reactant with an unfilled valence shell, showing the binding energy of the intermediate and the reaction exoergicity.[4]

Where both reactants $XH^+ + Y$ have filled valence shells, there is no chance of strong chemical bonding and no strongly bound intermediates. Intermediates such as $XH^+{\cdot}Y$ are bound by purely electrostatic forces — ion-induced dipole and ion-dipole — with binding energies of only a few tenths of an eV.[15] Unlike Figure 1, the hypersurface is not dominated by a single, deep basin. Instead the appropriate hypersurface is that for proton transfer from X to Y, because the minimum energy pathway for isotope exchange involves *internal proton transfer* $XH^+{\cdot}Y \leftrightarrow X{\cdot}YH^+$ within the reaction intermediate. The distinctive features of this hypersurface (Figure 2) are *two shallow basins*, corresponding to the two intermediates $XH^+{\cdot}Y$ and $X{\cdot}YH^+$.

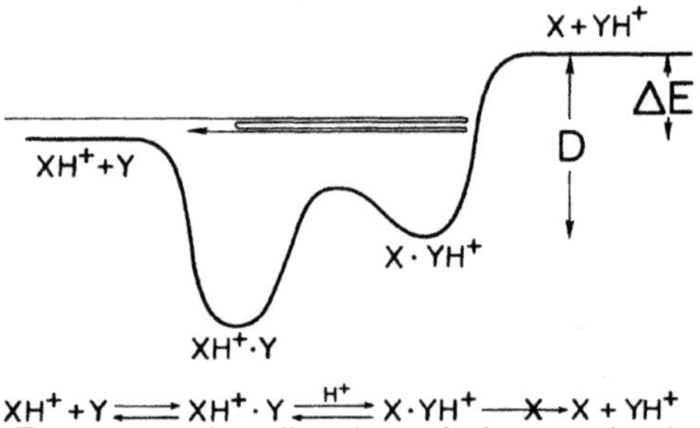

Figure 2. Representative slice through the reaction hypersurface for deuterium exchange between reactants with filled valence shells. D is the binding energy of the intermediate $X{\cdot}YH^+$. ΔE, the endoergicity of the proton transfer reaction $XH^+ + Y \rightarrow X + YH^+$, is the difference in proton affinity between Y and X. Exchange is possible if the reactants can reach the second well, i.e. if $\Delta E < D$. [Such diagrams have been used extensively by ion chemists to describe deuterium exchange reactions.[16]]

5. DEUTERIUM EXCHANGE

5.1 General Conditions

No barriers inhibit deuterium exchange for exoergic reactions where at least one reactant has an unfilled valence shell. The single deep well in the hypersurface — a consequence of the chemical bonding between the reactants — attracts the reactants to form an intermediate which subsequently dissociates either to give products or to give back reactants (Figure 1). This partitioning gives a negative temperature dependence to the exoergic reaction, such as (1) — the rate constant increasing with decreasing temperature until every collision results in reaction in the limit of absolute zero.

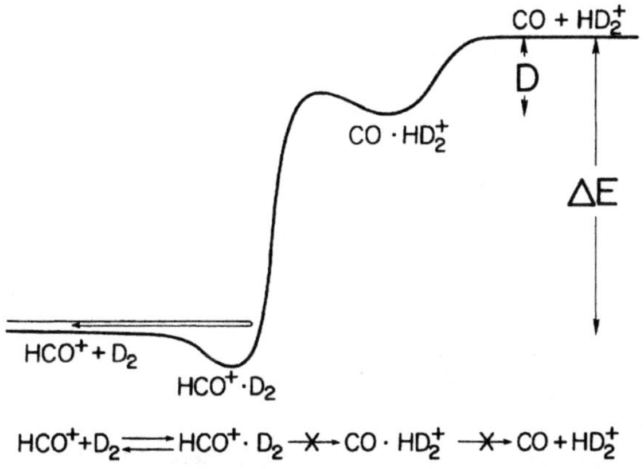

Figure 3. Representative slice through the reaction hypersurface for deuterium exchange between reactants with filled valence shells, for a system where exchange does not occur. For the system shown, $HCO^+ + HD$, $\Delta E \sim 1.8$ eV and $D \sim 0.5$ eV.[14,15] Because $\Delta E > D$, the second well is inaccessible, so that proton shuttling within the intermediate is not possible.

In contrast, barriers can inhibit deuterium exchange wherever both reactants show filled valence shells. For the reactants $XH^+ + Y$ to undergo exchange by the internal proton transfer or shuttling mechanism (Figure 2), the corresponding proton transfer reaction $XH^+ + Y \rightarrow X + YH^+$ reaction must be endoergic, i.e. $\Delta E > 0$. For deuterium exchange to be possible by this mechanism, the second basin, $X{\cdot}YH^+$, must be accessible to the reactants, i.e. $D > \Delta E$. Figure 3 shows a representative example where this condition is not met.

Reaction rate constants show the same negative temperature dependence for the double-basin mechanism (Figure 2)[17] as they do for the single-basin mechanism (Figure 1).[11]

5.2. Applications to Interstellar Chemistry

Using the guidelines developed above, we can now re-examine the deuterium exchange reactions that succeed (4,5) and fail (6).

Filled / unfilled:	HCO^+	+ D	⟶	DCO^+	+	H		(10)
Filled / filled:	H_3^+	+ HD	⟶	H_2D^+	+	H_2		(11)
Unfilled / filled:	CH_3^+	+ HD	⟶	CH_2D^+	+	H_2		(12)
Filled / filled:	HCO^+	+ HD	–x→	DCO^+	+	H_2		(13)
Filled / filled:	N_2H^+	+ HD	–x→	N_2D^+	+	H_2		(14)
Filled / filled:	NH_4^+	+ HD	–x→	NH_3D^+	+	H_2		(15)
Filled / filled:	H_3CO^+	+ HD	–x→	H_2DCO^+	+	H_2		(16)

In deuterium exchange reactions of the type $XH^+ + HD \rightarrow XD^+ + H_2$, where X is a stable molecule such as N_2, NH_3 etc., both reactants show filled shells. Consequently, energy barriers stop Reactions (13) - (16) because the proton affinity of H_2 is less than that of the other species X. The lowest value found for ΔE is 0.7 eV, the value for Reaction (14).[14] Even this exceeds the binding energy D of the intermediate $H_2D^+.X$, which is not more than ~ 0.5 eV.[15] Thus the condition $\Delta E > D$ consistently prevails so that access to the second basin — and deuterium exchange — are not possible (Figure 3). Exceptions arise where the reaction is symmetric, requiring that $\Delta E \sim 0$, that necessarily $\Delta E < D$ and therefore the absence of any barrier to deuterium exchange. Reaction (11) is one example of a symmetric reaction[13] and the reaction $D^+ + H_2 \rightarrow H^+ + HD$ is another.[18]

In contrast, for Reactions (10) and (12) there are no barriers to deuterium exchange. Each shows one reactant with an unfilled valence shell ; each forms an intermediate which is chemically bound and which features as a single deep well in the hypersurface. (For (10) see Figure 1 : for (12) the binding energy of the CH_5^+ -type intermediate is ~ 2 eV.[14])

Laboratory measurements support these conclusions consistently. Thus HCO^+ and N_2H^+ exchange readily with D but not with D_2 ;[11] both CH_3^+ and H_3^+ exchange with HD ;[13] and the predicted negative temperature dependence is found. In certain cases the laboratory measurements have only explored deuterium exchange with D_2 but not with HD. Our description shows that what is found for D_2 must apply equally to HD.

In conclusion, it may be appropriate to emphasize the importance of Reaction (1) as a pathway for deuterating interstellar molecules. It synthesizes in a single step deuterated species of the same complexity as the starting material ; and, as the deuterating agent, it uses deuterium atoms, the primary source of deuterium. Thus it would appear to provide possible routes for the synthesis of CCD [8] via $C_2H_2^+ + D \rightarrow C_2HD^+ + H$ and for the synthesis [6] of C_3HD via $C_3H^+ + D \rightarrow C_3D^+ + H$.[19]

ACKNOWLEDGEMENTS

M. H. thanks the U. S. Air Force for support — as an AFSC-URRP Visiting Professor at the Air Force Geophysics Laboratory (1984-1986) and under contract # RADC 4-40375.

REFERENCES

(1) Dalgarno, A. and Lepp, S. 1984, *Ap. J. (Letters)* **287**, L47.
(2) Crutcher, R. M. and Watson, W. D. 1985, in *Molecular Astrophysics*, eds. G. H. F. Diercksen, W. F. Huebner and P. W. Langhoff (Dordrecht: Reidel), p. 255.
(3) Smith, D. 1987, *Phil. Trans. Roy. Soc. Lond.* **A 323**, 269.

(4) Henchman, M., Paulson, J. F., Lindinger, W., Smith, D. and Adams, N. G., to be submitted to *Chem. Phys. Lett.*
(5) Wootten, A. 1987, in *Astrochemistry*, eds. M. S. Vardya and S. P. Tarafdar (Dordrecht: Reidel), p. 311.
(6) Bell, M. B., Feldman, P. A., Matthews, H. E. and Avery, L. W. 1986, *Ap. J. (Letters)* **311**, L89.
(7) Boesgaard, A. M. and Steigman, G. 1985, *Ann. Rev. Astron. Astrophys.* **23**, 319.
(8) Herbst, E., Adams, N. G., Smith, D. and DeFrees, D. J. 1987, *Ap. J.* **312**, 351.
(9) Smith, D. and Adams, N. G. 1984, in *Ionic Processes in the Gas Phase*, ed. M. A. Almoster Ferreira (Dordrecht: Reidel), p. 41.
(10) Henchman, M. J., Smith, D. and Adams, N. G. 1982, *Int. J. Mass Spectrom. Ion Phys.* **42**, 25.
(11) Adams, N. G. and Smith, D. 1985, *Ap. J. (Letters)* **294**, L63.
(12) Federer, W., Villinger, H., Tosi, P., Bassi, D., Ferguson, E. and Lindinger, W. 1985, in *Molecular Astrophysics*, eds. G. H. F. Diercksen, W. F. Huebner and P. W. Langhoff (Dordrecht: Reidel), p. 649.
(13) Smith, D., Adams, N. G. and Alge, E. 1982, *Ap. J.* **263**, 123.
(14) Wagman, D. D., Evans, W. H., Parker, V. B., Schumm, R. H., Halow, I., Bailey, S. M., Churney, K. L. and Nuttall, R. L. 1982, *J. Phys. Chem. Ref. Data* **11**, Suppl. 2, pp.390-391. Lias, S. G., Liebman, J. F. and Levin, R. D. 1984, *ibid.* **13**, 695.
(15) Keesee, R. G. and Castleman, A. W., Jr. 1986 *ibid.* **15**, 1011.
(16) Henchman, M. J., Smith, D. and Adams, N. G. *Proton-Transfer Reactions, Hydrogen-Exchange Reactions, and Proton-Transfer / Hydrogen-Exchange Reactions*, presented at the 28th ASMS Conference, New York City, 1980, paper WAMOC 13 ; Ausloos, P. and Lias, S. G. 1981, *J. Am. Chem. Soc.* **103**, 3641 ; Squires, R. R., DePuy, C. H. and Bierbaum, V. M. 1981, *ibid.* **103**, 4256.
(17) Meot-Ner, M., Lloyd, J. R., Hunter, E. P., Agosta, W. A. and Field, F. H. 1980, *J. Am. Chem. Soc.* **102**, 4672.
(18) Henchman, M. J., Adams, N. G. and Smith, D. 1981, *J. Chem. Phys.* **75**, 1201.
(19) Adams, N. G. and Smith, D. 1987, *Ap. J. (Letters)* **317**, L25.

DIFFUSE CLOUD CHEMISTRY

Ewine F. van Dishoeck
Princeton University Observatory
Princeton, NJ 08544

John H. Black
Steward Observatory, University of Arizona
Tucson, AZ 85721

ABSTRACT. The current status of models of diffuse interstellar clouds is reviewed. A detailed comparison of recent gas–phase steady–state models shows that both the physical conditions and the molecular abundances in diffuse clouds are still not fully understood. Alternative mechanisms are discussed and observational tests which may discriminate between the various models are suggested. Recent developments regarding the velocity structure of diffuse clouds are mentioned. Similarities and differences between the chemistries in diffuse clouds and those in translucent and high latitude clouds are pointed out.

I. INTRODUCTION

Diffuse interstellar clouds are usually defined as the regions of the interstellar gas in which the total continuous extinction by dust particles at visual wavelengths, A_V^{tot}, is less than 2 magnitudes. They have been studied since the beginning of this century by the absorption lines of the atoms and molecules in the clouds superposed on the spectra of bright background stars. Diffuse clouds are considered to be simpler astrophysical regions to model than dense clouds: they usually do not contain embedded heat sources, only the simplest diatomic molecules are found in diffuse clouds, and the ultraviolet radiation penetrates through them so that the time scales for most chemical processes are short enough to attain chemical equilibrium. Diffuse clouds may therefore provide the best framework in which to test the basic concepts of interstellar chemistry, and a good understanding of diffuse cloud chemistry is a prerequisite for having any confidence in the more complicated chemistry occurring in dense clouds.

Despite their relative simplicity, however, the structure and chemistry of diffuse interstellar clouds remain elusive. As summarized recently by Black (1987) and Crutcher and Watson (1985), many of the basic questions are still unanswered such as (1): What are the typical temperatures and densities in diffuse clouds? Are the densities of the order of a few hundred cm^{-3} or a few thousand cm^{-3}? (2): Can the abundances of most species indeed be described by quiescent steady–state gas–phase reactions, or do shocks play a crucial role in the formation of most species? (3): What is the role of the grains in the cloud? Can molecules other than H_2 be formed efficiently on grain surfaces and returned to the gas–phase? (4): What is the structure of diffuse clouds? Can they indeed be represented

T. J. Millar and D. A. Williams (eds.), Rate Coefficients in Astrochemistry, 209–237.

by homogeneous plane–parallel slabs? (5): What is the relation, if any, between diffuse clouds and dense clouds?

In this paper, the current status of answers to these questions will be reviewed. Emphasis will be placed on a detailed discussion of the most recent theoretical models and new observational data, with only limited reference to earlier work.

2. OBSERVATIONAL CONSTRAINTS

What observational constraints do we have on the chemistry in diffuse clouds? As Table 1 indicates, the list of molecules that have securely been identified in diffuse clouds is still small, and includes no polyatomic species. Most of these molecules are found only in the thicker diffuse clouds with $A_V^{tot} \geq 1$ mag. The list of molecules sought but not detected is considerably longer and includes such interesting species as NH, HCl, NaH, MgH, H_2O and C_3. No new molecules have been identified since the detection of C_2 in 1977 by Souza and Lutz. The discovery of new molecules in diffuse clouds is hampered by the fact that there is at present no sensitive high–resolution spectrograph available in space for searches at ultraviolet wavelengths, where most molecules have their strongest electronic transitions. Some claims of detections of the molecules NO (Pwa and Pottasch 1986), HCl and MgH^+ (Somerville 1987) have been made based on IUE observations. However, owing to the limited resolution and sensitivity of this instrument, these identifications are suspect, and need to be confirmed by independent measurements. Observations using ground based optical telescopes have recently been reinvigorated by great improvements in the sensitivity of detectors at visible and near–infrared wavelengths. This has led to sensitive searches for some specific molecules such as C_3 (Snow, Seab and Joseph 1988) and NaH and MgH (Czarny, Felenbok and Roueff 1987), but so far without success. The suggested detection of the CS^+ ion in diffuse cloud (Ferlet *et al.* 1983) has not been confirmed by subsequent observations (Ferlet *et al.* 1986). On the other hand, high–quality spectra at various wavelengths show numerous weak unidentified features with equivalent widths of less than 1 mÅ. Many of these features probably have an atmospheric origin. Some of them can be catalogued as new diffuse interstellar bands, but a few are sharp enough that they may be due to some as yet unidentified molecule. For example, recent spectra by Herbig (1987) toward some more reddened lines of sight show an interesting series of diffuse bands around 6800 Å with a spacing of about 35 cm^{-1}, which corresponds to a typical rotational spacing of a light hydride molecule.

In contrast with millimeter observations of molecules in dense clouds, the absorption line observations provide accurate column densities of the species, unless the lines are highly saturated. An additional advantage of the study of diffuse clouds compared with dense clouds is that direct information is available on the atomic abundances.

3. CHEMISTRY IN DIFFUSE CLOUDS

The first attempts to explain the observed abundances of molecules on the basis of non–thermodynamic equilibrium models were by Kramers and ter Haar (1946) and Bates and Spitzer (1951). These investigations considered in some detail the basic formation and destruction processes through radiative association, photodissociation, photoionization and dissociative recombination, but were unable to reproduce the observed abundances of the then known interstellar molecules CH and CH^+. Because of the difficulties encountered with the early gas–phase reaction schemes, subsequent models focussed on grain surface

TABLE 1. Observed Molecules in Diffuse Clouds[a]

Species	λ(Å)	Star	N(cm^{-2})	Ref.	Species	λ(Å)	Star	N(cm^{-2})	Ref.
H_2	912–1108	ζ Oph	4.2(20)	1	NH	3350	ζ Per	<6.3(11)	1
HD	912–1108	ζ Oph	2.1(14)	1	NH^+	2890	ζ Oph	<1.0(14)	2
H_3^+	37155	ζ Per	<1.0(15)	8	NH_2	5693	ζ Oph	<1.5(13)	9
CH	3880, 4300	ζ Oph	2.5(13)	1	N_2	958	δ Sco	<8.2(12)	2
CH^+	4232	ζ Oph	2.9(13)	1	NO	2262	ζ Oph	≤2.6(13)	4
OH	1222, 3080	ζ Oph	4.8(13)	1	NO^+	1313	ζ Oph	<1.1(14)	2
CO	1076–1545	ζ Oph	2.0(15)	1	HCl	1290	ζ Oph	<4.5(11)	1
C_2	8750	ζ Oph	2.6(13)	1	NaH	3991	ζ Oph	<1.3(10)	5
CN	3874	ζ Oph	2.9(12)	1	MgH	5187	ζ Oph	<1.9(9)	5
C_3	4050	ζ Oph	<5.4(10)	3	MgH^+	2806	ξ Per	<2.0(11)f^{-1}	2
CH_2	1397	ζ Oph	<3.1(14)	2	MgO	4998	...	...	...
H_2O	1114, 1240	ζ Oph	≤2.2(13)	1	AlH	2242	ζ Oph	<2.2(12)	2
H_2O^+	6147	ζ Oph	<1.9(12)	10	SiH	4119	ζ Oph	<2.6(12)	2
CO^+	4260	ζ Oph	<1.4(14)	1	SH	1257	o Per	<1.5(11)f^{-1}	2
CO_2	1089	ζ Oph	<3.0(12)	2	SH^+	3363	ζ Oph	<1.0(13)	6
CS	2577	ζ Oph	<9.6(12)	2	CaH	2717	ζ Oph	<9.3(10)f^{-1}	2
CS^+	6840	ζ Oph	<1.9(11)	7	SiO	1310	ζ Oph	<3.2(11)f^{-1}	2

Note: In this and following tables, the notation a(b) means a × 10^b.

[a] *Column densities are given for the ζ Oph cloud, unless more stringent limits are available for other clouds. References: (1) van Dishoeck and Black 1986a, but with $f_{20}(C_2)$=1.0 × 10^{-3}; (2) Snow 1980, but with updated oscillator strengths; (3) Snow et al. 1988; (4) Pwa and Pottasch 1986; (5) Czarny et al. 1988; (6) Millar and Hobbs 1988; (7) Ferlet et al. 1986; (8) Black et al. 1988; (9) G. H. Herbig (private communication); (10) Federman and Trauger, quoted by Federman 1987, but with oscillator strength listed by van Dishoeck and Black 1986a. See (1) and (2) for references to original papers.*

processes for the formation of molecules. Plausible models were made for the formation of the H_2 molecule on grains (Hollenbach and Salpeter 1971), but the production of other molecules appeared difficult to quantify due to uncertainties in the composition of the grain surfaces, the surface reactions that can occur, and the desorption processes. Around the same time, Solomon and Klemperer (1972) and Herbst and Klemperer (1973) realized that reactions between ions and neutral species could be very rapid at the low temperatures prevailing in interstellar clouds, and therefore reintroduced the gas–phase reaction schemes. These papers, together with contributions from Black and Dalgarno (1973), Dalgarno and Black (1976) and Watson (1978), have become the basis of the present interstellar gas–phase reaction networks.

The key processes for the formation of the simplest carbon–, oxygen– and nitrogen–bearing molecules have been discussed extensively before (Dalgarno and Black 1976; Watson 1978; Crutcher and Watson 1985; van Dishoeck 1988), and will be only briefly reiterated here. Since H and H_2 are so much more abundant in interstellar clouds than any other species, the dominant reactions usually involve hydrogen, whenever possible. Table 2 lists the current best estimates of the solar abundances of the various elements relative to hydrogen. Some of the heavier elements are depleted from the gas phase in interstellar clouds. In diffuse clouds, this depletion is very mild and tends to exceed a factor of four only for heavier metals like Ca, Ti, Mn, and Fe. The fraction of the solar abundance of element X in the gas phase is denoted by the depletion factor δ_X, with $\delta_X \leq 1$.

TABLE 2. Solar Elemental Abundances

Element	Abundance	Element	Abundance
H	1.00	Si	4.3(-5)
He	0.075	S	1.7(-5)
O	8.3(-4)	P	2.8(-7)
C	4.7(-4)	Cl	1.1(-7)
N	1.0(-4)	K	1.3(-7)
Na	2.1(-6)	Ca	2.2(-6)
Mg	4.2(-5)	Fe	4.3(-5)
Al	3.1(-6)		

3.1 Carbon chemistry

The carbon chemistry differs from that of oxygen and nitrogen in that the ionization potential of carbon is less than 13.6 eV, so that carbon can be ionized by the interstellar radiation field. Thus carbon exists mostly as C^+ in diffuse clouds. However, the reaction between C^+ and H_2 to form CH^+ is endothermic by about 0.4 eV, and does not proceed at low temperatures. The carbon chemistry is therefore thought to be initiated by the relatively slow radiative association reaction (Black and Dalgarno 1973)

$$C^+ + H_2 \rightarrow CH_2^+ + h\nu. \tag{1}$$

Once CH_2^+ is formed, subsequent ion–molecule reactions followed by dissociative recombination lead to the formation of simple molecules such as CH and C_2. The dominant destruction process of the neutral molecules is photodissociation.

3.2 Oxygen chemistry

Oxygen exists mostly in neutral atomic form in diffuse interstellar clouds. In this case, the ionization necessary to initiate the ion–molecule reactions is provided by cosmic rays which penetrate the clouds and ionize H and H_2 to form H^+ and H_2^+. H_2^+ subsequently reacts very rapidly with H_2 to form H_3^+. The rate at which atomic hydrogen is ionized is denoted by ζ_o. Since the ionization potential of O is accidentally close to that of H, the charge transfer reaction

$$O + H^+ \rightarrow O^+ + H \tag{2}$$

requires only a small energy input, and takes place efficiently at low temperatures. Once O^+ is formed, it reacts rapidly with H_2 to form OH^+. Alternatively, OH^+ can be formed by the fast reaction between H_3^+ and O. The formation of neutral oxygen–bearing molecules such as OH and H_2O results again from ion–molecule reactions followed by dissociative recombination. Subsequent reactions with C^+ lead to the formation of, for example, CO in diffuse clouds.

3.3 Nitrogen chemistry

The formation routes for nitrogen–bearing molecules in diffuse clouds are particularly uncertain. Most previous models have assumed that the exothermic reaction

$$N + H_3^+ \rightarrow NH_2^+ + H \tag{3}$$

is rapid at interstellar temperatures. The corresponding reaction to form NH^+ and H_2 is now known to be endoergic. Huntress (1977) pointed out that reaction (3) is a very unusual reaction in which two protons are transferred. Detailed theoretical calculations by Herbst *et al.* (1987) indicate indeed that large barriers exist along the reaction path, so that the reaction is likely to be very slow at low temperatures. The other initiating reaction

$$N^+ + H_2 \rightarrow NH^+ + H \tag{4}$$

has also recently been measured to be slow at low temperatures (Marquette *et al.* 1985*a*; Luine and Dunn 1985). Alternative gas-phase reactions through which nitrogen-containing molecules can be formed include neutral-neutral reactions of atomic nitrogen with small oxygen- and carbon-containing molecules such as OH, CH and C_2 to form CN and NO, and ion-molecule reactions of atomic nitrogen with small hydrocarbon ions to form e.g. H_2CN^+, which can then recombine to form CN and HCN.

3.4 Deuterium chemistry

Although the molecule HD, like H_2, can be formed on grain surfaces, it is produced more rapidly by the gas-phase reaction sequence

$$H^+ + D \rightleftarrows H + D^+ \tag{5}$$

$$D^+ + H_2 \rightleftarrows H^+ + HD. \tag{6}$$

Thus the abundance of HD is also sensitive to the cosmic-ray ionization rate ζ_o which produces H^+.

3.5 Chemistry of other species

The ionization potentials of Si, S and Cl are less than 13.6 eV, so that these species are mostly ionized in diffuse clouds. Analogous to the carbon chemistry, the reaction of Si^+ with H_2 to form SiH^+ is endothermic by about 1.2 eV, so that the formation of silicon-bearing molecules is most likely initiated by the radiative association reaction between Si^+ and H_2 to form SiH_2^+. Subsequent ion-molecule reactions will lead to the formation of molecules such as SiH.

The reaction of S^+ with H_2 to form SH^+ is highly endothermic as well by about 0.9 eV, but in this case the radiative association reaction to form SH_2^+ is also very slow. The reaction of H_3^+ with S to form SH^+ can lead to small amounts of sulfur hydrides. Reactions of S^+ with hydrocarbon molecules lead to the formation of sulfur-bearing species like CS.

Ionized chlorine can react with H_2 to form HCl^+, which subsequently reacts rapidly with H_2 to form H_2Cl^+. The dissociative recombination of H_2Cl^+ produces HCl, although the branching ratio is apparently small.

Metals such as magnesium and sodium also exist primarily as singly-charged atomic ions in diffuse clouds, and observations suggest that they are only mildly depleted from the gas phase. It is still uncertain whether these elements form significant amounts of gas-phase molecules. Reactions of Na^+ and Mg^+ with H_2 to form NaH^+ and MgH^+ are highly endothermic, and the radiative association reactions to form NaH_2^+ and MgH_2^+ are extremely slow. Thus any gas-phase formation route of the metal hydrides NaH and MgH is expected to be negligible. The only alternative formation mechanism of metal-containing molecules is by the recombination of ions on grain surfaces.

3.6 Grain–surface chemistry

The formation rate of a molecule on a grain surface is governed by the frequency of collisions between the atoms and the grains, the probability y_S that the atoms stick to the cold surface, the probability y_r that molecule formation occurs, and the probability y_d that the molecule is released from the surface back into the gas phase. Thus the rate r_f for a reaction of X with a grain g to form XY

$$\mathrm{X} + \mathrm{Y}:g \rightarrow \mathrm{XY} + g \qquad (7)$$

is given by (Williams 1984)

$$r_f \approx 3 \times 10^{-18} T^{1/2}\ n_H\ n(\mathrm{X})\ y_S\ y_r\ y_d \quad \mathrm{cm^{-3}s^{-1}} \qquad (8)$$

where T is the kinetic temperature and n_H the density of hydrogen nuclei in the cloud. The probabilities y_S, y_r and y_d are highly uncertain for any species except for the formation of H_2, for which all three factors are expected to be close to unity.

For carbon– and oxygen–bearing molecules, the rate of grain surface production is at least an order of magnitude less than that for gas–phase formation through reactions (1) and (2). However, grain surface reactions may be dominant in the formation of some nitrogen– or metal–bearing species, for which the gas–phase processes are extremely slow. The fact that the observed OH and CH abundances are at least two orders of magnitude higher than that of NH in diffuse clouds is still one of the strongest arguments against any significant grain surface formation of carbon– and oxygen–bearing molecules (Crutcher and Watson 1976). Also, the stringent upper limits on the abundances of the metal hydrides NaH and MgH suggest that grain surface formation of any species other than H_2 is inefficient (Czarny *et al.* 1987).

The effects of the possible presence of a significant amount of very small grains or large molecules on the chemistry in diffuse clouds will be discussed in §4.3.5.

3.7 Recent developments

Since many of the key chemical processes were identified more than a decade ago, most progress in our understanding of diffuse cloud chemistry has come from detailed laboratory and theoretical work on important reaction rates. A recent compilation of ion–molecule reaction rate coefficients has been given by Anicich and Huntress (1986). References to rate coefficients for other processes are given elsewhere in this volume. Some of the important developments for the chemistry in diffuse interstellar clouds are summarized below.

- Photodissociation and photoionization are the dominant destruction processes of the neutral species in diffuse clouds. However, many of the rates for these processes adopted in early models were based on educated guesses. For some species, more accurate rates are now available through detailed theoretical calculations or laboratory experiments. In most cases, the new rates are significantly larger than the previous estimates (van Dishoeck 1987*a*; van Dishoeck, this volume). Important examples include the OH and CO molecules. One exception is the H_3^+ ion, for which the previously adopted photodissociation rate was at least four orders of magnitude too large.
- Dissociative recombination is the dominant destruction process for many molecular ions. It has recently been found (Adams and Smith 1987; this volume) that the rate coefficients for some species are significantly smaller at low temperatures than suggested

by earlier measurements at room temperature. The most important development for diffuse cloud studies is the realization that the dissociative recombination of H_3^+

$$H_3^+ + e \rightarrow H_2 + H \text{ or } H + H + H \tag{9}$$

may be very slow at interstellar temperatures (Michels and Hobbs 1984; Adams and Smith 1984). The actual value of the rate coefficient for $H_3^+(v=0)$ is still uncertain. Adams and Smith (1987) suggest an upper limit $k_9 \leq 10^{-11}$ cm^3 s^{-1} at T=80 K, whereas the experiments of Mitchell (1987) and Amano (1987) indicate $k_9 \approx 10^{-8}$ and $k_9 \approx 2 \times 10^{-7}$ cm^3 s^{-1}, respectively. As will be shown below, it is important for the oxygen chemistry to know whether this rate coefficient is indeed smaller than 10^{-9} cm^3 s^{-1} at low T.

- Little information is still available on the branching ratios to the various products of dissociative recombination. Most models assume that either a single hydrogen atom is detached, or that H and H_2 are formed in equal amounts

$$\begin{aligned} XH_n^+ + e &\rightarrow XH_{n-1} + H \\ &\rightarrow XH_{n-2} + H_2. \end{aligned} \tag{10}$$

  Bates (1986; see also this volume), however, has argued that the product distribution may be significantly different. In particular, he has suggested that the dissociative recombination of H_3O^+ produces exclusively H_2O, and no OH.
- Although it was realized more than a decade ago that reactions between ions and molecules with a permanent dipole moment may be more rapid at low T than reactions between ions and non–polar molecules, this effect was not taken into account in early models. Important reactions in diffuse clouds whose rates will be affected at low T include (Marquette *et al.* 1985*b*; see also Rowe this volume)

$$C^+ + OH \rightarrow CO^+ + H; \ CO + H^+. \tag{11}$$

- As discussed above, reactions (3) and (4) which were previously thought to initiate the nitrogen chemistry are probably very slow at low temperature, and alternative formation schemes need to be investigated.

4. STEADY–STATE MODELS

4.1 Approaches

Since the early work of Bates and Spitzer (1951) and Solomon and Klemperer (1972), many steady–state models have been developed to describe the chemistry in diffuse clouds. Two classes of models can be distinguished. One class consists of models that try to reproduce the general trends in observed column densities for many lines–of–sight. Examples are found in the work of Glassgold and Langer (1974; 1976), Barsuhn and Walmsley (1977), Clavel, Viala and Bel (1978), Federman (1982), Federman, Danks and Lambert (1984), Mann and Williams (1985) and Viala (1986). Figure 1 illustrates the observed CH column densities as functions of H_2 column densities for a number of clouds, with the results of a typical low density and high density model included (Danks, Federman and Lambert 1984). Note that these models usually provide only the relative densities of the species at a given temperature and density, and that it is assumed that these can be equated to relative column densities. The observations of CH appear to be well bracketed by

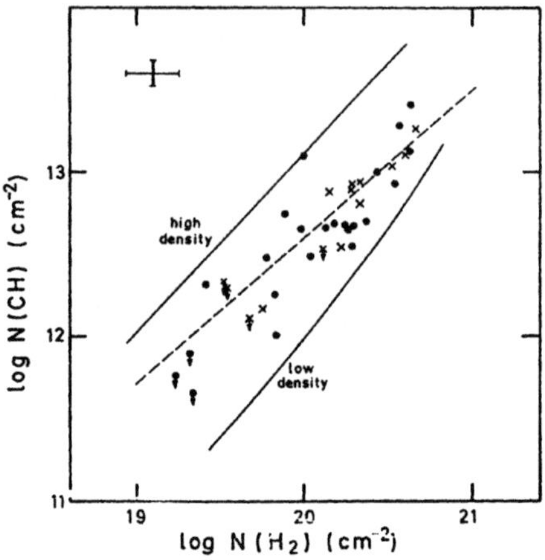

Figure 1. Observed N(CH) vs. N(H_2) for various lines of sight. The results of a typical low density model (n_H=150 cm^{-3}) and a high density model (n_H=2500 cm^{-3}) are included (from: Danks et al. 1984).

models with densities ranging from 200–2500 cm^{-3}. These methods are useful because they indicate that in first approximation, the adopted chemical network is correct and that the physical conditions in the various diffuse clouds probably differ by no more than an order of magnitude. Note that for such investigations to be of value, a homogeneous data set, preferably obtained with a single instrument and interpreted with a consistent set of molecular data, is a prerequisite.

In a few investigations the opposite approach has been taken, i.e., to reproduce the observations of a wide variety of species in individual clouds. Such comprehensive models of a few specific clouds are much more restrictive and may reveal best the current limitations of the steady-state models. Models of individual diffuse clouds have been developed by Black and Dalgarno (1977), Black, Hartquist and Dalgarno (1978), Federman and Glassgold (1980), van Dishoeck and Black (1986*a*) and Viala, Roueff and Abgrall (1988).

4.2 Physical conditions

Because the chemical reaction rates depend on the physical conditions, the first step in the modeling of a specific line of sight is the determination of the temperature T and density $n_H=n(\mathrm{H}) + 2n(\mathrm{H_2})$ in the cloud, as well as the strength of the ultraviolet radiation field incident on the cloud surface. In this review, all intensities will be given by a scaling factor I_{UV} where $I_{UV}=1$ refers to the intensity of the standard radiation field as given by Draine (1978). The physical parameters are usually constrained from observations of the excitation of various species. Most models adopt as the primary diagnostic the observed abundance and rotational population distribution of the H_2 molecule. As has been described in detail, e.g. by Jura (1975), Black and Dalgarno (1976, 1977) and van Dishoeck and Black (1986*a*), the populations of the high rotational levels $J \geq 5$ are sensitive primarily to the strength of the ultraviolet radiation field, whereas the population distribution over the lower levels reflects the temperature structure in the cloud. Note that strictly speaking

TABLE 3. Selected Observational Data for the ζ Per and ζ Oph Clouds[a]

Species	ζ Per	ζ Oph
H	(6.5±0.7)(20)	(5.2±0.2)(20)
H_2	(4.8±1.0)(20)	(4.2±0.3)(20)
H_2 J=5	(2.3±0.2)(14)	(4.3±0.5)(14)
HD	(3.8±1.4)(15)	(2.1±1.0)(14)
OH	(4.2±0.5)(13)	(4.8±0.5)(13)
CH	(2.0±0.3)(13)	(2.5±0.3)(13)
CH^+	(3.5±0.4)(12)	(2.9±0.3)(13)
CO	(6.1±3.0)(14)	(2.0±0.3)(15)

[a] *See van Dishoeck and Black (1986a) for references to observed values.*

the H_2 excitation constrains the strength of the radiation field only in the wavelength region where the H_2 ultraviolet pumping occurs, i.e., 912–1100 Å. The relative abundance of H with respect to H_2 depends on the combination of parameters $n_H y_f / I_{UV}$, where y_f is a scaling factor for the formation rate of H_2 on grain surfaces (van Dishoeck and Black 1986*a*). Thus n_H, I_{UV} and T can be constrained in principal from observations of H_2 and H alone. The inferred parameters have to be consistent with those suggested by other observations. In particular, the rotational excitation of the HD, C_2 and CO molecules, as well as the fine structure excitation of C, C^+ and O provide additional constraints on the physical conditions.

Compared with the early models of Black and Dalgarno (1977), there have been a few developments in the molecular data relevant to the excitation calculations. Although the collisional excitation rates for H + H_2 and H_2 + H_2 are still uncertain, more accurate values for the latter system have been provided by Schaefer (1985). The rotational excitation rates of CO by H_2 have recently been recomputed independently by Schinke *et al.* (1985) and Flower and Launay (1985), and show good agreement. Monteiro and Flower (1987) have reminded us of a neglected selection rule for atomic fine structure excitation, which suggests that the J=0→1 excitation of C and O by H_2 is in first order forbidden. Finally, Chambaud *et al.* (1988) have computed rotational excitation cross sections for C_2 by H_2.

Before discussing the detailed model results, it is instructive to compare briefly the observational data for two well studied lines of sight, namely those toward ζ Per and ζ Oph. Table 3 lists some selected observational data. The two clouds have very similar column densities of H and H_2, suggesting that $n_H y_f / I_{UV}$ has similar values in the two clouds. However, the excited rotational levels $J \geq 5$ are more populated in the ζ Oph cloud than in the ζ Per cloud, indicating that the ζ Oph cloud is exposed to more intense radiation. Assuming that the H_2 formation efficiency on grains is similar in the two clouds, this would imply that the density is higher in the ζ Oph cloud than in the ζ Per cloud by a factor of at least two. However, the rotational population distributions of the C_2 molecule are found to be very similar for the two lines of sight (Danks and Lambert 1983; van Dishoeck 1984). In fact, for all diffuse clouds for which both C_2 and H_2 observations exist, the C_2 rotational excitation appears to be nearly identical, whereas significant differences in H_2 excitation are found (van Dishoeck and Black 1986*a*). The C_2 excitation is sensitive to the parameters n/I_R, with n=n(H)+n(H_2) and I_R the scaling factor for the radiation field in the near-infrared part of the spectrum (van Dishoeck and Black 1982). Since the strength of the red part of the radiation field is not expected to vary significantly from place to place, the C_2 data suggest that the densities in diffuse clouds are very similar, in conflict

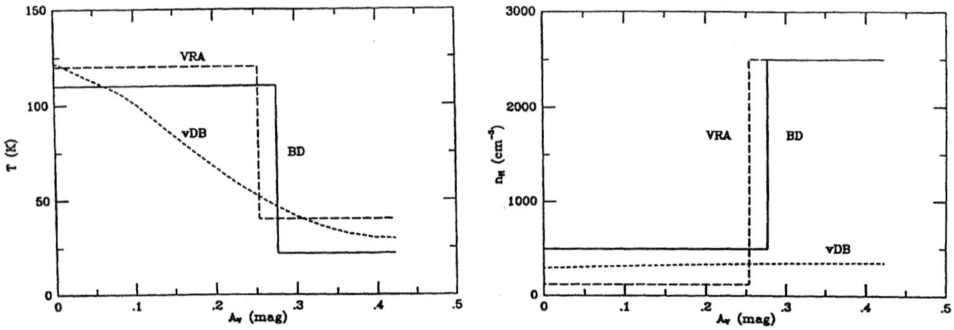

Figure 2. Temperature and density structure in the BD, vDB and VRA models of the ς Oph cloud.

with the densities inferred from the H and H_2 observations. Unless some other mechanism contributes to the population of J=5 in the ς Oph cloud, but not to that of J=5 in the ς Per cloud, or unless either the grain formation efficiency or the intensity of the radiation field in the red is substantially different in the two regions, the various diagnostics give inconsistent results for the densities. As to the observations of other species in the two clouds, the column densities of CH, C_2 and OH are nearly the same in the two regions, but the HD column density is apparently an order of magnitude larger in the ς Per cloud than in the ς Oph cloud, and the CO and CH^+ column densities somewhat lower. These differences may eventually provide significant clues to the structure of diffuse clouds, and illustrate the importance of studying more than one line of sight.

The analyses of the various diagnostic observations by different people have often led to conflicting results. In the extreme case of the ς Oph cloud, central temperatures between 20 and 60 K, central densities n_H between 200 and 2500 cm^{-3} and enhancement factors for the radiation field I_{UV}=2–6 have been inferred. The original two component plane-parallel model of Black and Dalgarno (1977; hereafter BD) had about 2/3 of the material at n_H=500 cm^{-3}, T=110 K with I_{UV}=2.5, and the remainder at n_H=2500 cm^{-3} with T=20 K. The structure is illustrated in Figure 2. However, the self-shielding in the H_2 ultraviolet lines through which the dissociation and excitation occurs was significantly underestimated in these models. Crutcher and Watson (1981) subsequently argued that the observed CO and HD rotational excitations imply a low density, $n_H \approx$ 200 cm^{-3}, throughout most of the cloud, whereas the absence of significant ^{13}CO fractionation suggested a rather high temperature, $T \approx$ 60 K. Roberge (1981) and van Dishoeck and Black (1986*a*; hereafter vDB) used a corrected treatment of the radiative transfer in the H_2 lines and showed that the original BD structure was inconsistent with observations. In particular, the intensity of the radiation field had to be increased and the density decreased to reproduce the H and H_2 measurements. vDB subsequently developed new models for the ς Oph cloud and 3 other diffuse clouds, in which continuous gradients of temperature and density have replaced an unrealistic two component structure. The clouds are in hydrostatic equilibrium and are

TABLE 4. Excitation of Various Species in the ζ Oph Cloud

Species	vDB Model G	VRA Model 6	VRA/vDB[a]	Observed[b]
H	5.1(20)	4.6(20)	4.5(20)	(5.2±0.2)(20)
H_2	4.2(20)	4.8(20)	4.8(20)	(4.2±0.3)(20)
H_2 J=0	2.9(20)	3.3(20)	3.5(20)	(2.9±0.3)(20)
1	1.3(20)	1.5(20)	1.3(20)	(1.3±0.2)(20)
2	8.3(17)	3.2(18)	1.1(18)	(4.0±1.6)(18)
3	2.2(16)	8.9(16)	2.6(16)	(1.7±1.2)(17)
4	1.5(15)	9.1(14)	1.3(15)	(5.3±2.2)(15)
5	4.5(14)	1.7(14)	3.6(14)	(4.3±0.5)(14)
6	5.3(13)	2.8(13)	4.6(13)	(4.9±0.5)(13)
7	3.8(13)	1.4(13)	3.0(13)	(3.6±0.4)(13)
C J=0	0.65[c]	...	0.46[c]	(0.63±0.06)[c]
1	0.29	...	0.39	(0.31±0.09)
2	0.06	...	0.15	(0.06±0.02)
CO J=3→2 0.87mm[d]	0.052	...	0.573	...
CN N=1→0 2.64mm[d]	0.0025	...	0.019	...

[a] *VRA temperature and density structure adopted in vDB programs with I_{UV}=4, grain model 2, y_f=1.5 and H_2 formation model φ=1;* [b] *See vDB for references to observed values;* [c] *Relative populations in various levels;* [d] *Predicted Rayleigh–Jeans radiation temperatures (T_R) in K.*

pressure bounded. The vDB models are not unique, but they typically have densities of a few hundred cm^{-3} at the edge with temperatures of 100–200 K and I_{UV}=2–6, depending on the adopted grain scattering properties. If the grain model 2 of Roberge, Dalgarno and Flannery (1981) is adopted, the scaling factor is at the high end of this range, whereas with the more forward scattering model 3, it can be rather smaller. The densities increase to 300–800 cm^{-3} at the center and the temperatures decrease to 20–30 K. The high temperature at the edge is needed to reproduce the observed H_2 populations in J=2 and 3, whereas the low temperature in the core is suggested by the H_2 J=0 to 1 population ratio. In addition, the observed J=0/J=2 population ratios of C_2 suggest low temperatures in the centers of diffuse clouds, in particular the ζ Oph cloud (van Dishoeck and Black 1986*b*). The central density is constrained mostly by the H/H_2 abundance ratio. The temperature and density structure in the preferred model G of vDB of the ζ Oph cloud is included in Figure 2.

Viala, Roueff and Abgrall (1988; hereafter VRA) have recently also reinvestigated the ζ Oph cloud through models which incorporate a detailed treatment of the excitation of H_2 and HD. These models do not have continuous variations of temperature and density but retain the BD "core + envelope" structure. Thus the number of free parameters in the VRA models is larger than that in the vDB models, since not only the temperature and density of each of the two regions, but also their relative sizes can be adjusted so as to best reproduce the data. The temperature and density structure in the preferred model 6 of VRA is included in Figure 2 for comparison. VRA share the conclusions of vDB that a single component model cannot reproduce the observational data toward ζ Oph; that a significant part of the cloud must have a low density to reproduce the observed H/H_2 ratio; that an enhanced temperature of about 100–200 K is required at the edge by the H_2 J=2–3 populations; and that an enhanced radiation field is necessary to reproduce the high

J population of H_2. The VRA models differ, however, in a few details. In particular, the density in the core of the VRA models of the ζ Oph cloud is not well constrained by the H_2 and H observations alone, and can vary between 1000 and 5000 cm^{-3}. The enhancement factor of the radiation field is 7.5 in the VRA models with respect to the radiation field of Gondhalekar *et al.* (1980), which corresponds to I_{UV}=5–6 with respected to the Draine (1978) field. The scaling factor for the radiation field in the VRA models was chosen so as to reproduce about half of the observed J=5 column density, and appears slightly higher than that found in the vDB models. A detailed comparison is complicated by differences in adopted grain scattering properties, which in the VRA models are probably intermediate between those of grain model 2 and model 3. Table 4 compares the computed H_2 excitation in the two models. As a test of the sensitivity of the results to the details of the calculations, the table also includes the results of a model in which the temperature and density structure of the VRA model has been used in the vDB programs, with all other parameters as close as possible to those adopted by VRA. The slight discrepancies between the computed H_2 population distributions obtained in the two methods are probably due to the different treatments of the self–shielding in the H_2 lines and the continuum attenuation by dust. Compared with observations, all models generally reproduce well the observed populations in the various levels, with the exception of the J=3 and 4 levels, for which the computed populations are too low by factors of a few.

It thus appears that the main difference between the vDB and VRA models is the central density, which may be significantly higher in VRA models. However, VRA did not yet consider the constraints on the temperature and density from the observed C_2 and CO rotational excitations and the atomic fine structure excitation. As discussed by vDB, the available data on these species all favor a rather low density of a few hundred cm^{-3}. These conclusions are not modified significantly if the new collisional cross section data for C_2 with H_2 (Chambaud *et al.* 1988) are used, nor if the selection rule for C J=0→1 excitation (Monteiro and Flower 1987) is taken into account. Table 4 includes the computed atomic carbon fine structure excitation in the various models, which is sensitive primarily to the density structure in the cloud. A high central density appears to be less compatible with the observational data than a low density. Le Bourlot *et al.* (1987) have recently investigated the excitation of C_2 in the VRA models using the new collisional excitation data, and found indeed that the C_2 excitation can only be reproduced with a high central density if the intensity of the radiation field in the red is enhanced as well. They also presented a new model with a lower central density of about 800 cm^{-3} which reproduces the observations equally well, and which is closer to the vDB models.

In summary, the various efforts to model the ζ Oph cloud yield central densities that are uncertain by at least a factor of two, although most of the currently available data favor a rather low density throughout the cloud. The density structure could be further constrained by observations of the 2.65 millimeter rotational line of CN (Black 1988) or the J=3→2 0.87 millimeter line of CO. Predicted radiation temperatures, T_R, for these lines in the various models are included in Table 4. These values of T_R are based upon the model values of central densities and temperatures and the observed column densities and linewidths. The intensity of the incident UV radiation field is also not well constrained. If other processes contribute to the population of the high–J levels of H_2, the inferred scaling factors provide only upper limits to the strength of the radiation field. Eventual observations of vibrationally excited H_2 would provide a better probe of I_{UV} (see also discussion in §5). Additional information on the strength of the radiation field may come from studies of the ionization balance of the atomic species, although the observational data are still uncertain. The uncertainties in the physical conditions appear to be smaller for other lines of sight, such as the ζ Per cloud.

TABLE 5. Comparison of Model and Observed Column Densities

	ς Per		ς Oph			
Species	vDB[a] Model F	Observed[a]	vDB[a] Model G	VRA[b] Model 6	VRA/vDB[c]	Observed[a]
δ_C	0.50		0.52	0.28	0.28	
δ_D	1.10		0.30	0.40	0.40	
ς_o (s^{-1}) ...	6.0(-17)		8.0(-17)	5.1(-17)	5.1(-17)	
k_1 ($cm^3\ s^{-1}$)	7.0(-16)		7.0(-16)	2.6(-16)	2.6(-16)	
C	2.9(15)	(3.3±0.4)(15)	2.8(15)	3.7(15)	5.1(15)	(3.2±0.6)(15)
C^+	3.8(17)	(3.0±1.0)(17)	3.2(17)	1.8(17)	1.8(17)	(1.0±0.5)(17):
CH	1.9(13)	(2.0±0.3)(13)	2.0(13)	2.1(13)	2.2(13)	(2.5±0.3)(13)
C_2	2.0(13)	(1.9±0.2)(13)	2.2(13)	2.5(13)	5.8(13)	(2.5±0.2)(13)
OH	4.7(13)	(4.2±0.5)(13)	4.4(13)	4.7(13)	2.8(13)	(4.8±0.5)(13)
H_2O	5.2(12)[d]	...	4.9(12)[d]	7.2(12)	5.0(12)[d]	≤2.2(13)
HD	5.8(15)	(3.8±1.4)(15)	2.6(14)	6.0(14)	1.6(14)	(2.1±1.0)(14)
CO	8.9(13)	(6.1±3.0)(14)	8.2(13)	...	2.0(14)	(2.0±0.3)(15)
CH^+	2.4(11)	(3.5±0.4)(12)	2.8(11)	2.3(10)	7.9(10)	(2.9±0.3)(13)
CN	4.4(11)	(3.0±0.3)(12)	4.1(11)	3.8(11)	3.0(12)	(2.9±0.3)(12)
NH	2.3(10)	<6.3(11)	2.2(10)	5.4(11)	2.5(10)	<7.5(12)

[a] *From vDB, but with the new CO photodissociation rates, improved ionization balance of metals, and* $k_3=10^{-12}$ *cm*^{3} *s*$^{-1}$; [b] *VRA model 6 with* $k_9=2\times10^{-7}T^{-1/2}$ *cm*^{3} *s*$^{-1}$; [c] *VRA temperature and density structure adopted in vDB programs cf. Table 4; all other parameters similar to model G;* [d] *Assuming a branching ratio for dissociative recombination of* H_3O^+ *to form* H_2O *of 0.9.*

4.3 Chemistry

4.3.1 General results

Given the physical structure of a cloud, the chemical network can be solved at each depth and column densities or line intensities can be calculated for direct comparison with observations. The uncertainties in the physical parameters directly affect the confidence with which the chemistry networks can be tested. In particular, the dominant destruction process of neutral molecules in diffuse clouds is photodissociation, for which the rates scale directly with the intensity of the radiation field.

In Table 5, the column densities of various species in the vDB models are compared with observational data for the ς Oph and ς Per clouds. At first sight, the agreement appears to be very good for most species. However, this result is deceptive, since for each of these species there is at least one unknown reaction rate or parameter that can be adjusted *ad hoc* to obtain the best agreement with observations. For example, the CH abundance depends directly on the rate coefficient of reaction (1) for the radiative association between C^+ and H_2. Since the main destruction route of CH by photodissociation is now fairly well understood (van Dishoeck 1987*b*), a value $k_1 \approx 7 \times 10^{-16}$ cm^3 s^{-1} appears necessary to reproduce the observations. This value is consistent with the theoretical estimates $k_1 \approx 10^{-16} - 10^{-15}$ cm^3 s^{-1} by Herbst (1982), but is close to the measured upper limit $k_1 < 1.5 \times 10^{-15}$ cm^3 s^{-1} at 13 K (Luine and Dunn 1985). Note however that this experimental upper limit pertains to the reaction of C^+ with normal hydrogen having an ortho:para ratio of 3:1, whereas the interstellar models require the rate for an ortho:para ratio of about 1:2. Since the radiative association with ortho–H_2 is expected to be slower than that with para–H_2, the upper limit under interstellar conditions may be higher.

The abundance of C_2 is directly related to that of CH. In this case, certain branching ratios have to be adjusted to get the best agreement with observations. In particular, the production of C_2 is favored by a large branching ratio to CH_2 in the dissociative recombination of CH_3^+.

The abundances of all oxygen–bearing molecules are directly proportional to the cosmic ray ionization rate ζ_o. In the models of vDB, a fairly high ionization rate, $\zeta_o \approx (7\pm3)\times10^{-17}$ s^{-1} is needed to reproduce the observed OH abundances. With this rate, the computed abundances of H_2O^+ and H_2O are consistent with the measured upper limits. Note that the model OH results are not sensitive to the branching ratios for the dissociative recombination of H_3O^+, since H_2O rapidly photodissociates to form OH. The H_2O abundance is, of course, significantly increased if the H_3O^+ recombination results mostly in H_2O. The vDB models adopt a dissociative recombination rate of H_3^+ of $k_9 = 10^{-10}$ cm^3 s^{-1}. With this low rate, the H_3^+ abundance is large and the main formation route of OH is through the O + H_3^+ reaction. If k_9 were as large as 10^{-8} cm^3 s^{-1}, however, the H_3^+ abundance would be significantly lower, and the value of ζ_o would have to be increased even further. On the other hand, the analysis assumes that all the observed OH is formed by steady–state gas–phase reactions. As will be discussed in §5, OH can also be produced in shock–heated gas, in which case the inferred values of ζ_o should be regarded as upper limits. However, at least for the ζ Oph cloud, the amount of shock–produced OH seems to be at most 20% of the total OH abundance (Crutcher 1979). A direct test of the inferred cosmic ray ionization rate would be provided by observations of the H_3^+ ion, since its abundance is also directly proportional to ζ_o. If the dissociative recombination of H_3^+ is indeed as slow as $10^{-9} - 10^{-10}$ cm^3 s^{-1}, the predicted H_3^+ column densities are large enough that its absorption lines at near–infrared wavelengths would become detectable.

Another measure of the cosmic ray ionization rate would be provided by the observed HD abundances, if the overall deuterium abundance [D]/[H] in interstellar clouds were known. Alternatively, the values of ζ_o derived from the oxygen chemistry can be used to infer [D]/[H] from the measured HD column densities. The derived deuterium abundances in the models of vDB are in the range $(0.5\text{–}2.0)\times10^{-5}$ for the various clouds, consistent with other estimates of the deuterium abundance in the interstellar medium, [D]/[H]=$(1.5\pm 1.0)\times10^{-5}$ (Vidal–Madjar and Gry 1984). The models of the ζ Per cloud favor the upper part of this range, whereas those of the ζ Oph cloud give somewhat lower values, due to the order of magnitude lower HD column density. A similar deuterium abundance for the ζ Oph cloud has been obtained by VRA.

In Table 5, the results for the carbon–bearing species in the ζ Oph cloud are compared with those obtained by VRA. Since the published VRA models do not yet take into account the recent developments in the CO photodissociation rate, their CO result is not included in the table. To investigate the influence of the temperature and density structure of the cloud, Table 5 also contains the abundances that are obtained with the vDB chemistry network if the physical structure of VRA for the ζ Oph cloud is adopted.

Even with the current flexibility in some of the reaction rate coefficients, the steady–state models fail to reproduce the observed column densities of a number of species, most notably those of CH^+, CO and CN. The vDB models also give C^+/C column density ratios that are somewhat large compared with the (uncertain) observed ratio. In the following some of these discrepancies will be discussed in more detail.

4.3.2 The CO problem

As discussed by van Dishoeck (this volume), the photodissociation processes of CO are finally well understood through the detailed laboratory work of Letzelter *et al.* (1987) and Yoshino *et al.* (1988). The unshielded CO photodissociation rate of 2×10^{-10} s^{-1} is a factor of four larger than the rate adopted by vDB, and a factor of 50 larger than the rate used by BD and VRA. The CO photodissociation rate decreases rapidly with depth inside the cloud, due to self-shielding and shielding by lines of H and H_2 (Glassgold, Huggins and Langer 1985; van Dishoeck and Black 1988*a*; Viala *et al.* 1988). The new results confirm the conclusion of vDB that the current diffuse cloud models give CO column densities that are an order of magnitude too low compared with observations. Similar discrepancies occur if the new photodissociation rate is used with the VRA structure of the ζ Oph cloud.

The main formation route of CO in diffuse clouds is through the reaction of C^+ and OH, and it is difficult to find a significantly more efficient route. The discrepancy must therefore be due to the large destruction rate. The photodissociation rate of CO could be lowered if the intensity of the radiation field were smaller. As discussed above, the intensity of the radiation field in the models is constrained by the observed high-J population of H_2, under the assumption that ultraviolet pumping at 912–1100 Å is the dominant population mechanism of $J \geq 5$ (see also §5). The photodissociation of CO occurs in the same wavelength region, with $\geq 75\%$ of the rate due to transitions at $\lambda < 1000$ Å. As pointed out by van Dishoeck and Black (1988*a*), the shape of the radiation field at $\lambda < 1000$ Å is not well determined. If the radiation field at $\lambda < 1000$ Å were significantly smaller than thought previously, the CO photodissociation rate could be decreased significantly without affecting the H_2 $J \geq 5$ excitation by more than about 50%. Such modifications of the shape of the ultraviolet radiation field have been explored by van Dishoeck and Black (1988*a*) and CO column densities up to a few times 10^{14} cm^{-2} may be obtained. Although these modifications may not be sufficient to remove the discrepancies for all lines of sight, they can significantly reduce them.

4.3.3 The CN and NH problems

A similar problem is encountered for the CN molecule. If the unshielded CN photodissociation rate is indeed 10^{-10} s^{-1} or larger (cf. van Dishoeck, this volume), the models produce too little CN by up to a factor of a few, if the rates for both reactions (3) and (4) are taken to be negligibly small at interstellar temperatures. In that case, CN is formed mostly by the neutral–neutral reactions of C_2 and CH with N (Federman *et al.* 1984; Federman and Lambert 1988), and by reactions of hydrocarbon ions with N. However, these reactions appear insufficient to reproduce the observed CN column densities, if the models are constrained not to exceed the observed CH and C_2 column densities. Any additional formation route of CN has to be such that it does not produce NH abundances in excess of the observed upper limits. With the new estimate of the NH photodissociation rate (Kirby and Goldfield 1988) and the low rates for reactions (3) and (4), the computed NH abundances are well below these upper limits.

Grain surface formation of NH_3 followed by reactions with C^+ could enhance the production of CN in diffuse clouds. However, the rapid photodissociation of NH_3 also produces NH, and estimates of the efficiency of this process result in NH abundances that are close to the current upper limits for some clouds (Mann and Williams 1984; 1985). More sensitive searches for NH in diffuse clouds would put significant constraints on any grain surface production of molecules.

The problem of the CN abundance in diffuse clouds could be alleviated by adopting a less intense radiation field. In particular, the CN photodissociation also occurs mostly at short wavelengths, $\lambda < 1000$ Å. If the modified radiation field suggested by van Dishoeck and Black (1988*a*) based on the CO results is used, the CN column densities can be increased to about 10^{12} cm^{-2}.

4.3.4 The CH⁺ problem

The large observed abundance of CH^+ in diffuse clouds has been a puzzle for almost 50 years. As discussed, for example, by Dalgarno (1976), and most recently by Black (1988), steady-state models fail to reproduce the observed column densities of CH^+ by at least one order of magnitude. Since the reaction between C^+ and H_2 to form CH^+ cannot proceed at low temperatures, the formation of CH^+ occurs mostly through photoionization of CH and reactions of C with H_3^+. CH^+ is efficiently destroyed by photodissociation, by reactions with H_2 to form CH_2^+, and by H to form C^+ and H_2. Alternative formation channels for CH^+ have been proposed. In particular, Stecher and Williams (1974) suggested that the exothermic reaction between C^+ and vibrationally excited molecular hydrogen, H_2^*, could produce significant amounts of CH^+. This process has recently been included in the vDB models, which calculate in detail the amount of vibrationally excited H_2 due to ultraviolet pumping. However, the resulting CH^+ column densities are increased by only 20% or less, due to the fact that the concentration of vibrationally excited H_2 is large only in a narrow boundary layer of the cloud. Other theories include sublimation of CH_4 from dust grains followed by photoionization and photodissociation (Bates and Spitzer 1951), and chemical reactions in gas heated by starlight in regions close to young hot stars (White 1984). However, none of these mechanisms can account quantitatively for the observed CH^+ abundances. Shock production of CH^+ will be discussed briefly in §5.

4.3.5 Large molecule chemistry

It has been suggested that large molecules (LMs) such as Polycyclic Aromatic Hydrocarbons (PAHs) may be present in substantial abundances in interstellar clouds (Duley and Williams 1981; Léger and Puget 1984; Omont 1986; Allamandola, Tielens and Barker 1987). From the observed strengths of the infrared features that have been attributed to them, it has been estimated that they may contain up to 10% of the available carbon in a cloud. The LMs can be either in neutral form, or they may be positively or negatively ionized. The distribution among the different forms is determined by the balance between the photoionization, photodissociation and photodetachment rates, and the rates of recombination and attachment with electrons. In diffuse clouds, the LMs are mostly neutral or positively ionized at the edges. Deeper into the cloud, the abundance of LM^+ decreases rapidly, and the abundance of LM^- becomes comparable to that of neutral LM.

The presence of LMs in diffuse clouds can change the chemistry both quantitatively and qualitatively through the reactions

$$\begin{aligned} \mathrm{LM} + \mathrm{X}^+ &\rightarrow \mathrm{LM}^+ + \mathrm{X} \\ \mathrm{LM}^- + \mathrm{X}^+ &\rightarrow \mathrm{LM} + \mathrm{X} \end{aligned} \tag{12}$$

where X^+ can be any ion such as C^+, a metal ion or a molecular ion. Thus the main effect of the LMs will be to neutralize ions. Note that in this respect the specific identification of the LMs as PAHs is not crucial: the same chemical effects will occur for any other large molecule or small grain which is capable of neutralizing ions.

A simplified LM chemistry, characteristic of that of PAHs, has been added to the vDB models (Lepp *et al.* 1988). The aim of this study was to explore whether useful

TABLE 6. Effects of Large Molecules on Diffuse Cloud Chemistry[a]

Cloud	Model	x_{LM}	δ_C	δ_D	C	C^+	HD
ς Per	F	0	0.62	1.1	3.8(15)	4.7(17)	3.2(15)
		1(-7)	0.40	1.1	3.7(15)	3.0(17)	3.7(15)
		1(-6)	0.40	1.1	1.5(16)	2.9(17)	3.7(14)
			Observed:		(3.3±0.4)(15)	(3±1)(17)	(3.8±1.4)(15)
ς Oph	G	0	0.67	0.30	4.3(15)	4.2(17)	2.6(14)
		6(-7)	0.21	0.87	3.2(15)	1.3(17)	2.2(14)
		1(-6)	0.23	1.00	5.5(15)	1.4(17)	1.8(14)
			Observed:		(3.2±0.6)(15)	(1±0.5)(17):	(2.1±1.0)(14)

[a] *From Lepp et al. (1988).*

estimates of —or upper limits to— the fractional abundance of LMs in diffuse clouds can be derived from comparison with observations. It appears that the strongest constraints are provided by the ratios of successive ionization stages of atomic systems that do not react with hydrogen. Lepp *et al.* have presented ionization ratios for various species with LM abundances $x_{LM} = n(\mathrm{LM})/n_H$ ranging from 0 to 10^{-5}. As Table 6 shows, the observed C^+/C column density ratios in the ς Per and ς Oph clouds can be well reproduced if LMs are present with abundances of 10^{-7} and 6×10^{-7}, respectively. The presence of LMs or some other neutralizing agent also improves the comparison between observations and models for other species such as Mg, S, Ca and Fe. LM abundances larger than a few times 10^{-6} appear to be excluded by the observations. Thus, about 1–7 % of the carbon may be in the form of 50–atom LMs in diffuse clouds. About 20–40 % of the carbon is in the gas–phase, whereas the remaining 60–70% resides in larger grains.

The presence of LMs with abundances of a few times 10^{-7} has little effect on the abundances of small molecules such as CH, OH, CO and CN. However, as Table 6 shows, the LMs strongly affect the amounts of HD, and consequently the inferred deuterium abundances in diffuse clouds. The decrease in the HD abundance is caused mainly by the large reduction of both the H^+ and D^+ concentrations through reactions (12), which initiate the formation of HD through reactions (5) and (6). As discussed above, the HD abundance depends on two disposable parameters: the deuterium abundance δ_D, where δ_D=1 refers to [D]/[H]=1.5×10^{-5}, and the cosmic ray ionization rate ς_o. If ς_o is chosen to be constrained by the observed OH column densities, then the order of magnitude smaller HD column density observed toward ς Oph compared with ς Per would imply a four times smaller deuterium abundance for the former cloud, in the absence of LMs. However, if LMs are included at the levels suggested by the atomic ionization balances, the inferred deuterium abundances for the two clouds are virtually the same, [D]/[H]=$(1.5 \pm 0.5) \times 10^{-5}$.

4.4 Time–dependent and evolutionary models

Can time–dependent or evolutionary effects account for some of the weaknesses of the steady–state models? The time scales for all chemical processes at the edges of diffuse clouds are so short, $t \approx 10^2 - 10^3$ yr, that chemical equilibrium is rapidly attained on typical dynamical time scales of $10^5 - 10^6$ yr. Even at the centers of diffuse clouds, the time scales for most processes are less than 10^5 yr. The only exception is the H/H_2 chemistry, for which the time to reach chemical equilibrium in the center of a diffuse cloud like ς Oph is about 5×10^6 yr. Time dependent chemistry has been investigated by Allen and Robinson (1976) and Iglesias (1977), but such studies have ignored the effects of photodissociation and depth structure. Wagenblast and Hartquist (1988) have treated the time dependence of the H/H_2 abundance and the H_2 rotational excitation in diffuse clouds recently in detail. The results depend on the history of the cloud, and in particular on the initial H/H_2 ratio. If diffuse clouds evolved significantly on time scales less than a few million years, some relics of their history might appear in the form of non–equilibrium abundances in their centers. Most potentially observable effects of time dependence, however, are expected to be small.

A special case is presented by the cloud toward ζ Oph, since this star is known to be a runaway star (Blaauw 1961). If the inferred enhancement of the ultraviolet radiation field were provided mostly by the star itself, then the cloud has been exposed to this radiation for a comparatively short time. The star is presumably no closer to the neutral cloud than the projected radius of the ionization region around it, which is about 10 pc. At its space velocity of 39 km s^{-1}, ζ Oph could move a comparable distance in 3×10^5 yr. This time scale is still long compared with that for CO photodissociation, even at the center of the cloud, so that no significant effects are expected.

Evolutionary models in which also the physical parameters evolve with time have been developed by Gerola and Glassgold (1978) and Tarafdar *et al.* (1985). In this picture, dense clouds are the result of gravitational contraction of an initially diffuse cloud. The clouds spend most of their lifetimes in the diffuse state, and dense cores are formed on a time scale of a few times 10^6 yr. The time scales for the chemical processes in the diffuse state are small, and the dynamical effects even serve to shorten the approach to chemical equilibrium. No significant differences with steady–state models are therefore expected in the modeling of the diffuse stage of a contracting cloud.

5. SHOCK MODELS

5.1 Models

Although the steady–state models are fairly successful in reproducing the observed abundances of most simple species in diffuse clouds—albeit with some *ad hoc* assumptions about reaction rate coefficients and other physical parameters—they still fail to reproduce the large observed column densities of CH^+. Elitzur and Watson (1978, 1980) suggested that substantial amounts of CH^+ could be produced in a narrow shock–heated layer of a cloud, where the temperature is high enough that the endothermic reaction

$$C^+ + H_2 \rightarrow CH^+ + H - 4640\ \mathrm{K} \tag{13}$$

proceeds rapidly. In addition to CH^+, significant amounts of other species can be formed in the shock–heated gas, most noticeably OH through the reaction

$$O + H_2 \rightarrow OH + H - 2980\ \mathrm{K}. \tag{14}$$

Furthermore, collisional excitation processes at high temperatures can contribute significantly to the populations of the rotational levels of H_2. Thus the shock models have to be consistent with a variety of observations.

The original Elitzur and Watson (1980) one–fluid shock models were capable of reproducing the observed CH^+ column densities, but tended to produce too much OH and rotationally excited H_2 compared with observations. However, if magnetic fields are included in the models, the shock may have a multi–fluid character when the fractional ionization is sufficiently low. In that case, a comparatively large warm region exists in which the ionized and neutral species have different flow speeds and temperatures. This systematic velocity difference between the ions and the neutrals can contribute part of the 0.4 eV energy required for reaction (13) to occur. On the other hand, reaction (14) will proceed less rapidly in magnetohydrodynamic (MHD) shocks compared with non-magnetic shocks, because the maximum temperature is lower in the MHD shocks and because no velocity difference occurs between the two neutral species to assist the reaction.

MHD shock models of diffuse clouds have been developed by Draine (1986), Draine and Katz (1986*a*, *b*), and Pineau des Forêts *et al.* (1986), and will be discussed in detail in the chapter by Flower in this volume. Both studies conclude that substantial amounts of CH^+ can be produced in MHD shock models with reasonable assumptions about the shock velocity and preshock density. The two studies differ, however, on the question whether the same models are in harmony with observed rotational populations of H_2 and column density of OH. The models of Draine and Katz suggest that the H_2 J=3, 4 and 5 levels are populated significantly in the shock, but that the amount of shock produced OH is small. The shock-production of H_2 J=3 and 4 is consistent with the fact that the steady-state models generally produce too little population in these levels compared with observations. If indeed part of the J=5 population results from the shock, the scaling factor for the radiation field could be lowered, if no other data on high-J levels were available. However, the current MHD shock models are not capable of producing large populations in the higher J=6 and 7 levels. The observed populations of these levels toward ζ Oph still require a significant enhancement of the radiation field. The models of Pineau des Forêts *et al.* (1986) give column densities of OH and rotationally-excited H_2 in excess of observations.

5.2 Shock vs steady-state models: observational tests

Various observational tests of the relative importance of shock chemistry compared with steady-state chemistry in diffuse clouds are possible:

- Although the shock models produce large populations in the H_2 J=3–5 levels, they result in negligible amounts of vibrationally excited H_2. On the other hand, the ultraviolet pumping mechanism in the steady-state models naturally gives populations of H_2 in $v \geq 1$ that are characterized by excitation temperatures T_{ex} >5000 K. The vDB models have predicted equivalent widths for the Lyman lines originating in $v'' > 0$ that should be detectable with the high resolution ultraviolet spectrometer on board the Hubble Space Telescope. Failure to detect these lines at the predicted level would be a clear indication that the intensity of the radiation field has been overestimated in the models, and that a significant fraction of the observed rotational excitation in $v''=0$ is produced in shock-heated gas. Some ultraviolet pumping of both the shock-excited and quiescent H_2 by the average background starlight must occur, however, in any case.
- A common problem of the current MHD shock models is that they all require preshock H_2/H abundance ratios close to unity in a gas with a density of only 20 cm^{-3}. Such ratios can be obtained only if the total column density of H_2 in the preshock gas is of the order of 10^{18} cm^{-2} or more. In that case, one would expect the preshock gas to show strong lines of atomic species like Na. Toward ζ Oph, these lines are not observed.
- A strong argument in favor of shock production of CH^+ has come from the observed correlation between the CH^+ column densities and the H_2 column densities in levels J=3–5 (Frisch and Jura 1980; Lambert and Danks 1986). However, it also appears that the CH^+ column densities continue to increase for more reddened lines of sight (Black 1985; Lambert and Danks 1986). This fact remains difficult to reconcile with shock

production of CH^+. New, more accurate data on CH^+ abundances in thicker diffuse clouds are needed.

- The shock models make specific predictions about the velocities and velocity dispersions of the lines. In particular, they suggest that the CH^+ absorptions should be displaced from those of the cool postshock gas by typically 1–2 km s^{-1}. Differences of this order in observed CH and CH^+ line positions have been claimed e.g. by Federman (1982). However, the magnitude of these differences has recently become smaller due to a more accurate determination of the rest wavelength of the CH^+ line (Carrington and Ramsay 1982), and measurable velocity differences in excess of 1 km s^{-1} are found for only 50 % of the lines of sight (Federman 1987). Although small velocity differences could be explained by projection effects, more accurate measurements for a larger number of clouds are warranted. The observed velocity dispersions of CH^+ are reasonably well reproduced in the models.
- The MHD models also produce significant amounts of CH in the shocks. This shock-produced CH is predicted to be displaced from the quiescent CH by several km s^{-1} and should thus be identifiable as a weak absorption feature on the shoulder of the strong quiescent CH line. The new high quality spectra of Palazzi, Mandolesi and Crane (1988) provide significant upper limits on the amounts of shocked CH toward ζ Oph.
- CO is one of the few molecules that can also be observed by its emission lines at millimeter wavelengths in diffuse clouds. The order of magnitude higher spectral resolution of these measurements shows CO line widths <0.5 km s^{-1} in the ζ Oph cloud, which indicates that the molecule exists in gas with a kinetic temperature T <100 K: a hot component of shocked CO is apparently excluded (Langer *et al.* 1987, see Figure 3 in §7). CO millimeter observations for other diffuse clouds would be valuable.

6. ISOTOPIC ABUNDANCES

Several important observations of the abundances of the ^{13}C varieties of molecules in diffuse clouds have recently become available. Hawkins and Jura (1987) determined accurate $^{12}CH^+/^{13}CH^+$ column density ratios for 4 lines of sight. The inferred ratio (43±4) was found to be remarkably uniform in the different directions, and is a factor of two lower than the terrestrial isotope ratio of 89. If CH^+ is indeed formed in a warm region without fractionation, the measured $^{12}CH^+/^{13}CH^+$ ratio should be directly representative of the $[^{12}C]/[^{13}C]$ abundance ratio of carbon in all forms in the interstellar medium.

Accurate information on the ^{13}CO column density in diffuse clouds can be obtained from observations of the ultraviolet absorption lines of CO. Except for lines arising in (0,0) bands, the ^{13}CO features are well separated from the ^{12}CO lines and can easily be distinguished on spectra with resolving powers $\lambda/\Delta\lambda \geq 2 \times 10^4$, such as provided by the *Copernicus* satellite. Wannier, Penzias and Jenkins (1982) inferred $^{12}CO/^{13}CO \approx (55 \pm 11)$ for the ζ Oph cloud.

The very weak line of ^{13}CN has recently been discovered in the high signal-to-noise spectra of Crane *et al.* (1986) toward ζ Oph (Crane 1986). The inferred ratio $^{12}CN/^{13}CN$ (50+13, -10).

These observations thus lead to the remarkable conclusion that within the measurement errors the $^{12}CH^+/^{13}CH^+$, $^{12}CO/^{13}CO$ and $^{12}CN/^{13}CN$ abundance ratios are the same in the ζ Oph cloud, and probably equal to the $[^{12}C]/[^{13}C]$ abundance. This indicates either that no isotope specific processes are significant in the ζ Oph cloud for any of the species, or that these effects accidentally cancel out in the ζ Oph cloud. Since enhancement of

^{13}CO can occur at low temperatures through the reaction

$$^{13}C^{+} + {}^{12}CO \rightleftarrows {}^{12}C^{+} + {}^{13}CO + 35\ K \tag{15}$$

the former conclusion would imply that the (CO column–averaged) temperature in the ζ Oph cloud must be more than 50 K (Crutcher and Watson 1981), in possible conflict with other diagnostics of the temperature (van Dishoeck and Black 1986*b*). On the other hand, the enhancement of ^{13}CO at low T may be counteracted by the isotope selective photodissociation of ^{13}CO (Bally and Langer 1982; Glassgold *et al.* 1985). Indeed, van Dishoeck and Black (1988*a*) find with the new CO spectroscopic data that the ^{13}CO photodissociation rate inside the ζ Oph cloud is about a factor of two larger than that of ^{12}CO, so that the two effects tend to cancel. In the case of ^{13}CN, a temperature–sensitive exchange reaction analogous to reaction (15) with an energy defect $\Delta E/k = 31$ K is expected to operate. The lack of apparent fractionation in the ζ Oph cloud may simply reflect the inability of the exchange reaction to compete with rapid photodissociation of a molecule that is not effectively shielded in a cloud of low density.

The deuterium abundance as derived from observations of the HD molecule has been discussed already in §4. No information is available on other isotopic species.

7. VELOCITY STRUCTURE

How well do we understand the structure of diffuse clouds? Can they really be represented by homogeneous plane–parallel slabs? All models of diffuse clouds developed so far have been based primarily on optical absorption line observations for which the resolving power is typically no better than $\lambda/\Delta\lambda \approx 10^5$, corresponding to about 3 km s^{-1}. At this resolution, none of the molecular absorption lines shows evidence for multiple components. Information on smaller scale structures can be obtained from observations at millimeter and centimeter wavelengths. Although such observations of weak lines in diffuse clouds are very difficult owing to the small column densities and low excitations, they benefit from spectrometers conventionally designed to provide resolution of <0.1 km s^{-1}. Early observations of Liszt (1979) of the CO 2.1 mm and CH 9 cm line profiles toward ζ Oph suggested small velocity dispersions for the main component in that direction. Higher quality spectra of the CO J=1→0 and 2→1 line emission toward ζ Oph have recently been obtained by Langer, Glassgold and Wilson (1987) and Crutcher and Federman (1987). The spectrum of Langer *et al.* is reproduced in Figure 3 and suggests that the gas toward ζ Oph is located in 4 distinct clumps, whose velocities differ by less than 1 km s^{-1} and whose line widths ($\Delta V \leq 0.5$ km s^{-1}) indicate temperatures less than 100 K. It would be of great interest to obtain CO millimeter observations for other diffuse clouds to determine how common such complicated velocity structures are.

Crane *et al.* (1986) and Palazzi *et al.* (1988) have recently obtained very high signal–to–noise optical absorption line observations of both CH and CN at $\lambda/\Delta\lambda \approx 10^5$ toward ζ Oph. The inferred Doppler parameters for the CH and CN profiles after correction for instrumental broadening are b = (1.27±0.02) and (0.88±0.02) km s^{-1}. Quite surprisingly, the ratio of these two line widths is (1.44 ±0.04), which is exactly the ratio that is expected for pure thermal broadening in a gas of 1200 K. This temperature is clearly much higher than that suggested by the other diagnostic data discussed above. The alternative possibility is that a complicated velocity distribution of cold gas along the line of sight causes the CH and CN lines to be broadened in just such a way as to mimic high–temperature thermal broadening. Black and van Dishoeck (1988) have recently investigated this possibility using the velocity structure for the ζ Oph cloud suggested by the CO observations

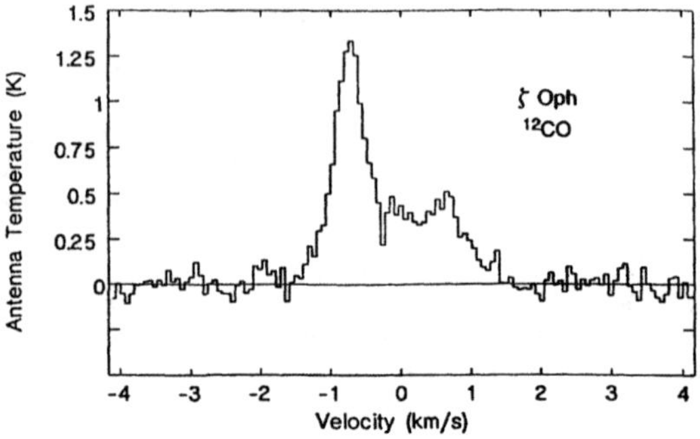

Figure 3. Observed CO J=1→0 profile toward ς Oph (from: Langer et al. 1987).

of Langer *et al.* It appears that at a resolving power of about 10^5, the high-temperature thermal broadening model and the low-temperature complex velocity structure model are virtually indistinguishable. Only at resolving powers $\lambda/\Delta\lambda \approx 4 \times 10^5$ does the velocity structure become apparent. The observed differences in the CH and CN line widths may reflect slightly different distributions of these molecules over the 4 components.

Unresolved velocity structure in diffuse clouds is also indicated by the high-resolution, $\lambda/\Delta\lambda \approx 2 \times 10^5$, ultraviolet rocket observations by Jenkins *et al.* (1988) toward π Sco. The observed profiles of the H_2 J=2–5 lines strongly suggest the presence of multiple components.

If these small scale structures turn out to be common in diffuse clouds, what is their origin and what will the effect be on the models? The clumpiness will certainly increase the surface area of the cloud, so that all photorates are expected to be effectively increased. Thus the inferred scaling factors for the radiation field for each of the clumps are expected to be less. However, it still remains to be investigated to what extent the various clumps shield each other; the radiative transfer in both continuum and lines through the various clumps will be a non-trivial exercise, and will depend on the adopted geometry. An additional problem is that most of the observational data have been interpreted assuming a single unresolved component along the line of sight, so that the column densities inferred from highly-saturated lines may be quite uncertain.

8. HEATING AND COOLING PROCESSES IN DIFFUSE CLOUDS

The various diagnostic species suggest temperatures at the edges of diffuse clouds of the order of 100 K or more at densities of about 100 cm^{-3}. The question arises whether such high temperatures can be maintained by the currently known heating processes. The main heat input in diffuse clouds comes from the photoelectric effect, while most of the cooling is provided by the fine-structure excitation of C^+ As discussed by Roberge (1981) and vDB, the current estimates of the efficiency of the photoelectric effect are insufficient to maintain temperatures as high as 100 K at the edges of diffuse clouds. In the centers of diffuse clouds where $T \approx 20$ 30 K, the heating and cooling processes appear to balance within a factor of two. An additional heat source may be provided by the photoionization of LMs (d'Hendecourt and Léger 1987), but large LM abundances are required to remove

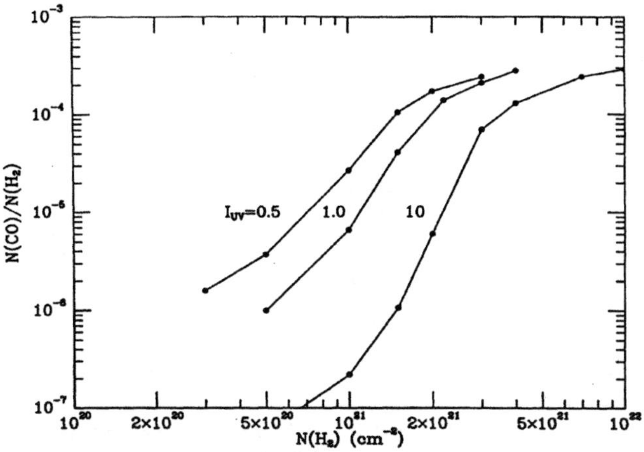

Figure 4. Computed CO/H_2 column density ratio as a function of total H_2 column density for translucent clouds exposed to different strengths of the radiation field (from: van Dishoeck and Black 1988a).

the discrepancies. The presence of LMs can have another effect on the thermal balance: LMs enhance the effective rate of recombination of C^+ and thus allow smaller abundances of gas phase carbon —the principal coolant— to be in harmony with absorption line observations of C (Lepp *et al.* 1988; see §4.3.5). The presence of a warm outer zone is based mostly on the high observed H_2 $J=2$ and 3 populations. As discussed in §5, these levels can also be populated in a shock-heated layer of the cloud. In particular, Table 2 of Draine (1986) illustrates that about 25% of the $J=2$ abundance can be produced in shocks. In that case, the inferred temperatures and heating requirements at the edges of the quiescent steady-state part of the cloud could be lowered. Even the narrowest line widths observed in diffuse clouds exceed the thermal Doppler widths expected at the inferred temperatures. The dissipation of these superthermal motions (loosely called "turbulence") is a heating mechanism. Although it appears that the global source of this turbulence can be explained in an energetic sense (Black and van Dishoeck 1988), turbulent dissipation is not a significant heating source in regions like the ζ Oph cloud. Taking the best-fitting vDB Model G of the ζ Oph cloud as an example, we find a ratio of total heating and cooling rates that varies from 0.3 near the boundary to 2.0 at the center. Inclusion of LMs at the level of 5% of the carbon with corresponding reductions in the abundances of gas phase coolants lowers the cooling rates by 10–20 %.

9. TRANSLUCENT MOLECULAR CLOUDS

All of the models and observations discussed so far have been concerned with the classical diffuse clouds that have $A_V^{tot} \approx 1$ mag. These have been studied mostly by absorption line observations against young, hot O and B stars. In recent years, a growing body of data has become available for more reddened lines-of-sight with $A_V^{tot} \approx 2$–5 mag. These clouds are of interest because they provide the bridge between diffuse and dense interstellar clouds, and

TABLE 7. Computed Column Densities for the HD 169454 Cloud[a]

Species	Model 1	Model 2	Observations
H_2	1.5(21)	2.2(21)	(1–2)(21)
H	6.3(20)	3.3(20)	>2(19)
C	1.1(16)	2.1(16)	...
C^+	1.4(17)	9.8(16)	...
CH	4.7(13)	7.6(13)	(4.6±0.8)(13)
C_2	9.9(13)	1.6(14)	(7.3±1.4)(13)
CN	1.4(13)	4.9(13)	(5.6±0.9)(13)
CO	2.0(16)	1.0(17)	(1–9)(16)
OH	3.5(14)	2.1(15)	...
CH_2	2.2(14)	4.0(14)	...
C_2H	2.7(13)	5.2(13)	...
C_3	5.7(9)	1.5(10)	...
A_V (mag)	2.25	2.95	3.5

[a] *From Jannuzi et al. (1988). Models 1 and 2 have* n_H^0*=500 and 700* cm^{-3}, T_0*=20 and 15 K, and* I_{UV}*=2 and 1, respectively, with* δ_C*=0.1.*

may give insight into the chemical processes occurring in both regions. They are denoted as "translucent" clouds, to indicate that photoprocesses play an important role in the chemistry throughout the cloud, even though the photo rates diminish rapidly toward the center. Although the translucent clouds are usually taken to refer to isolated small clouds, they may also represent the outer edges of dense molecular clouds. Translucent clouds have the virtue that they can be studied observationally not only by millimeter emission lines, but also by absorption line techniques, provided that a suitable background star is available. A number of such translucent clouds, or small molecular clouds, have been detected by searches for absorption lines toward highly-reddened stars (Hobbs, Black and van Dishoeck 1983; Lutz and Crutcher 1983; van Dishoeck and de Zeeuw 1984; Crutcher 1985; Crutcher and Chu 1985; Cardelli and Wallerstein 1986; Gredel and Münch 1986; Federman and Lambert 1988; van Dishoeck and Black 1988*b*), and occasionally by accident through CO millimeter observations (Knapp and Bowers 1988). Compared with the classical diffuse clouds, the column densities of CH, C_2 and CN are larger by up to an order of magnitude, whereas the CO column density is larger by several orders of magnitude. A detailed study of the rapid increase in CO abundance with increasing cloud thickness using the new spectroscopic data on the CO photodissociation has been made by van Dishoeck and Black (1988*a*) and is illustrated in Figure 4.

A good example of a translucent cloud is provided by the small cloud in front of the star HD 169454 investigated recently by Jannuzi *et al.* (1988). The CO column density in this cloud lies in the interesting regime of parameter space where it starts to account for a large fraction of the available carbon. Specific models for the HD 169454 cloud are collected in Table 7, where they are compared with observations. The temperature and density in the models are fairly well constrained by the observed rotational excitation of C_2. The modeling is further simplified by the fact that no observations on the rotational excitation of H_2 are available, so that the scaling factor I_{UV} for the ultraviolet radiation field can be chosen to be close to unity. Another difference with the classical diffuse clouds is that no information is available on the column densities of atomic or ionized carbon, so that the carbon depletion factor δ_C can be treated as a free parameter as well. The models can satisfactorily reproduce the observations, provided that the depletion of carbon is fairly large, $\delta_C \approx 0.1$.

TABLE 8. Computed Column Densities for High-Latitude Clouds[a]

Species	Model H3	Model H5	Observations
H_2	1.0(21)	2.0(21)	...
H	6.9(19)	7.8(19)	...
C	5.9(15)	1.1(16)	...
C^+	7.8(16)	7.2(16)	...
CO	1.7(16)	1.1(17)	6.4(16)[b]
CH	6.5(13)	8.3(13)	3(13)[c]
C_2	1.2(14)	1.3(14)	...
CH^+	5.0(10)	5.1(10)	6(12)[c]
CN	1.4(13)	2.8(13)	...
OH	2.6(14)	2.1(15)	9(14)[b]
H_2CO	2.8(9)	1.5(10)	4.6(12)[b]
A_V (mag)	1.3	2.6	0.7[b]

[a] *From van Dishoeck and Black (1988a, b);* [b] *Average properties from Magnani et al. (1988); abundances for individual clouds vary considerably; the extinction may be underestimated by a factor of 2;* [c] *From de Vries and van Dishoeck (1988) for one line of sight.*

Another interesting example of a translucent cloud is that toward HD 29647. This cloud has been studied extensively by Crutcher (1985) and has recently been modeled by Nercessian, Benayoun, and Viala (1988) and van Dishoeck and Black (1988*a*, *b*). More detailed observations of small polyatomic molecules such as C_2H and C_3H_2, and molecules containing second row elements like CS, in individual translucent clouds will provide significant tests of the gas-phase chemistry appropriate for their formation.

10. HIGH-LATITUDE CLOUDS

Another class of diffuse or translucent clouds is formed by the high-latitude molecular clouds detected by Magnani, Blitz and Mundy (1985) through CO millimeter observations. These clouds have visual extinctions $A_V^{tot} \approx 1$ mag similar to those of the classical diffuse clouds such as the ς Oph and ς Per clouds, yet their CO column densities and CO/H_2 ratios are at least an order of magnitude higher (Lada and Blitz 1988). As discussed by van Dishoeck and Black (1988*a*) and illustrated in Figure 4, such high CO column densities for low A_V can be produced within the steady-state framework only if the clouds are exposed to an ultraviolet radiation field that is smaller than the average background radiation field by a factor of 2–4. A reduction in the intensity by a factor of two is not implausible because the early type stars in the galactic plane illuminate the clouds from one side only (Draine 1978).

Magnani, Blitz and Wouterloot (1988) have recently investigated the abundances of the OH and H_2CO molecules in a number of high latitude clouds through observations at centimeter wavelengths. Within the large observational uncertainties, the OH column densities are in the range $10^{14} - 10^{15}$ cm^{-2}, an order of magnitude larger than those found for the classical diffuse clouds, while the H_2CO column densities are in the range $10^{12} - 10^{13}$ cm^{-2}. H_2CO has not yet been detected in any classical diffuse cloud.

Absorption line observations of high latitude clouds are complicated by the fact that almost no bright early-type stars are available as background light sources. However, with the improved sensitivity of the detectors, absorption line observations toward less bright background A and F stars are possible, provided that the spectral resolution is high enough to separate the narrow interstellar lines from the stellar features. Observations of absorption lines of interstellar Na have been performed by Hobbs, Blitz and Magnani (1986), Hobbs *et al.* (1988) and de Vries and van Dishoeck (1988). de Vries and van Dishoeck have also detected absorption lines of interstellar CH and CH^+ for lines of sight with strong interstellar Na features. The observed CH column densities are of the order of 10^{13} cm^{-2}, and are not significantly larger than those found in the classical diffuse clouds. CH^+ has been detected in one high latitude cloud, and its column density of 6×10^{12} cm^{-2} is small compared with that found in the ζ Oph cloud.

Steady-state models appropriate to high-latitude clouds have been developed by van Dishoeck and Black (1988*a*, *b*). The modeling is again simplified by the fact that few constraints on the density, temperature and strength of the radiation field are available. Two examples of models with n_H=500 cm^{-3}, T=40 K and I_{UV}=0.5, but with different total H_2 column densities, are presented in Table 8. Column densities of CO in excess of 10^{16} cm^{-2} are readily obtained in such models. The abundances of other species depend strongly on the adopted depletion factor of carbon. In particular, if the depletion of carbon is taken to be similar to that found in diffuse clouds, δ_C=0.4, and if k_1= 7×10^{-16} cm^3 s^{-1} is retained, the computed CH abundances are large compared with observations, whereas those of OH are somewhat low. On the other hand, if carbon is more depleted, the OH column density is increased significantly, because its removal rate through the reaction with C^+ is decreased. The depletion factor $\delta_C \approx 0.1$ adopted in the models in Table 8 appears to give CH and OH column densities that are consistent with the observations of high latitude clouds. Observations of other carbon-bearing molecules such as C_2H and C_3H_2 may provide additional constraints on the gas-phase carbon abundance in these clouds. The H_2CO column densities in the models are at least an order of magnitude below observations. However, the amount of H_2CO can be increased if an additional source of gas-phase CH_4 is added to the chemistry. Such a source may be provided by reactions of C and C^+ with LMs to form CH_4 (cf. Lepp *et al.* 1988), or by the disruption or evaporation of grain mantles. If 5–10% of the carbon were in the form of LMs in high-latitude clouds, the H_2CO column density in model H5 is increased to about 10^{12} cm^{-2}. Far-infrared emission from some high-latitude clouds may also be attesting to a significant abundance of LMs or small grains (cf. Weiland *et al.* 1986). Note that similar reactions of O and N with LMs would produce significant amounts of H_2O and NH_3 in high-latitude clouds. Searches for species such as NH or metal hydrides would provide additional indications whether grain processes or LMs play an important role in the chemistry in these clouds. Observations of sulfur-bearing molecules like CS may provide insight into their formation. Further searches for CH^+ might assess the importance of shock-induced processes.

11. CONCLUDING REMARKS

The discussion of the current status of steady-state models of diffuse clouds raises the question whether any significant new insights have emerged since the first detailed models by Black and Dalgarno (1977). A number of physical and chemical processes that enter the models are now better known, but the case of CO illustrates that the better rates often only complicate the interpretation. Future models should focus more on differences in measured abundances for some of the clouds, rather than on similarities. Many of the reaction rate coefficients that enter the simple networks are still not well determined. As a result, there is -strictly speaking- not even one molecule in diffuse clouds for which we can be *fully* certain that the adopted gas-phase reaction scheme is correct. A better determination of the C^+ + H_2 radiative association rate is of paramount importance. The discrepant results for the dissociative recombination rate for H_3^+ (v=0) need to be solved urgently. Measurements of the branching ratios for dissociative recombination of simple ions such as CH_3^+ and CH_5^+, and calculations of the photodissociation rates of simple molecules such as CH_2, C_2H and CH_3 are crucial as well.

Although the steady-state models can most likely account for many of the observed features of diffuse clouds, additional shocked layers appear necessary to reproduce the measured CH^+ abundances and part of the H_2 rotational excitation. It is still not known how ubiquitous shocks are in diffuse clouds, and to what extent they contribute to the formation of other molecules. It would be of interest to find a diffuse cloud which does not show any shock signatures, but still has substantial molecular abundances. Grain surface production of molecules other than H_2 in the classical diffuse clouds is evidently very inefficient, although large molecules or small grains may play a role in the ionization balance.

The study of diffuse clouds has suffered from a lack of new observational data over the last decade. However, new impetus for the modeling of the clouds has come from the recent high resolution CO millimeter observations, and from high quality optical data. Together with more detailed studies of chemically related regions such as the translucent and high-latitude clouds through optical absorption line observations and millimeter emission line measurements, substantial progress can be made in our understanding of the physical and chemical structure of diffuse clouds.

ACKNOWLEDGMENTS

This work was supported by NSF grant RII 86-20342 to Princeton University, and by NASA astrophysical theory grant NAGW-763 to the University of Arizona.

REFERENCES

Adams, N.G. and Smith, D. 1984, *Ap. J. (Letters)*, **284**, L13.
Adams, N.G. and Smith, D. 1987, in IAU Symposium **120**, *Astrochemistry*, eds. M.S. Vardya and S.P. Tarafdar (Reidel, Dordrecht), p. 1.
Allamandola, L.J., Tielens, A.G.G.M., and Barker, J.R. 1987, in *Interstellar Processes*, eds. D.J. Hollenbach and H.A. Thronson (Reidel, Dordrecht), p. 471.
Allen, M. and Robinson, G.W. 1976, *Ap. J.*, **207**, 745.
Amano, T. 1987, preprint.
Anicich, V.G. and Huntress, W.T. 1986, *Ap. J. Suppl.*, **62**, 553.
Bally, J. and Langer, W.D. 1982, *Ap. J.*, **255**, 143; erratum: **261**, 747.
Barsuhn, J. and Walmsley, C.M. 1977, *Astr. Ap.*, **54**, 345.
Bates, D.R. 1986, *Ap. J. (Letters)*, **306**, L45.
Bates, D.R. and Spitzer, L. 1951, *Ap. J.*, **113**, 441.
Blaauw, A. 1961, *Bull. Astr. Inst. Neth.*, **15**, 265.
Black, J.H. 1985, in *Molecular Astrophysics*, eds. G.H.F. Diercksen *et al.*, NATO ASI Series **157** (Reidel, Dordrecht), p. 215.
Black, J.H. 1987, in IAU Symposium **120**, *Astrochemistry*, eds. M.S. Vardya and S.P. Tarafdar (Reidel, Dordrecht), p. 217.
Black, J.H. 1988, *Adv. Atom. Mol. Phys.*, in press.
Black, J.H. and Dalgarno, A. 1973, *Astrophys. Letters*, **15**, 79.
Black, J.H. and Dalgarno, A. 1976, *Ap. J.*, **203**, 132.
Black, J.H. and Dalgarno, A. 1977, *Ap. J. Suppl.*, **34**, 405 (BD).
Black, J.H., Hartquist, T.W., and Dalgarno, A. 1978, *Ap. J.*, **224**, 448.
Black, J.H. and van Dishoeck, E.F. 1988, *Ap. J.*, submitted.
Black, J.H., Willner, S.P., van Dishoeck, E.F., Woods, R.C., and Churchwell, E.B. 1988, in preparation.
Cardelli, J.A. and Wallerstein, G. 1986, *Ap. J.*, **302**, 492.
Carrington, A. and Ramsay, D.A. 1982, *Physica Scripta*, **25**, 272.
Chambaud, G., Lavendy, H., Levy, B., Robbe, J.M., and Roueff, E. 1988, *Chem. Phys.*, submitted.

Clavel, J., Viala, Y.P., and Bel, N. 1978, *Astr. Ap.*, **65**, 435.
Crane, P. 1986, *The Messenger* **45**, 1.
Crane, P., Hegyi, D.J., Mandolesi, N., and Danks, A.C. 1986, *Ap. J.*, **309**, 822.
Crutcher, R.M. 1979, *Ap. J.*, **231**, L151.
Crutcher, R.M. 1985, *Ap. J.*, **288**, 604.
Crutcher, R.M. and Chu, Y.-H. 1985, *Ap. J.*, **290**, 251.
Crutcher, R.M. and Federman, S.R. 1987, *Ap. J. (Letters)*, **316**, L71.
Crutcher, R.M. and Watson, W.D. 1976, *Ap. J.*, **209**, 778.
Crutcher, R.M. and Watson, W.D. 1981, *Ap. J.*, **244**, 855.
Crutcher, R.M. and Watson, W.D. 1985, in *Molecular Astrophysics*, eds. G.H.F. Diercksen *et al.*, NATO ASI Series **157** (Reidel, Dordrecht), p. 255.
Czarny, J., Felenbok, P., and Roueff, E. 1987, *Astr. Ap.*, **188**, 155.
Dalgarno, A. 1976, in *Atomic Processes and Applications*, eds. P.G. Burke and B.L. Moiseiwitsch (North Holland, Amsterdam), Chap. 5.
Dalgarno, A. and Black, J.H. 1976, *Rep. Prog. Phys.*, **39**, 573.
Danks, A.C., Federman, S.R., and Lambert, D.L. 1984, *Astr. Ap.*, **130**, 62.
Danks, A.C. and Lambert, D.L. 1983, *Astr. Ap.*, **124**, 188.
d'Hendecourt, L.B. and Léger, A. 1987, *Astr. Ap.*, **180**, L9.
de Vries, C.P. and van Dishoeck, E.F. 1988, in preparation.
Draine, B.T. 1978, *Ap. J. Suppl.*, **36**, 595.
Draine, B.T. 1986, *Ap. J.*, **310**, 408.
Draine, B.T. and Katz, N.S. 1986*a*, *Ap. J.*, **306**, 655.
Draine, B.T. and Katz, N.S. 1986*b*, *Ap. J.*, **310**, 392.
Duley, W.W. and Williams, D.A. 1981, *M. N. R. A. S.*, **196**, 269.
Elitzur, M. and Watson, W.D. 1978, *Ap. J.*, **222**, L141.
Elitzur, M. and Watson, W.D. 1980, *Ap. J.*, **236**, 172.
Federman, S.R. 1982, *Ap. J.*, **257**, 125.
Federman, S.R. 1987, in IAU Symposium **120**, *Astrochemistry*, eds. M.S. Vardya and S.P. Tarafdar (Reidel, Dordrecht), p. 123.
Federman, S.R., Danks, A.C., and Lambert, D.L. 1984, *Ap. J.*, **287**, 219.
Federman, S.R. and Glassgold, A.E. 1980, *Astr. Ap.*, **89**, 113.
Federman, S.R. and Lambert, D.L. 1988, *Ap. J.*, in press.
Ferlet, R., Roueff, E., Czarny, J., and Felenbok, P. 1986, *Astr. Ap.*, **168**, 259.
Ferlet, R., Roueff, E., Horani, M., and Rostas, J. 1983, *Astr. Ap.*, **125**, L5.
Flower, D.R. and Launay, J.M. 1985, *M. N. R. A. S.*, **214**, 271.
Frisch, P.C., and Jura, M. 1980, *Ap. J.*, **242**, 560.
Gerola, H. and Glassgold, A.E. 1978, *Ap. J. Suppl.*, **37**, 1.
Glassgold, A.E., Huggins, P.J., and Langer, W.D. 1985, *Ap. J.*, **290**, 615.
Glassgold, A.E. and Langer, W.D. 1974, *Ap. J.*, **193**, 73.
Glassgold, A.E. and Langer, W.D. 1976, *Ap. J.*, **206**, 85.
Gondhalekar, P.M., Phillips, A.P., and Wilson, R. 1980, *Astr. Ap.*, **85**, 272.
Gredel, R. and Münch, G. 1986, *Astr. Ap.*, **154**, 336.
Hawkins, I. and Jura, M. 1987, *Ap. J.*, **317**, 926.
Herbig, G.H. 1987, preprint.
Herbst, E. 1982, *Ap. J.*, **252**, 810.
Herbst, E., DeFrees, D.J., and McLean, A.D. 1987, *Ap. J.*, **321**, 898.
Herbst, E. and Klemperer, W. 1973, *Ap. J.*, **185**, 505.
Hobbs, L.M., Black, J.H., and van Dishoeck, E.F. 1983, *Ap. J. (Letters)*, **271**, L95.
Hobbs, L.M., Blitz, L., and Magnani, L. 1986, *Ap. J. (Letters)*, **306**, L109.
Hobbs, L.M., Blitz, L., Penprase, B., Magnani, L., and Welty, D. 1988, *Ap. J.*, in press.
Hollenbach, D. and Salpeter, E.E. 1971, *Ap. J.*, **163**, 155.
Huntress, W.T. 1977, *Ap. J. Suppl.*, **33**, 495.
Iglesias, E. 1977, *Ap. J.*, **218**, 697.
Jannuzi, B.T., Black, J.H., Lada, C.J., and van Dishoeck, E.F. 1988, *Ap. J.*, submitted.
Jenkins, E.B. *et al.* 1988, in preparation.
Jura, M. 1975, *Ap. J.*, **197**, 581.
Kirby, K and Goldfield, E. 1988, in preparation.
Knapp, G.R. and Bowers, P.F. 1988, *Ap. J.*, submitted.
Kramers, H.A. and ter Haar, D. 1946, *Bull. Astr. Inst. Neth.*, **10**, nr 371.
Lada, E.A. and Blitz, L. 1988, *Ap. J. (Letters)*, in press.
Lambert, D.L. and Danks, A.C. 1986, *Ap. J.*, **303**, 401.
Langer, W.D., Glassgold, A.E., and Wilson, R.W. 1987, *Ap. J.*, **322**, 450.
Le Bourlot, J., Roueff, E., and Viala, Y. 1987, *Astr. Ap.*, **188**, 137.

Léger, A. and Puget, J.L. 1984, *Astr. Ap.*, **137**, L5.
Lepp, S., Dalgarno, A., van Dishoeck, E.F., and Black, J.H. 1988, *Ap. J.*, in press.
Letzelter, C., Eidelsberg, M., Rostas, F., Breton, J., and Thieblemont, B. 1987, *Chem. Phys.*, **114**, 273.
Liszt, H.S. 1979, *Ap. J. (Letters)*, **233**, L147.
Luine, J.A. and Dunn, G.H. 1985, *Ap. J. (Letters)*, **299**, L67.
Lutz, B.L. and Crutcher, R.M. 1983, *Ap. J. (Letters)*, **271**, L101.
Magnani, L., Blitz, L., and Mundy, L. 1985, *Ap. J.*, **295**, 402.
Magnani, L., Blitz, L., and Wouterloot, J.G.A. 1988, *Ap. J.*, in press.
Mann, A.P.C. and Williams, D.A. 1984, *M. N. R. A. S.*, **209**, 33.
Mann, A.P.C. and Williams, D.A. 1985, *M. N. R. A. S.*, **214**, 279.
Marquette, J.B., Rowe, B.R., Dupeyrat, G., and Roueff, E. 1985*a*, *Astr. Ap.*, **147**, 115.
Marquette, J.B., Rowe, B.R., Dupeyrat, G., Poissant, G., and Rebrion, C. 1985*b*, *Chem. Phys. Lett.*, **122**, 431.
Michels, H.H. and Hobbs, R.H. 1984, *Ap. J. (Letters)*, **286**, L27.
Millar, T.J. and Hobbs, L.M. 1988, *M. N. R. A. S.*, in press.
Mitchell, J.B.A. 1987, *Phys. Rev.*, in press.
Monteiro, T.S. and Flower, D.R. 1987, *M. N. R. A. S.*, **228**, 101.
Nercessian, E., Benayoun, J. J., and Viala, Y.P. 1988, *Astr. Ap.*, in press.
Omont, A. 1986, *Astr. Ap.*, **164**, 159.
Palazzi, E., Mandolesi, N., and Crane, P. 1988, *Ap. J.*, in press.
Pineau des Forêts, G., Flower, D.R., Hartquist, T.W., and Dalgarno, A. 1986, *M. N. R. A. S.*, **220**, 801.
Pwa, T.H. and Pottasch, S.R. 1986, *Astr. Ap.*, **164**, 116.
Roberge, W.G. 1981, Ph. D. thesis, Harvard University.
Roberge, W.G., Dalgarno, A., and Flannery, B.P. 1981, *Ap. J.*, **243**, 817.
Schaefer, J. 1985, in *Molecular Astrophysics*, eds. G.H.F. Diercksen *et al.*, NATO ASI Series **157** (Reidel, Dordrecht), p. 497.
Schinke, R., Engel, V., Buck, U., Meyer, H., and Diercksen, G.H.F. 1985, *Ap. J.*, **299**, 939.
Snow, T.P. 1980, in IAU Symposium **87**, *Interstellar Molecules*, ed. B.H. Andrew (Reidel, Dordrecht), p. 247.
Snow, T.P., Seab, C.G., and Joseph, C.L. 1988, preprint.
Solomon, P.M. and Klemperer, W. 1972, *Ap. J.*, **178**, 389.
Somerville, W.B. 1987, in IAU Symposium **120**, *Astrochemistry*, eds. M.S. Vardya and S.P. Tarafdar (Reidel, Dordrecht), p. 133.
Souza, S.P. and Lutz, B.L. 1977, *Ap. J.*, **216**, L49.
Stecher, T.P. and Williams, D.A. 1974, *M. N. R. A. S.*, **168**, 51P.
Tarafdar, S.P., Prasad, S.S., Huntress, W.T., Villere, K.R., and Black, D.C. 1985, *Ap. J.*, **289**, 220.
van Dishoeck, E.F. 1984, Ph. D. thesis, University of Leiden.
van Dishoeck, E.F. 1987*a*, in IAU Symposium **120**, *Astrochemistry*, eds. M.S. Vardya and S.P. Tarafdar (Reidel, Dordrecht), p. 51.
van Dishoeck, E.F. 1987*b*, *J. Chem. Phys.*, **86**, 196.
van Dishoeck, E.F. 1988, to appear in *Millimetre and Submillimetre Astronomy*, eds. R.D. Wolstencroft and W.B. Burton (Reidel, Dordrecht).
van Dishoeck, E.F. and Black, J.H. 1982, *Ap. J.*, **258**, 533.
van Dishoeck, E.F. and Black, J.H. 1986*a*, *Ap. J. Suppl.*, **62**, 109 (vDB).
van Dishoeck, E.F. and Black, J.H. 1986*b*, *Ap. J.*, **307**, 332.
van Dishoeck, E.F. and Black, J.H. 1988*a*, *Ap. J.*, submitted.
van Dishoeck, E.F. and Black, J.H. 1988*b*, in preparation.
van Dishoeck, E.F. and de Zeeuw, T. 1984, *M. N. R. A. S.*, **206**, 383.
Viala, Y.P. 1986, *Astr. Ap. Suppl.*, **64**, 391.
Viala, Y.P., Letzelter, C., Eidelsberg, M., and Rostas, F. 1988, *Astr. Ap.*, in press.
Viala, Y.P., Roueff, E., and Abgrall, H. 1988, *Astr. Ap.*, in press (VRA).
Vidal-Madjar, A. and Gry, C. 1984, *Astr. Ap.*, **138**, 285.
Wagenblast, R. and Hartquist, T.W. 1988, priv. comm.
Wannier, P.G., Penzias, A.A., and Jenkins, E.B. 1982, *Ap. J.*, **254**, 100.
Watson, W.D. 1978, *Ann. Rev. Astr. Astrophys.*, **16**, 585.
Weiland, J.L., Blitz, L., Dwek, E., Hauser, M.G., Magnani, L., and Rickard, L. J. 1986, *Ap. J. (Letters)*, **306**, L101.
White, R.E. 1984, *Ap. J.*, **284**, 695.
Williams, D.A. 1984, in *Galactic and Extragalactic Infrared Spectroscopy*, eds. M.F. Kessler and J.P. Phillips (Reidel, Dordrecht), p. 59.
Yoshino, K., Stark, G., Smith, P.L., Parkinson, W.H., and Ito, K. 1988, *J. de Physique* in press; and in prep.

DENSE INTERSTELLAR CLOUD CHEMISTRY

Eric Herbst
Department of Physics
Duke University
Durham, NC 27706
USA

ABSTRACT. Significant advances in our understanding of the gas phase chemistry of dense interstellar clouds have occurred in the last few years. These advances include the delineation of the reaction pathways by which complex molecules are produced, a preliminary understanding of the varied chemistries existing in and near regions of star formation, an appreciation that photodissociation caused by internally generated photons is important, and a realization that PAH's affect the chemistry of less complex species. However, all of these advances and indeed all of our understanding of the chemistry of dense interstellar clouds is dependent on knowledge of rate coefficients for a variety of processes that occur under the unusual conditions present in these sources. Our knowledge of some types of reactive processes is sufficiently poor that model results dependent on these proceses are often of questionable validity. Highest on the list of poorly understood processes are dissociative recombination and neutral-neutral reactions. In the former type of reaction, the neutral product branching ratios are not known definitively, while in the latter it is unclear whether or not reactions involving radicals are inhibited at low temperature by small activation energy barriers. Further progress in understanding the chemistry of dense clouds depends on significant laboratory and/or theoretical advances in the treatment of these and other processes.

1. INTRODUCTION

Much of our understanding of the chemistry of dense interstellar clouds derives from chemical models of these sources. A chemical model of an interstellar cloud is simply the solution of coupled differential equations that contain the time derivatives of the concentrations of individual molecular species. Each individual differential equation relates the time dependence of a specific molecular concentration to its formation and depletion rates. For example, consider the formation and depletion of a molecular ion labelled C^+:

$$A^+ + B \text{ -----> } C^+ + D \tag{1}$$

$$C^+ + D \text{ -----> } E^+ + F . \tag{2}$$

The species is produced via one chemical reaction and destroyed by a second reaction. For

T. J. Millar and D. A. Williams (eds.), Rate Coefficients in Astrochemistry, 239–262.

this simple case, the differential or kinetic equation governing the concentration of C^+ as a function of time is

$$d[C^+]/dt = k_1[A^+][B] - k_2[C^+][D] \quad (3)$$

where the symbol [] refers to concentration and the k_i's are the so-called rate coefficients for the formation and depletion processes. To produce a model, one must write similar equations for each species in the model and then solve these coupled equations with certain constraints to yield concentrations as functions of time.

There is a hierarchy of complexity of chemical models. The simplest are the steady-state models, in which the time derivatives of the molecular concentrations are all set equal to zero. In the above case, this leads to the algebraic equation

$$[C^+]_{ss} = k_1[A^+]_{ss}[B]_{ss} / \{k_2 [D]_{ss}\} \quad (4)$$

where the subscript ss stands for steady state. Solution of coupled algebraic equations is far simpler than solution of coupled differential equations, so that it is no surprise that the first detailed, inclusive models of dense interstellar clouds were steady state in nature (Herbst and Klemperer 1973; Mitchell, Ginsburg, and Kunz 1978). However, one can argue for using a steady-state approach even in these days of increased computational power. The argument is that our knowledge of the history of physical conditions in dense clouds is minimal and any time-dependent approach must incorporate this history of physical conditions.

The early models were on the whole quite successful in reproducing observed abundances for a variety of the smaller interstellar molecules although there were and are some important discrepancies even among small species such as the abundances of atomic neutral carbon and molecular oxygen (Prasad *et al.* 1987). More recent models have also used the steady-state concept. Herbst (1983), Suzuki (1983), and Millar and Freeman (1984a,b) have studied the chemistry of small hydrocarbons and related compounds in dark clouds using this approach. Tielens and Hollenbach (1985) have undertaken a model calculation that simulates cloud edges where external photodissociation is important. Dalgarno and co-workers have performed a variety of steady-state model calculations in which topics such as internal, cosmic ray-induced photodissociation (Sternberg, Dalgarno, and Lepp 1987) and the chemical effect of PAH's (Lepp and Dalgarno 1987) are investigated. The steady-state concept has also been used extensively in incomplete model work in which a small set of reactions is decoupled from the remainder and used to investigate a particular chemical topic of interest. For example, the abundance ratio of a deuterated species to the normal species can be investigated in this manner (for a recent study of CCD/CCH, see Herbst *et al.* 1987).

Despite its successes, the steady-state concept suffers from a difficulty. Even if one can ignore changes in physical conditions, the time-scale for gas phase species sticking to grains is shorter by a considerable factor than the time-scale for reaching steady state so that some sort of grain desorption process (e.g., see Leger, Jura, and Omont 1985) is a requirement if the steady-state condition is ever to be achieved.

The next more complex type of model is the pseudo-time-dependent approach, pioneered by Prasad and Huntress (1980a,b) and later utilized by Langer and co-workers (Graedel, Langer, and Frerking 1982; Langer *et al.* 1984), Herbst, Leung and co-workers (Leung, Herbst, and Huebner 1984; Herbst and Leung 1986a,b), Millar and co-workers (Millar and Nejad 1985; Millar, Leung, and Herbst 1987; Millar *et al.* 1987), and Brown and Rice (1986a,b). In addition to these studies, which are concerned with gas phase processes, mention should be made of the model of d'Hendecourt *et al.* (1985) in which both gas phase and surface processes are included. In pseudo-time-dependent models, chemical concentrations evolve from initial values under fixed physical conditions. A common set of initial values derives from the measured fractional abundances (concentration of species divided by either total gas density or H_2 concentration) of the diffuse cloud ζ Oph, in which atomic abundances dominate except for hydrogen where there is a significant amount of the molecular form. Note also that the dominant form of carbon is the singly ionized atom - C^+. Pseudo time-dependent models show in general that for molecules of some degree of complexity, calculated fractional abundances peak at an early time (≈ 3(5) yrs.) before declining as steady state is reached (≈ 1(7) yrs.) Smaller molecules tend in general to show less of a temporal effect with abundances at early time being closer to steady-state abundances. Another key result of pseudo-time-dependent models is that large abundances of neutral atomic carbon are present at early time. Pseudo-time-dependent models represent an improvement over steady-state models because the chemistry can be investigated at eras before steady state is reached. Early time abundances are particularly interesting because it is important to realize that large abundances of complex molecules can be produced fairly quickly.

The most complex type of model is a time-dependent approach in which physical as well as chemical conditions evolve. Time-dependent methods involving both diffuse and dense clouds have been reviewed by Prasad *et al*. (1987) within the last year. Since no major models have appeared since the time of that review, the subject of time-dependent models will not be raised at length here. A brief history of the method is in order, however, especially as regards calculations involving dense interstellar clouds. The simplest approach is to allow the cloud initially under diffuse conditions to undergo a free-fall or delayed free-fall collapse concomitantly with chemical evolution. Some early treatments of this variety include those of Kiguchi *et al.* (1974) and Suzuki *et al.* (1976). A later unpublished treatment in this vein by myself and Chun Leung includes some complex molecules. Given the nature of the collapse, what transpires in these models is that the clouds spend a large percentage of their time under diffuse conditions before undergoing rapid collapse. The efficient formation of large molecules at early time does not occur because at so-called early time the cloud is still diffuse and pervaded by stellar radiation. Given the amount of matter under dense cloud conditions in the galaxy, it is not clear that this simple time dependence is at all realistic. More detailed treatments treat collapse hydrodynamically and include density and temperature variations in both space and time. Such a calculation was first undertaken by Gerola and Glassgold (1978), but has been brought to fruition with a much more complex chemistry by Prasad, Tarafdar, Villere, and co-workers (Tarafdar *et al.* 1985; Prasad *et al.* 1987). The model of these scientists is sufficiently complex to require supercomputer implementation. After a certain amount of time which is quite model dependent, evolving clouds in this picture develop a core-envelope structure. Like the simpler free-fall models, the hydrodynamic models also show that clouds spend most of their time in the diffuse state before collapse becomes rapid. Dense clouds are not predicted to reach steady state chemically so that the early-time calculated abundances of the simpler pseudo-time-dependent models may be quite important if cloud collapse can be halted long enough at typical dense cloud densities.

All of these types of chemical models - steady-state, pseudo-time-dependent, and fully time-dependent - are themselves dependent on rate coefficients which are often highly uncertain and which need to be determined by experimental and/or theoretical techniques before the models can be put on a sound basis. The approach in this review will be to demonstrate this assertion by repeated references to it in the course of discussions of a significant number of models and less detailed chemical treatments that have been published recently. Basic reviews of some of the gas phase processes utilized in dense interstellar cloud models have been given recently (Herbst 1985; 1987a; Winnewisser and Herbst 1987). A general discussion of gas phase pathways to complex molecules is given by Bohme in this volume.

2. FORMATION OF COMPLEX MOLECULES

In recent years, several different pseudo-time-dependent models have been performed to elucidate the gas phase pathways to the formation of complex molecules in dense clouds. The models of Millar and Nejad (1985) and of Leung, Herbst, and Huebner (1984) (as updated by Herbst and Leung 1986a,b) show strong disagreements in the calculated abundances of complex molecules. This is shown most graphically in Figure 1 where calculated results for the fractional abundance of HC_3N as a function of time are depicted for physical conditions that pertain to dark clouds such as TMC-1. For all of the model results shown in Figure 1, the abundance at early time is far in excess of the steady-state value. This is due primarily to the high abundance of neutral atomic carbon calculated to be present at early time; the neutral carbon can be "fixed" into complex organic species (Herbst 1983,1985) although the reactions between C and other species thought to be involved in this process have not been studied in the laboratory. The actual measured fractional abundance of HC_3N in TMC-1 is 6(-9) (Irvine, Goldsmith, and Hjalmarson 1987); this is in reasonable agreement with the Millar and Nejad (1985) model at steady state but with the Herbst and Leung (1986a,b) models at early time.

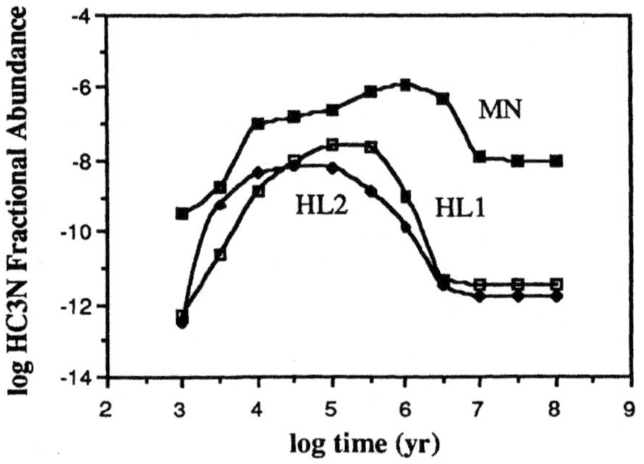

Figure 1. A plot of calculated fractional abundances for HC_3N versus time for three different models of TMC-1. MN refers to Millar and Nejad (1985), whereas HL1 and HL2 refer to Herbst and Leung (1986a,b) respectively.

Before a discussion of the discrepancy between these calculations, it is of interest to ask how secure is the calculated difference between early-time and steady-state results. The low abundance of HC_3N at steady state is due to the difficulty of forming this molecule in an environment in which almost all of the available carbon is predicted to be in the form of CO. However, the dominant process that converts neutral atomic carbon into CO is the neutral-neutral reaction

$$C + O_2 \text{ -----> } CO + O . \qquad (5)$$

This reaction has only been studied in the laboratory at room temperature via flash photolysis of carbon suboxide in a fairly complex system (Braun *et al.* 1969). If it possesses a small amount of activation energy, it will not proceed at low temperature and the amount of atomic carbon at steady state will be much larger than currently calculated. A larger abundance of C I will lead to larger abundances of complex molecules. It is therefore crucial that this reaction be studied at lower temperatures. Abundances at steady state for complex molecules that do not contain oxygen can also be increased if the elemental abundance of carbon is assumed to exceed that of oxygen in the gas phase (Langer *et al.* 1984; Watt 1985; Herbst and Leung 1986a).

Let us now return to the discrepancies among the various pseudo-time-dependent models. The two model results for HC_3N in Figure 1 labelled HL1 and HL2 refer to Herbst and Leung (1986a,b) respectively. The major difference between these two models is the use by Herbst and Leung (1986b) of the rapid rate coefficients for ion-polar neutral reactions as advocated by Adams, Smith, and Clary (1985). According to a number of theoretical treatments (see, e.g. Bates 1982, Clary 1988) that utilize the long-range attraction between ion and permanent dipole (and ignore shorter range effects) and some experimental evidence as well (Rowe 1988), the rate coefficients for reactions between ions and polar neutrals possess an inverse temperature dependence. For reactions between light ions such as H_3^+ and strongly polar neutrals such as HC_3N, the expected rate coefficient at 10 K can be as large as 1(-7) $cm^3\ s^{-1}$. Although there is both theoretical (Herbst 1986) and some recent experimental evidence (Clary 1988; Rowe 1988) that not all ion-polar reactions will behave in this manner, the evidence is strongly suggestive that most will. In fact, Herbst (1986) argues that ion-polar reactions that obey the long-range theoretical models at room temperature will obey these theories at lower temperatures. The effect on dense cloud interstellar chemistry is normally not great; polar neutrals such as HC_3N are reduced in abundance when large ion-polar rate coefficients are used by factors normally under one order of magnitude. In addition, as can be seen in Figure 1, the early-time abundances are reached somewhat earlier. The abundances of selected molecules are strongly affected by the use of large ion-polar neutral rate coefficients, however. In particular, protonated ions of polar neutrals have strongly enhanced abundances in relation to the neutrals (Millar *et al.* 1985)

As can be seen in Figure 1, the discrepancy between the HL1 and HL2 models is small compared with the several order-of-magnitude discrepancy between these models and that of Millar and Nejad (MN, 1985). The case of HC_3N is not unique; many small organic molecules in these models are also calculated by MN to have much higher abundances than calculated by HL1/2. Millar, Leung, and Herbst (1987) have studied these discrepancies in some detail and resolved them. They are not due to numerical errors in either calculation nor to differences in elemental abundances chosen. Rather they are due to a philosophical difference. HL1/2 tend to estimate reaction rates and products for gas phase reactions involving larger species based on analogous laboratory results for smaller species if they exist. MN tend to neglect these unstudied processes if they hinder the growth of organic molecules. The inclusion of a small number of unstudied reactions present in the HL1/2 models into the MN model results in good agreement for most species. In order of importance, the types of reactions which must be included in MN are

(i) $O + C_n$ -----> Products

(ii) $O + C_nH_m{}^+$ -----> Products

(iii) $AH^+ + e$ -----> channels other than $A + H$.

Reaction types (i) and (ii) are the more important. These are destructive processes in which reactive oxygen atoms hinder chain growth. Very little is known about reaction type (i) and HL1/2 preferred to include these because of what they perceived to be the reactive nature of the C_n species. However, the experimental evidence regarding activation energy for neutral-neutral reactions involving oxygen atoms and hydrocarbons is mixed (Cvetanovic 1987) and, in addition, there may be spin correlation problems that slow these processes down. It is to be noted that neither MN nor HL1/2 let the C_nH radicals react rapidly with atomic oxygen except for the case of CCH. Reaction type (ii) is of somewhat lesser importance in causing discrepancies between HL1/2 and MN; laboratory studies have only been conducted for hydrocarbon ions with one or two carbon atoms. These reactions proceed at room temperature but with rate coefficients typically an order of magnitude below the collisional rate. (A useful compendium including these and other measured ion-molecule reactions has been published by Anicich and Huntress 1986). What happens at lower temperatures is not well understood at present and more work certainly needs to be accomplished.

For selected dissociative recombination reactions involving protonated molecular ions AH^+, HL1/2 include more neutral product channels than MN, who on occasion only include the A + H channel. Including only this one channel means that protonation of the neutral species A by any of the abundant ions $H_3{}^+$, HCO^+, or H_3O^+ is, because of 100% recycling, not a destruction step and therefore leads to enhanced abundances for A. The question of the proper neutral branching ratios for dissociative recombination reactions is a vexing one; theoretical ideas on this subject are discussed by Bates and Herbst in this volume and their implications for calculated molecular abundances in this article in section 4. However, the astrochemical community eagerly awaits some definitive experimental studies.

For the particular case of HC_3N, there is one additional difference between MN and HL1/2; MN include the unstudied neutral-neutral reaction

$$N + C_3H_2 \text{ -----> } HC_3N + H \quad (6)$$

which is not included in HL1/2 because of the possibility of activation energy. This reaction dominates the formation of HC_3N in the MN model.

In summary, almost all of the differences between the models of MN and HL1/2 can be resolved by inclusion of (or in one case exclusion of) reactions in the MN model that have not yet been studied in the laboratory. It is somewhat frightening that these unmeasured rate coefficients can lead to several order-of-magnitude differences in calculated abundances!

2.1. A New Model

Despite the obvious problem of calculating the abundances of organic molecules in dense interstellar clouds uncovered by the discrepancies between the HL1/2 and MN models, Herbst and Leung (1987) are finishing a new pseudo-time-dependent model calculation in which significantly more complex species are included than heretofore. The new neutral species, listed in Table I, contain hydrocarbons through nine carbon atoms in complexity, and include cyanoacetylenes through HC_9N. No attempt has been made to

TABLE I NEW NEUTRAL MOLECULES ADDED TO HL MODEL

C_5	C_5H	C_5H_2	C_7N	HC_7N
C_6	C_6H	C_6H_2		
C_7	C_7H	C_7H_2	C_9N	HC_9N
C_8	C_8H	C_8H_2		
C_9	C_9H	C_9H_2		
CH_3C_4H		CH_3C_3N		
CH_3C_6H		CH_3C_5N		
		CH_3C_7N		

consider metastable isomers as independent species nor has much thought been given to the likely structure/s of the new radicals. The ion-molecule processes thought to be responsible for the production of complex species have in the main not yet been studied in the laboratory but are extensions of schemes discussed previously (see, e.g. Winnewisser and Herbst 1987 or Bohme 1988). Some important work on the hydrogenation of hydrocarbon ions has been done by McElvaney, Dunlap, and O'Keefe (1987) and on the chemistry of the complex hydrocarbon C_4H_2 by Dheandhanoo *et al.* (1986). Based on the work of these authors, one can envisage how more complex hydrocarbons might be synthesized although there are of course large gaps in our knowledge. A possible synthetic pathway from previously studied four-carbon species to C_6H follows:

$$C^+ + C_4H_2 \text{ -----> } C_5H^+ + H \quad ***** \tag{7}$$

$$C^+ + C_4H \text{ -----> } C_5^+ + H \tag{8}$$

$$C_5^+ + H_2 \text{ -----> } C_5H^+ + H \quad ***** \tag{9}$$

$$C_5H^+ + H_2 \text{ -----> no reaction} \quad ***** \tag{10}$$

$$C_5H^+ \; + \; C \; \text{----->} \; C_6^+ \; + \; H \tag{11}$$

$$C_6^+ \; + \; H_2 \; \text{----->} \; C_6H^+ \; + \; H \quad ***** \tag{12}$$

$$C_6H^+ \; + \; H_2 \; \text{----->} \; C_6H_2^+ \; + \; H \quad ***** \tag{13}$$

$$C_6H_2^+ \; + \; e^- \; \text{----->} \; C_6H \; + \; H, \tag{14}$$

where asterisks are used to signify that the reaction has been studied in the laboratory. The type of synthesis above is based on insertion via C^+ and C; its difficulty lies in hydrogenating the more complex ions once they are formed. Note, for example, that C_5H^+ cannot add more hydrogen atoms via normal ion-molecule reactions and the possibility of association reactions has not been investigated. The neutral species C_5H cannot be formed from C_5H^+ efficiently so that another type of synthesis is needed. This second type involves so-called condensation reactions between smaller hydrocarbon ions and neutrals; e.g.

$$CH_3^+ \; + \; C_4H \; \text{----->} \; C_5H_2^+ \; + \; H_2. \tag{15}$$

The synthesis of cyanoacetylenes is thought to proceed most rapidly via reactions involving nitrogen atoms and hydrocarbon ions (see for example Millar and Freeman 1984a; Winnewisser and Herbst 1987); some supportive laboratory evidence has been obtained by Federer *et al.* (1986) but far more is needed, especially involving the very unsaturated hydrocarbon ions important in dense clouds. A likely pathway to the synthesis of HC_7N via nitrogen atoms is

$$N \; + \; C_7H_2^+ \; \text{----->} \; HC_7N^+ \; + \; H \tag{16}$$

$$HC_7N^+ \; + \; H_2 \; \text{----->} \; H_2C_7N^+ \; + \; H \tag{17}$$

$$N \; + \; C_7H_3^+ \; \text{----->} \; H_2C_7N^+ \; + \; H \tag{18}$$

$$H_2C_7N^+ \; + \; e^- \; \text{----->} \; HC_7N \; + \; H \tag{19}$$

since some analogous ion-molecule reactions are known to proceed for the HC_3N case.

Radiative association reactions may be critical in the formation of somewhat more saturated hydrocarbons and nitrogen-containing species but there is little laboratory evidence for even the analogous ternary association reactions. One reaction for which there is ternary evidence (Bohme and Raksit 1985) is the radiative association

$$CH_3^+ \; + \; HC_3N \; \text{----->} \; C_4H_4N^+ \; + \; h\nu \tag{20}$$

which, if it proceeds efficiently, leads to CH_3C_3N via dissociative recombination.

The new results of Herbst and Leung are similar to their previous results in that complex molecules are only produced in reasonable abundance at early times rather than at steady state. In Table II some early-time results for a cloud resembling TMC-1 are listed along with observational results for TMC-1 taken from the compendia of Irvine, Goldsmith, and Hjalmarson (1987) and Cernicharo *et al.* (1987). The model results shown are obtained using rapid ion-polar neutral rate coefficients. Although preliminary, these results show that gas phase reactions can produce large amounts of complex molecules, even in the "pessimistic" limit of Herbst and Leung although there does appear to be a problem with the cyanoacetylenes, for which the calculated abundances become low

TABLE II COMPARISON OF HL MODEL AND OBSERVATION IN TMC-1

	Fractional Abundance With Respect to H_2 (Early Time)	
Molecule	Calculated	Observed
HC_3N	6(-9)	6(-9)
HC_5N	4(-10)	3(-9)
HC_7N	7(-11)	1(-9)
HC_9N	1(-11)	3(-10)
C_5H	3(-9)	4(-10)
C_6H	3(-9)	1(-9)
CH_3C_3N	1(-10)	5(-10)
CH_3C_4H	1(-9)	2(-9)

by over an order of magnitude for HC_9N. Whether this discrepancy is important or not depends on whether the pessimistic limit of Herbst and Leung is correct. If MN are closer to the truth and O atom reactions can be neglected in the main, then the abundances of complex molecules will certainly be boosted. Another possibility is the importance of neutral-neutral reactions such as process (6). More laboratory work is needed.

2.2. Some Additional Considerations

The pseudo-time-dependent models discussed above neglect the phenomenon of cosmic ray-induced photodestruction. As originally discussed by Prasad and Tarafdar (1983), cosmic rays generate energetic secondary electrons, which excite molecular hydrogen from the ground electronic state into higher states. The subsequent relaxation of excited H_2 produces ultra-violet radiation of sufficiently low wave lengths to dissociate many molecules. The work of Prasad and Tarafdar (1983) was directed at a possible explanation of the observed large neutral atomic carbon abundance in selected dense clouds via photodissociation of CO. How does the inclusion of cosmic ray-induced photodestruction processes affect models of dense interstellar clouds generally? Certainly, if enough ultra-violet photons are produced, the abundances of complex molecules will be seriously diminished. However, at small fluxes of ultra-violet radiation, it is not clear what will happen. In one of their pseudo-time-dependent models, Herbst and Leung (1986a) considered the effect of external photons on a dense cloud with a small visual extinction of $A_V = 5$. These scientists found that even with such a low value for the visual extinction, the effect of including photodissociation processes was small. Calculated abundances of complex molecules were slightly decreased at early time but slightly increased at steady state. This increase is presumably due to the photodissociation of CO into atomic C and O and the subsequent synthesis of complex molecules from additional atomic C. However, the photodissociation rate for CO used in the HL model was probably too high since it was assumed that at least some photodissociation proceeded via a continuum rather than via lines, which are easily self-shielded (see the discussion by van Dishoeck in this volume).

In a recent series of papers, Lepp, Dalgarno, and co-workers have attempted to determine the effect of cosmic ray-induced photodestruction on steady-state models of dense clouds. Calculation of the individual photodestruction rates is a prerequisite for such models and is not simple for the case of CO in which photodissociation is now known to proceed through lines and the coincidence between emission lines of H_2 and absorption lines must be considered. Even for molecules that undergo continuous photodissociation, cross sections and/or products are often unknown (Herbst and Leung 1986a). A final uncertain parameter in the rate calculations is the albedo of the grains.

Sternberg, Dalgarno, and Lepp (1987) first investigated the effect on dense cloud chemistry when cosmic ray-induced photodestruction is included for a variety of species other than CO. One would expect in this case that molecular abundances would decrease and that is what was found. For example, the calculated abundance of C_3H_2 decreased by approximately two orders of magnitude in a model with grains of high albedo. In a more recent paper, Gredel, Lepp, and Dalgarno (1987) have reinvestigated the effect of including the cosmic ray induced photodestruction of CO. These scientists show that at steady state the abundance of atomic carbon is increased far more dramatically than was originally estimated by Prasad and Tarafdar (1983). For a cloud at 30 K and with CO/H_2 of 1.5(-4), the C/CO ratio rises from 5(-6) to 5(-3). The latter number is an order of magnitude higher than the C/CO ratio achieved at steady state in the model of Herbst and Leung (1986a) with $A_V = 5$ and should result in increased calculated abundances of complex molecules at steady state although it does not appear that the distinction between early time and steady state will be erased. Millar (1987) is currently performing pseudo-time-dependent model calculations with internal photodissociation rate coefficients including that for CO. A proper inclusion of cosmic ray-induced photodestruction into dense cloud models awaits a large body of accurate rates.

Another factor increasing the already large uncertainty in calculated abundances of "complex" molecules is the possible existence of truly complex molecules. In particular, if species as complex as PAH's are present in reasonable abundance, they will affect the dense cloud chemistry to a considerable extent (Omont 1986). Lepp and Dalgarno (1987) have investigated the effect of PAH's or similar complex molecules on ordinary dense cloud chemistry under steady-state conditions. According to these authors, some of the principal processess involving PAH's can be written as:

$$M^+ + PAH \longrightarrow PAH^+ + M \quad \text{charge transfer} \quad (21)$$

$$e^- + PAH \longrightarrow PAH^- + h\nu \quad \text{radiative attachment} \quad (22)$$

$$PAH^+ + PAH^- \longrightarrow PAH \quad \text{neutralization} \quad (23)$$

$$M^+ + PAH^- \text{-----}> M + PAH \quad \text{neutralization} \quad . \qquad (24)$$

where M^+ represents an ordinary positive ion (atomic or molecular). Lepp and Dalgarno (1987) assume that charge transfer can occur at the Langevin rate and that radiative attachment and neutralization can be exceedingly efficient. (The additional possibility of dissociative neutralization is considered by Bates and Herbst in this volume.) Some of the results of Lepp and Dalgarno (1987) are shown in Table III below for an assumed

TABLE III INFLUENCE OF PAH'S ON STEADY-STATE CHEMISTRY

	Calculated Fractional Abundances	
Species	$f_{PAH} = 0$	$f_{PAH} = 2(-7)$
PAH^-		6.8(-8)
PAH		1.3(-7)
PAH^+		1.2(-9)
e^-	4.4(-8)	3.8(-9)
C	6.2(-10)	2.2(-7)
C_2H	1.0(-10)	2.0(-9)
C_3H_2	4.5(-12)	6.0(-10)

fractional PAH abundance f_{PAH} of 2(-7). It can be seen that the rapid radiative attachment leads to a large fractional abundance for the negative ion PAH^-, a fact originally recognized by Omont (1986); indeed this species is more than an order of magnitude more abundant than e^-. The fact that the dominant carrier of negative charge becomes PAH^- promotes the importance of $M^+ + PAH^-$ reactions compared with ion-electron recombination processes. In particular, if M^+ is an atomic ion such as C^+, recombination with electrons is particularly slow since it occurs radiatively whereas neutralization with PAH^- is rapid. The result is a large increase in the products of such reactions; e.g., neutral carbon. As can be seen in Table III, C increases by 2.5 orders of magnitude. This large increase in turn promotes the formation of "complex" organic molecules such as C_3H_2. Although Lepp and Dalgarno (1987) only studied the effect of PAH's in a small steady-state model, it would be of considerable interest to generalize their treatment to larger pseudo-time-dependent models involving more complex molecules such as the recent model by Herbst and Leung. Perhaps the inclusion of *both* cosmic ray-induced photodestruction and PAH chemistry will lead to obliteration of much of the distinction between steady state and early time.

3. THE CHEMISTRY OF NEWLY DETECTED SPECIES

Many new interstellar molecules have been detected recently, chiefly at the IRAM and Nobeyama facilities, and new gas phase syntheses have been proposed to explain their abundances. The syntheses of the newly observed radicals C_5H and C_6H have already been discussed. Below we consider syntheses for the cyclic molecules C_3H and C_3H_2, the organo-sulfur species CCS and CCCS, and PN.

3.1. Cyclic Molecules

The basic problem in synthesizing cyclic molecules via gas phase reactions is determining which reactions actually lead to what chemists call "ring closure." Since it is assumed that dissociative recombination reactions do not lead to major structural changes of the heavy neutral fragments, the ionic precursors of cyclic neutrals are presumably cyclic ions. For C_3H and C_3H_2 in particular, the cyclic precursor $C_3H_3^+$ has been proposed (Thaddeus, Vrtilek, and Gottlieb 1985; Yamamoto, Saito, Ohishi *et al.* 1987; Adams and Smith 1987). This ion is known to be more stable than its non-cyclic isomer. Recently ion-molecule groups have managed to determine the structure/s of the product $C_3H_3^+$ ion in several reactions by testing its chemical reactivity and found that the cyclic isomer can indeed be formed from non-cyclic precursors. The most important interstellar production route of this ion is probably the association reaction

$$C_3H^+ + H_2 \text{ -----> } c\text{-}C_3H_3^+ \tag{25}$$

which has been studied by Adams and Smith (1987) under three-body conditions and shown to produce the cyclic isomer on approximately 50% of association collisions. In interstellar clouds, the mechanism for reaction (25) is radiative, but radiative association has not been studied extensively in the laboratory. Since reaction (25) is thought to be the dominant production route to $C_3H_3^+$ and since previous models predict large C_3H_2 abundances without specifying its structure, it appears that gas phase models can account for c - C_3H_2 (Herbst, Adams, and Smith 1984). Other ion-molecule reactions leading to cyclic $C_3H_3^+$ of somewhat lesser interstellar importance are (Adams and Smith 1987; Bohme 1986)

$$CH_3^+ + C_2H_2 \text{ -----> } c\text{-}C_3H_3^+ + H_2 \tag{26}$$

$$C_3H^+ + CH_4 \text{ -----> } c\text{-}C_3H_3^+ + CH_2. \tag{27}$$

Although the production of c - C_3H_2 from c - $C_3H_3^+$ is presumed to occur in all current theories of the products of dissociative recombination reactions, the production of c - C_3H from this precursor ion is on shakier grounds since the recent theory of Bates does not normally lead to predictions that two hydrogen atoms are removed (Bates 1986, 1987). In this theory, the cyclic $C_3H_2^+$ ion, if it exists, would be a more likely precursor to c-C_3H. Adams and Smith (1987) have determined two ion-molecule reactions that produce cyclic $C_3H_2^+$. They are

$$C_3H^+ + H_2 \text{ -----> } c\text{-}C_3H_2^+ + H \tag{28}$$

and

$$C^+ + C_2H_4 \text{ -----> } c - C_3H_2^+ \ (10\%) + H_2 \ . \qquad (29)$$

The first reaction is slightly endothermic (Herbst, Adams, and Smith 1984) and cannot proceed under normal interstellar conditions if the reactants are truly thermalized. However, it appears that only a small amount of collision energy (≈ 25 meV) is needed to turn the reaction on so that if the C_3H^+ ion is formed with significant translational energy, reaction (28) may well compete with elastic cooling collisions between this ion and H_2 to produce a sufficient quantity of c - $C_3H_2^+$. An analogous situation involving a slightly endothermic reaction in the synthesis of ammonia is discussed in Section 6. However, there is insufficient information in the case of reaction (28) to provide a definitive answer. Reaction (29) leads to cyclic $C_3H_2^+$ on only 10% of reactive collisions and does not appear to be fast enough to produce the required amount of c - $C_3H_2^+$. According to theoretical models of Herbst and Leung, the dominant formation mechanism for $C_3H_2^+$ involves reactions between atomic carbon and hydrocarbon ions such as $C_2H_3^+$ and $C_2H_4^+$. It would be of interest to measure whether or not ring closure occurs for these systems. Unfortunately, no reactions between carbon atoms and molecular ions have ever been studied in the laboratory, to the best of our knowledge. Other ion-molecule reactions that produce cyclic ions are sure to be discovered in the near future.

3.2. Organo-sulfur Species

The unusual linear radicals CCS and CCCS, recently detected in the interstellar medium (Yamamoto, Saito, Kawaguchi *et al.* 1987; Saito *et al.* 1987), are surprisingly abundant compared with their oxygen counterparts despite the fact that there is less sulfur than oxygen. One explanation for this has been supplied by Smith *et al.* (1987) who postulate that the atomic ion S^+, abundant because it does not react with H_2 unlike its oxygen counterpart O^+, leads to these species by reactions with assorted hydrocarbons. In particular, the well-studied reaction

$$S^+ + C_2H_2 \text{ -----> } HC_2S^+ + H \qquad (30)$$

leads to an ion precursor of CCS:

$$HC_2S^+ + e \text{ -----> } CCS + H \qquad (31)$$

Smith *et al.* (1987) estimate that this process produces a fractional abundance for CCS of 1(-9) at early time which is only somewhat lower than the observed value of 8(-9) in TMC-1. The calculated number is dependent on poorly understood parameters such as the elemental abundance of sulfur in the gas phase.

It is not clear whether a similar mechanism can produce CCCS since the analog of reaction (30)

$$S^+ + C_3H_2 \text{ -----> } HC_3S^+ + H \qquad (32)$$

has not been studied in the laboratory. Unless there is a significant amount of *non* cyclic C_3H_2 in dense interstellar clouds, it is necessary to use the cyclic form of this hydrocarbon in which case it is unclear whether ring opening can occur. The reaction between S^+ and

CH_3CCH has been studied in the laboratory by Smith *et al.* (1987) who find that the sulfur insertion channel

$$S^+ + CH_3CCH \text{ -----> } H_3C_3S^+ + H \qquad (33)$$

occurs on 60% of the reactive collisions. However, it is not clear that the ion formed can lead to the much less saturated CCCS via dissociative recombination.

Smith *et al.* (1987) studied a number of other reactions between S^+ and larger hydrocarbons and found that sulfur insertion is often a minor or non-existent channel and that charge exchange is often a dominant one. However, these scientists did not study hydrocarbons as unsaturated as those found in dense interstellar clouds (viz. C_4H, C_4H_2) since they are difficult to produce in the laboratory. Unless these unsaturated species react differently from their more saturated relatives, it does not appear that the S^+ - hydrocarbon route will lead to a rich organo - sulfur chemistry in dense clouds. An alternative mechanism to be studied involves reactions between sulfur atoms and hydrocarbon ions.

3.3. PN

Recently observed unambiguously in the interstellar medium (Turner and Bally 1987; Ziurys 1987), PN has been thought to be produced via a non-standard chemistry due to the work of Thorne *et al.* (1984) who showed that this species cannot be produced via ion-molecule reactions. Instead these authors found that PO is produced efficiently, a result in some conflict with negative observations. However, as pointed out by Millar, Bennett, and Herbst (1987), it is entirely possible that PO is destroyed by a neutral-neutral reaction with atomic nitrogen

$$PO + N \text{ -----> } PN + O \text{ and/or } NO + P \qquad (34)$$

which has two possible sets of exothermic products, PN + O and NO + P. The PN channel is analogous to the products of the well-studied rapid reaction between NO and N:

$$NO + N \text{ -----> } N_2 + O \qquad (35)$$

If PO and N react rapidly as well and if PN is formed on even a small fraction of the reactive collisions, then the interstellar abundance of PN can easily be accounted for according to the pseudo-time-dependent calculations of Millar, Bennett, and Herbst (1987). Indeed, with an elemental abundance 100x below the standard cosmic abundance of phosphorus and a branching ratio of 25 % for PN + O production in reaction (34), these authors calculate a fractional abundance for PN of 1(-9) at both early time and steady state for a cloud with gas density 2(5) cm^{-3} and temperature 50 K. The observed fractional abundance for PN is 1(-10). Of course, reaction (34) is a neutral-neutral reaction which may possess a small activation energy, which might make it too slow to be of importance in cold clouds but more important in warmer ones. Interestingly enough, PN appears to be concentrated towards warmer regions of giant molecular clouds; negative observations in dark clouds yield an upper limit to the PN fractional abundances an order of magnitude below the observed value in warm regions (Turner *et al.* 1987). Besides possible activation energy for process (34), another explanation of this phenomenon is that only in the warmer regions is enough phosphorus driven off the grains to lead to the production of PN.

4. THE PROBLEM OF DISSOCIATIVE RECOMBINATION

In most recent models of dense clouds, scientists have been guided by the ideas of Green and Herbst (1979) in estimating the neutral product branching ratios of dissociative recombination reactions between polyatomic positive ions and electrons. A previous statistical approach by Herbst (1978) shares some of the predictions of the Green and Herbst (1979) work but is impractical for use in large model calculations. In the simplified approach of Green and Herbst (1979), the idea of Bates (1950) is used that when the electron attaches itself to the ionic species the neutral molecule must fragment rapidly to avoid reionization. Since pairs of fragments in which one of the neutrals is a hydrogen atom separate most rapidly, it was assumed that these are the dominant channels. It was also assumed that different hydrogen atoms can be ejected with equal efficiency. Were this idea accepted totally by modelers, there would be no destruction of neutral molecules via reaction with protonated ions such as HCO^+ since protonation of the neutral species would be followed by dissociative recombination which would lead back to the initial neutral (or to an isomer if a different hydrogen were ejected). To avoid this somewhat artificial method of boosting the calculated abundances of neutral species, most modelers have allowed for alternative neutral products in dissociative recombination reactions. A standard approach is to allow for the ejection of more than one hydrogen atom at a time: viz.,

$$AH_2^+ + e^- \longrightarrow AH + H; \; A + 2H\,(H_2) \qquad (36)$$

if the process is exothermic; such channels are also allowed in the earlier statistical theory of Herbst (1978).

Bates (1986, 1987) has recently advocated a very different method of estimating the likely products of dissociative recombination reactions (see the discussion by Bates and Herbst in this volume). In this view, a polyatomic ion is considered to be formed by localized valence bonds between atoms of well-described valence. The bonds involving the charged atom or atoms are converted into anti-bonds by the incoming electron and then rupture. Normally only one bond can be ruptured at a time, ruling out processes that lead to the A + 2H (H_2) channel of reaction (36) unless they occur subsequently to the initial bond rupture. However, unlike previous approaches, even strong bonds between heavy atoms are susceptible to rupture if the atoms forming the bond contain positive charge. Therefore the approach of Bates is somewhat akin to what modelers have assumed; namely that polyatomic ions will often fragment into sets of different neutral products when reacting with electrons. On the other hand, in the Bates approach, the fragments will in general include smaller species than heretofore considered.

Presented with such differences in predicted neutral products of dissociative recombination reactions, modelers face a severe difficulty. The few experimental measurements of dissociative recombination neutral product branching ratios conducted heretofore have involved vibrationally excited polyatomic ions. Although these experiments on H_2O^+ (Vallee *et al.* 1987) and H_3^+ (Mitchell *et al.* 1983) are not in agreement with the predictions of Bates, there is strong theoretical evidence that vibrationally excited ions behave differently from cold ions. Experiments with cold ions are planned by several groups in the next year.

How sensitive are calculated abundances to dissociative recombination branching ratios? In order to answer this question, Millar *et al.* (1987) have attempted to use the Bates approach to estimate branching ratios for a wide variety of polyatomic ions, to incorporate the results into a pseudo-time-dependent model of dense clouds, and to compare calculated abundances with previously determined values. To determine dissociative recombination branching ratios, these scientists had first to ascertain the positive charge distributions on an assortment of polyatomic ions. This was achieved by two methods: *ab initio* calculations for ions of known structure and simple valence bond considerations for other ions. The *ab initio* quantal calculations ran into some difficulties because different basis sets and different methods of defining the charge on a given atom lead to different results (Millar *et al.* 1987). However, it was determined that if one added up the calculated charges on heavy atoms and the hydrogen atoms surrounding them to obtain so-called "group charges", then one got fairly consistent results from one basis set to another. For example, the ion protonated methyl amine, whose simple valence bond structure is as follows:

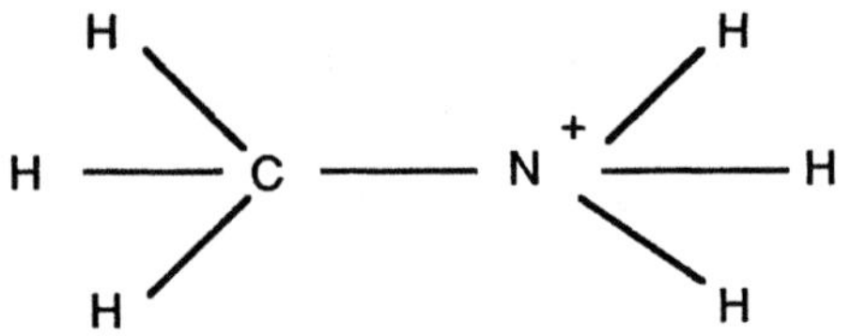

is calculated to have 1/4-1/3 of its positive charge on the methyl end. From the valence bond point of view, in which the positive charge is totally on N, one can only form CH_3NH_2 + H and CH_3 + NH_3 upon electron impact. The spreading out of the positive charge compared with the simple valence bond view leads to more fragment channels since the CH bonds can also be attacked upon electron impact. In this instance one can then expect further fragmentation if the large neutral species formed initially is unstable. Thus, initial formation of CH_2NH_3 + H might subsequently lead to CH_2 + NH_3 + H.

Although the adding of individual charges is useful to determine branching ratios for dissociative recombination reactions involving certain ions, it is perhaps too restrictive an approach for others. Consider the case of $HCNH^+$:

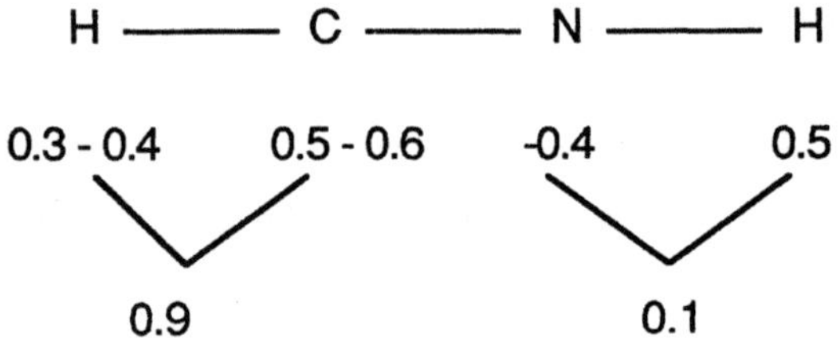

for which a limited range of *ab initio* individual atomic charges are shown as well as the overall group charges on the heavy atom-hydrogen atom combinations. If one only used these latter quantities, one would conclude that almost all of the positive charge is located on the HC portion of the molecule, so that the major neutral fragments are H + CNH and

HC + NH (although it is uncertain that this channel is exothermic). This analysis, leading to the lack of a rapid formation route to interstellar HCN, is perhaps too restrictive because despite their uncertainty the individual calculated charges on the NH end of the molecule show a significant dipole and a large positive charge on H. Electron attack on the NH bond would lead to HCN + H. In general, what Millar *et al.* (1987) have done is to give channels in which neutral products are formed based on the so-called group charges equal weights and to give product channels in which additional neutral products are formed based on atomic charges only (viz. HCN + H) somewhat lower weights (typically 1/2).

The detailed results of the Millar *et al.* (1987) model are quite complex and indicate some significant changes from earlier model results. The overall level of agreement between theory and observation is unchanged, however. It seems that a very important result using the Bates approach is that only one set of products occurs in the reaction between $H_3O^+ + e^-$:

$$H_3O^+ + e^- \longrightarrow H_2O + H \quad . \tag{37}$$

As also discussed by Sternberg, Dalgarno, and Lepp (1987), this results in a large increase in the calculated water abundance, especially at steady state, and a large decrease in the calculated OH abundance to below observed values. However, the OH abundance can be increased by including cosmic ray-induced photodissociation of H_2O in the model (Sternberg, Dalgarno, and Lepp (1987), at least at steady state.

As for the calculated abundances of more complex species, one would expect the additional fragmentation channels in the Bates approach to inhibit the growth of large species and to reduce their abundances. In general, this is found to be the case but only at steady state is the effect a large one. At early time, the synthetic power of the ion-molecule chemistry appears to be sufficient to withstand this assault, as can be seen in Table IV below. Despite the apparent insensitivity of some species to neutral product branching ratios at early time, it is still vitally important that the correct ratios be determined.

TABLE IV SENSITIVITY OF MOLECULAR ABUNDANCES TO CHANGES IN DISSOCIATIVE RECOMBINATION BRANCHING RATIOS

	New Abundance / Old Abundance	
Species	Early Time	Steady State
C_3	0.65	0.004
C_3H	0.95	0.05
C_3H_2	0.94	0.40
C_4	0.47	0.01
C_4H	0.50	0.56
HC_3N	0.65	8.3

As opposed to the products of dissociative recombination reactions, the rate coefficients of these processes have been studied extensively by a variety of techniques and are generally well known (see, e.g., Adams, Smith, and Alge 1984). There is one important exception to this trend, however, and it involves the reaction between H_3^+ and electrons. As discussed by Bates and Herbst in this volume, a rapid dissociative recombination rate coefficient under thermal conditions requires a curve crossing between the potential surface of the parent ion and a repulsive curve of the product neutrals somewhere near the equilibrium position of the ion. According to theory (Michels and Hobbs 1984), such a crossing does not occur for H_3^+ and therefore this ion should not react rapidly with electrons unless it is located in an excited vibrational state whose wave function is appreciable near the crossing with the repulsive surface. Previous experiments involving the merged beam and stationary afterglow techniques showed the reaction to be rapid but were possibly vitiated by the presence of H_5^+ or some vibrationally excited H_3^+. More recent experiments by Smith and Adams (1984) using the FALP technique obtained a null result for this reaction; this result was originally interpreted to signify an upper limit of 2(-8) cm^3 s^{-1} at 95 K, more than an order of magnitude smaller than normal rates. More recent merged beam work (Mitchell 1987) has reduced the amount of vibrationally excited H_3^+ in their experiment and obtained a rate coefficient at about this value. However Smith (1988) now argues that his null result implies a much lower rate coefficient ($\approx$ 1(-11) cm^3 s^{-1}) than he has heretofore published. This continuing discrepancy has been added to by another experimental result by Amano and Nakanaga (1987) who have studied the spectroscopy of the ions HCO^+ and HOC^+ in a discharge of H_2 and CO and feel that their results are best explained if a rapid rate coefficient for $H_3^+ + e^-$ ($\geq$ 1(-7) cm^3 s^{-1}) is used. These authors maintain that their H_3^+ is not vibrationally excited.

5. CHEMICAL INHOMOGENEITY

The preceding discussion perhaps give the erroneous impression that dense interstellar clouds are homogeneous entities the chemistry of which can all be explained using one set of either constant or varying physical conditions. However, the most advanced evolutionary models (Tarafdar *et al.* 1985) now include spatial variations in density and temperature. That nature is far more complex than this however, can be seen immediately in the case of Orion, where the evolution has been punctuated by stellar and protostellar formation. A detailed discussion of the heterogeneity in this source near the Kleinmann-Low position is found in Blake *et al.* (1987) based on results from the Owens Valley millimeter-wave spectral survey. A similar discussion can be found in Irvine, Goldsmith, and Hjalmarson (1987). There is a relatively quiescent cloud or "extended ridge", characterized by a kinetic temperature of 50-60 K and a density of $\approx$1(5) cm^{-3}. This moderately high density region is located in a much larger and more rarefied molecular cloud and is what the previously discussed chemical models should be used for. In addition to this region, there is a spatially confined or "compact" ridge which is somewhat warmer (100-200 K) and is characterized by high abundances of large oxygen-rich molecules such as CH_3OH, $HCOOCH_3$, and CH_3OCH_3. It appears that this source is at the intersection of two other sources - the plateau source and the quiescent or ambient cloud ("extended ridge"). The plateau source is traversed by shock waves related to outflow from the embedded infra-red source IRc 2. There are wide differences between abundances in the plateau source and abundances elsewhere; in particular molecules containing silicon and sulfur have abundances raised by factors of more than 100. Finally there is the so-called "hot core" region which is heated to temperatures of perhaps 300 K by radiative heating from IRc 2. The hot core is also quite dense - n > 1(6) cm^{-3}. In the hot core are enhanced abundances of relatively saturated molecules such as C_2H_5CN as well as copious amounts of NH_3.

The latter three regions require separate models to explain their chemistry. An initial study of the oxygen-rich compact ridge chemistry has been made by Blake *et al.* (1987). These scientists suggest that the high abundances of oxygen-containing organic species are caused by special conditions; a conclusion born out by calculations of Herbst (1987b) who shows that normal ion-molecule processes under the "extended ridge" conditions are insufficiently rapid. In the approach of Blake *et al.* (1987), the oxygen enrichment is initiated by an outflow of water from the plateau source. The large abundance of water eventually cools to the point where it will radiatively associate efficiently with ions in the quiescent cloud such as CH_3^+:

$$CH_3^+ + H_2O \text{ -----> } CH_3OH_2^+ + h\nu \, . \qquad (38)$$

The protonated methanol can dissociatively recombine to form methanol or react with methanol to form protonated dimethyl ether:

$$CH_3OH_2^+ + CH_3OH \text{ -----> } (CH_3)_2OH^+ + H_2O, \qquad (39)$$

a precursor of dimethyl ether. Reaction (39) has been studied in the laboratory and must compete with an association reaction. Another possible route leads to protonated methyl formate:

$$CH_3OH_2^+ + H_2CO \text{ -----> } H_2COOCH_3^+ + H_2 \qquad (40)$$

A simple steady-state analysis appears to be sufficient to explain the observed abundances of most of the oxygen-containing organic species although the model overpredicts HCOOH.

The chemistry of the plateau source is probably even more complicated. Shock models have been used, especially to treat the sulfur chemistry (e.g. Hartquist, Oppenheimer, and Dalgarno 1980) although Blake *et al.* (1987) deduce that grain disruption, shock chemistry, grain mantle desorption, and normal gas phase chemistry all play a role in this source. The chemistry of the hot core source is somewhat simpler and has been the focus of a detailed time-dependent model by Brown, Charnley, and Millar (1987), which is discussed in this volume. The temperature in the hot core is high enough to desorb material from grain mantles and it is thought that the desorption of material previously hydrogenated on the grains is responsible for much of the uniqueness of this source. In addition, the high density leads to a low fractional ionization, which increases the time scale for ion-molecule processes.

Overall, the chemical complexity observed in Orion is bewildering but should not be cause for despair. It is true, however, that progress in its understanding is utterly dependent on knowledge of the rates of the myriad chemical processes that are occurring.

6. NON-THERMAL EFFECTS

For certain processes, it is important to take non-thermal effects into account. Excluding the case of shock chemistry where non-thermal effects are known to be strong (Graff and Dalgarno 1987), there are still cases in the ambient cloud medium where they should not be neglected. One case that has been well treated in the literature is the process of radiative association. Here it turns out that if the internal modes of the reactants are subthermally excited, as often happens for polar reactants, the temperature dependence and absolute value of the rate coefficient are quite different from the simple thermal case. A discussion is given in this volume by Bates and Herbst. What concerns us here is the existence of translationally excited products from one reaction and their ability to use this energy to undergo subsequent reactions that are slightly endothermic. Although there are several important cases, the best studied endothermic reaction involves N^+ and H_2:

$$N^+ + H_2 \longrightarrow NH^+ + H . \qquad (41)$$

This reaction is a crucial first step in the synthesis of ammonia, since the alternative first step:

$$N + H_3^+ \longrightarrow NH_2^+ + H \qquad (42)$$

has been calculated to possess considerable activation energy (Herbst, DeFrees, and McLean 1987). Although reaction (41) has been studied by a number of investigators, the low temperature CRESU technique (Rowe 1988 and references therein) probably leads to our best understanding of it at low temperatures. It appears that the zero-point endothermicity of reaction is 19 meV (Marquette 1987). However, rotational energy is quite efficient at powering this reaction and the amount of rotational energy available to H_2 as a function of temperature must be determined carefully. Hydrogen consists of two independent rotational stacks of energy levels - ortho H_2 (odd rotational quantum number J) and para H_2 (even rotational quantum number J) which, due to nuclear spin restrictions, do not convert under normal conditions. At low temperatures, ortho H_2 relaxes to its lowest (J=1) state whereas para H_2 relaxes to its lowest (J=0) state. Since the ortho H_2 still possesses rotational excitation, it reacts more rapidly with N^+ than does para H_2. Under interstellar conditions, however, where ion-molecule reactions can interconvert ortho and para forms, thermal equilibrium prevails and at low temperatures essentially all H_2 is in its J=0 state. Therefore laboratory and interstellar rate coefficients for process (41) will differ at low temperatures unless pure para H_2 is used in the laboratory. The complexity of this process has been sorted out via the CRESU technique and it now appears that reaction (41) is too slow to lead to the production of sufficient amounts of interstellar ammonia at temperatures under 20 K *if the N^+ is truly thermalized.*

It is known however that the N^+ that reacts via process (41) in interstellar clouds is not thermal. It is produced in the main via the reaction

$$N_2 + He^+ \longrightarrow N^{+*} + H + He . \qquad (43)$$

Assuming this reaction to occur as a long-range charge exchange that produces a short-lived N_2^+ which then dissociates, the N^+ ion departs with a significant amount of translational energy - 147 meV. When converted into relative translational energy between N^{+*} and H_2, this energy becomes 18.5 meV or 215 K, an amount very close to the zero-point endothermicity of process (41). As first discussed by Adams, Smith, and Millar (1984), this additional translational energy can speed up reaction (41). However, process (41) is not the only process that translationally excited N^{+*} can undergo with H_2; N^{+*} can also cool via elastic collisions with H_2:

$$N^{+*} + H_2 \text{ -----> } N^{+} + H_2 \qquad (44)$$

and the elastic scattering reduces the ability of N^+ to undergo subsequent reaction with H_2 since it possesses less energy with which to surmount the reaction endothermicity. The competition between elastic scattering and reaction has been treated by Yee, Lepp, and Dalgarno (1987) and by Herbst, DeFrees, and McLean (1987) although both treatments suffer to some extent because the definitive CRESU results on the reaction endothermicity and influence of H_2 rotational energy were not available. This competition has been looked at anew by Galloway and Herbst (1987) using the CRESU results and treating reaction (41) by the phase space approach since CRESU results are not available for arbitrary N^+ collision energies. Since the phase space calculations reproduce the measured rate coefficients from the CRESU experiments, one can use them with confidence. The tentative result of Galloway and Herbst (1987) is that for a 10 K cloud, the non-thermal mechanism allows roughly 11% of N^+ ions to react with H_2 before being cooled by it. This results in an ammonia abundance which is only a factor of ≈3 or so below what is observed in TMC-1.

As stated above, the endothermic reaction between N^+ and H_2 is not the only such slightly endothermic reaction of importance to the chemistry of dense interstellar clouds. Reaction (28) involving C_3H^+ and H_2 is another example. Unfortunately, for this and other reactions, insufficient information is known about the amount of translational energy possessed by the ion to permit a determination of the importance of the reaction.

7. SUMMARY

Although a significant amount of research has been accomplished in the field of dense interstellar cloud chemistry in recent years, progress has been limited by a lack of knowledge of important classes of gas phase reactions. Only when more information is available will models be able to determine the true importance and limitations of gas phase chemistry. The most critical types of reactions are discussed individually below:

(i) dissociative recombination reactions. Although in almost all cases the absolute rate coefficients of these processes are not in dispute, the neutral fragments have not been measured under thermal conditions nor have theoretical treatments arrived at unambiguous answers. It is critical that these neutral products be determined. The most important reactions involve the ion H_3O^+, in which the existence of a channel leading to OH is in doubt, and reactions involving complex organic ions, in which the extent of fragmentation into smaller neutral entities is poorly understood.

(ii) neutral-neutral reactions. In contrast to the ion-molecule case in which much progress has been made in studying reactions at low temperatures, there is little if any such work on rapid neutral-neutral reactions, such as occur between two radicals or between an atom and a radical. If these reactions possess even small activation energy barriers, they will not be rapid under low temperature interstellar conditions. Yet model calculations currently make use of these processes and they are quite important to the chemistry. For example, the reaction between C and O_2 is the major destroyer of atomic carbon under steady-state conditions.

(iii) atom-ion reactions. Reactions involving the atoms N, C, and O and a variety of molecular ions are critical to dense cloud chemistry. No reactions involving C have been studied and only a few involving N and O have been studied. These latter two atoms appear to react slowly with ions at room temperature and it is not clear how to extrapolate the results to lower temperatures. The O atom reactions are important in hindering the growth of molecular complexity whereas the C atom reactions are important in reinforcing this growth. The initial C atom reaction studied in the laboratory should be the one involving H_3^+, which is supposed to start a long chain of synthetic processes.

(iv) radiative association reactions. Discussed by Bates and Herbst in this volume, radiative association reactions are important in promoting molecular growth in dense clouds. Yet very little experimental information is available and theoretical treatments of the rate coefficients suffer from some severe difficulties.

Great strides have been made in the field of interstellar chemistry in the last fifteen years. Much more progress remains to be made and this progress will require a better understanding of all of the types of reactions discussed here.

ACKNOWLEDGMENTS

Support of this work by the National Science Foundation (U.S.) and the Science and Engineering Research Council (U.K.) is gratefully acknowledged.

REFERENCES

Adams, N. G. and Smith, D. 1987, *Ap. J. (Letters)*, **317**, L25.
Adams, N. G., Smith, D., and Alge, E. 1984, *J. Chem. Phys.*, **81**, 1778.
Adams, N. G., Smith, D., and Clary, D. C. 1985, *Ap. J. (Letters)*, **296**, L31.
Adams, N. G., Smith, D., and Millar, T. J. 1984, *M. N. R. A. S.*, **211**, 857.
Amano, T. and Nakanaga, T. 1987, *Ap. J.* (in press).
Anicich, V. G. and Huntress, W. T., Jr. 1986, *Ap. J. Suppl.*, **62**, 553.
Bates, D. R. 1950, *Phys. Rev.*, **78**, 492.
Bates, D. R. 1982, *Proc. Roy. Soc. London A*, **384**, 289.
Bates, D. R. 1986, *Ap. J. (Letters)*, **306**, L45.
Bates, D. R. 1987, in *Modern Applications of Atomic and Molecular Physics*, ed. A. E. Kingston (London: Plenum).
Blake, G. A., Sutton, E. C., Masson, C. R., and Phillips, T. G. 1987, *Ap. J.*, **315**, 621.
Bohme, D. K. 1986, *Nature*, **319**, 473.
Bohme, D. K. 1988, this volume.
Bohme, D. K. and Raksit, A. B. 1985, *M. N. R. A. S.*, **213**, 717.
Braun, W., Bass, A. M., Davis, D. D., and Simmons, J. D. 1969, *Proc. Roy. Soc. London A*, **305**, 107.
Brown, P. D., Charnley, S. B., and Millar, T. J. 1987, *M. N. R. A. S.* (in press).
Brown, R. D. and Rice, E. H. N. 1986a, *M. N. R. A. S.*, **223**, 405.
Brown, R. D. and Rice, E. H. N. 1986b, *M. N. R. A. S.*, **223**, 429.
Cernicharo, J., Guélin, M., Menten, K. M., and Walmsley, C. M. 1987, *Astr. Ap.*, **181**, L1.

Clary, D. C. 1988, this volume.
Cvetanovic, R. J. 1987, *J. Phys. Chem. Ref. Data*, **16**, 264.
Dheandhanoo, S., Forte, L., Fox, A., and Bohme, D. K. 1986, *Can. J. Chem.*, **64**, 641.
d'Hendecourt, L. B., Allamandola, L. J., and Greenberg, J. M. 1985, *Astr. Ap.*, **152**, 130.
Federer, W., Villinger, H., Lindinger, W., and Ferguson, E. E. 1986, *Chem. Phys. Letters*, **123**, 12.
Galloway, E. and Herbst, E. 1987, in preparation.
Gerola, H. and Glassgold, A. E. 1978, *Ap. J. Suppl.*, **37**, 1.
Graedel, T. E., Langer, W. D., and Frerking, M. A. 1982, *Ap. J. Suppl.*, **48**, 321.
Graff, M. M. and Dalgarno, A. 1987, *Ap. J.*, **317**, 432.
Gredel, R., Lepp, S., and Dalgarno, A. 1987, *Ap. J. (Letters)*, in press.
Green, S. and Herbst, E. 1979, *Ap. J.*, **229**, 121.
Hartquist, T. W., Oppenheimer, M., and Dalgarno, A. 1980, *Ap. J.*, **236**, 182.
Herbst, E. 1978, *Ap. J.*, **222**, 508.
Herbst, E. 1983, *Ap. J. Suppl.*, **53**, 41.
Herbst, E. 1985, in *Molecular Astrophysics*, eds. G. H. F. Diercksen *et al.* (Dordrecht: Reidel), p.237.
Herbst, E. 1986, *Ap. J.*, **306**, 667.
Herbst, E. 1987a, in *Interstellar Processes*, eds. D. J. Hollenbach and H. A. Thronson, Jr. (Dordrecht: Reidel), p.611.
Herbst, E. 1987b, *Ap. J.*, **313**, 867.
Herbst, E., Adams, N. G., and Smith, D. 1984, *Ap. J.*, **285**, 618.
Herbst, E., Adams, N. G., Smith, D., and DeFrees, D. J. 1987, *Ap. J.*, **312**, 351.
Herbst, E., DeFrees, D. J., and McLean, A. D. 1987, *Ap. J.*, **321**, 898.
Herbst, E. and Klemperer, W. 1973, *Ap. J.*, **185**, 505.
Herbst, E. and Leung, C. M. 1986a, *M. N. R. A. S.*, **222**, 689.
Herbst, E. and Leung, C. M. 1986b, *Ap. J.*, **310**, 378.
Herbst, E. and Leung, C. M. 1987, in preparation.
Irvine, W. M., Goldsmith, P. F., and Hjalmarson, Å 1987, in *Interstellar Processes*, eds. D. J. Hollenbach and H. A. Thronson, Jr. (Dordrecht: Reidel), p.561.
Kiguchi, M., Suzuki, H., Sata, K., Miki, S., Tominatsu, A., and Nakagawa, Y. 1974, *Publ. Astr. Soc. Japan*, **26**, 499.
Langer, W. D., Graedel, T. E., Frerking, M. A., and Armentrout, P. B. 1984, *Ap. J.*, **277**, 581.
Leger, A., Jura, M., and Omont, A. 1985, *Astr. Ap.*, **144**, 147.
Lepp, S. and Dalgarno, A. 1987, *Ap. J.* (in press).
Leung, C. M., Herbst, E., and Huebner, W. F. 1984, *Ap. J. Suppl.*, **56**, 231.
Marquette, J. B. 1987, private communication.
McElvaney, S. W., Dunlap, B. I., and O'Keefe, A. 1987, *J. Chem. Phys.*, **86**, 715.
Michels, H. H. and Hobbs, R. H. 1984, *Ap. J. (Letters)*, **286**, L27.
Millar, T. J. 1987, private communication.
Millar, T. J., Adams, N. G., Smith, D., and Clary, D. C. 1985, *M. N. R. A. S.*, **216**, 1025.
Millar, T. J., Bennett, A., and Herbst, E. 1987, *M. N. R. A. S.* (in press).
Millar, T. J., De Frees, D. J., McLean, A. D., and Herbst, E. 1987, *Astr. Ap.* (in press).
Millar, T. J. and Freeman, A. 1984a, *M. N. R. A. S.*, **207**, 405.
Millar, T. J. and Freeman, A. 1984b, *M. N. R. A. S.*, **207**, 425.

Millar, T. J., Leung, C. M., and Herbst, E. 1987, *Astr. Ap.*, **183**, 109.
Millar, T. J. and Nejad, L. A. M. 1985, *M. N. R. A. S.*, **217**, 507.
Mitchell, G. F., Ginsburg, J. L., and Kuntz, P. J. 1978, *Ap. J. Suppl.*, **38**, 39.
Mitchell, J. B. A. 1987, in a talk given at the 5th International Swarm Seminar, Birmingham, U. K., 29-31 July.
Mitchell, J. B. A., Forand, J. L., Ng, C. T., Levac, D. P., Mitchell, R. E., Mul, P. M., Claeys, W., Sen, A., and McGowan, J. Wm. 1983, *Phys. Rev. Letters*, **51**, 885.
Omont, A. 1986, *Astr. Ap.*, **164**, 159.
Prasad, S. S. and Huntress, W. T., Jr. 1980a, *Ap. J. Suppl.*, **43**, 1.
Prasad, S. S. and Huntress, W. T., Jr. 1980b, *Ap. J.*, **239**, 151.
Prasad, S. S. and Tarafdar, S. P. 1983, *Ap. J.*, **267**, 603.
Prasad, S. S., Tarafdar, S. P., Villere, K. R., and Huntress, W. T., Jr. 1987, in *Interstellar Processes*, eds. D. J. Hollenbach and H. A. Thronson, Jr. (Dordrecht: Reidel), p. 631.
Rowe, B. R. 1988, this volume.
Saito, S., Kawaguchi, K., Yamamoto, S., Ohishi, M., Suzuki, H., and Kaifu, N. 1987, *Ap. J. (Letters)*, **317**, L115.
Smith, D. 1988, this volume.
Smith, D. and Adams, N. G. 1984, *Ap. J. (Letters)*, **284**, L13.
Smith, D., Adams, N. G., Giles, K., and Herbst, E. 1987, submitted to *Astr. Ap.*
Sternberg, A., Dalgarno, A., and Lepp. S. 1987, *Ap. J.*, **320**, 676.
Suzuki, H. 1983, *Ap. J.*, **272**, 579.
Suzuki, H., Miki, S., Sata, K., Kiguchi, M., and Nakagawa, Y. 1976, *Progr. Theor. Phys. (Japan)*, **56**, 1111.
Tarafdar, S. P., Prasad, S. S., Huntress, W. T., Jr., Villere, K. R., and Black, D. C. 1985, *Ap. J.*, **289**, 220.
Thaddeus, P., Vrtilek, J. M., and Gottlieb, C. A. 1985, *Ap. J. (Letters)*, **299**, L63.
Thorne, L. R., Anicich, V. G., Prasad, S. S., and Huntress, W. T., Jr. 1984, *Ap. J.*, **280**, 139.
Tielens, A. G. G. M. and Hollenbach, D. 1985, *Ap. J.*, **291**, 722.
Turner, B. E. and Bally, J. 1987, *Ap. J. (Letters)*, **321**, L75.
Turner, B. E., Bally, J., Amano, T., Lee, S., and Feldman, P. A. 1987, a poster presented at *Molecular Clouds in the Milky Way and External Galaxies*, a conference at U. Massachusetts (Amherst), November 2-4.
Vallée, F., Gomet, J. C., Rowe, B. R., Quéffelec, J. L., and Morlais, M. 1987, in *Astrochemistry (Internal Astronomical Union Symposium 120)*, eds. M. S. Vardya and S. P. Tarafdar (Dordrecht: Reidel), p.29.
Watt, G. D. 1985, *M. N. R. A. S.*, **212**, 93.
Winnewisser, G. and Herbst, E. 1987, *Topics in Current Chemistry*, **139**, 119.
Yamamoto, S., Saito, S., Kawaguchi, K., Kaifu, N., Suzuki, H., and Ohishi, M. 1987, *Ap. J. (Letters)*, **317**, L119.
Yamamoto, S., Saito, S., Ohishi, M., Suzuki, H., Ishikawa, S.-I., Kaifu, N., and Murakami, A. 1987, *Ap. J. (Letters)*, **322**, L58.
Yee, J. H., Lepp, S., and Dalgarno, A. 1987, *M. N. R. A. S.*, **227**, 461.
Ziurys, L. M. 1987, *Ap. J. (Letters)*, **321**, L81.

HOT MOLECULAR CORES - A CASE FOR ACCRETION

Paul D. Brown, S.B. Charnley and T.J. Millar
Department of Mathematics
UMIST
P O Box 88
Manchester M60 1QD, UK

ABSTRACT. Models of dense interstellar clouds should include accretion of gas-phase species onto dust grains. By modelling the chemistry of a hot molecular core we demonstrate that accretion does occur and that reactions on the surface with hydrogen and deuterium atoms are important. Observations of hot cores show enhanced abundances of ammonia, methanol and deuterium-bearing species. These enhancements cannot be reproduced by gas-phase chemistry alone. We explain the abundances by allowing accretion to occur as a cold cloud undergoes an isothermal free-fall collapse. Atoms and radicals are saturated with hydrogen on the surface, and the processed mantle material is then returned to the gas-phase as a result of grain heating caused by the formation of a nearby star. Support is thereby given to the idea that mantle material is returned to the gas-phase sporadically, on a timescale of a million years, as a result of star formation, shocks etc., rather than by a continually acting desorption mechanism.

1. INTRODUCTION

Chemical models of dense interstellar clouds have grown in complexity in recent years (see Herbst, this volume). However, since molecules are widely observed in the gas phase, many modellers have chosen to ignore the effects of accretion. Gas phase atoms and radicals, upon collision with cold dust grains present in the interstellar medium, are expected to stick to the dust forming an icy mantle and thus leaving the gas phase. This is thought to occur on a timescale comparable to the expected lifetime of the clouds. Certain models have incorporated continually acting, efficient desorption mechanisms (d'Hendecourt et al. 1985); however, because of the uncertainties surrounding such processes, they have not been widely accepted. The recently discovered "hot core" regions provide a good case for study and indicate that substantial accretion onto grains has occurred, giving support to the idea that sporadic release mechanisms such as star formation, shocks,

T. J. Millar and D. A. Williams (eds.), Rate Coefficients in Astrochemistry, 263–270.

supernova flashes and so on, are responsible for returning material to the gas phase.

In 1983 Pauls et al. detected emission from metastable transitions of NH_3 in the Orion-KL region, and concluded that it came from a hot, dense region (T_{KIN} = 150 K, n = 10^7 cm^{-3}). They calculated the fractional abundance of ammonia to be ~ 10^{-5}, two orders of magnitude larger than the typical value in cold, dense clouds, and suggested that the enhancement arose from the evaporation of molecular mantles on interstellar grains.

Since then several other "hot core" sources have been discovered, among them W51, NGC 7538, W31c - all with characteristic enhancements of NH_3, CH_3OH and HDO. Hot cores may be formed in several ways, but that investigated here is of the type believed to be represented by the Orion Hot Core; a clump of dense gas left-over from the formation of a nearby star, and now heated by that star. The so-called "compact ridge" in Orion is a region of similar temperature and density but is thought to have formed by the interaction of the wind from IRc2 with the surrounding cloud.

The hot core region has been modelled by following the chemical evolution of a cloud as it undergoes an isothermal, free-fall collapse. The collapse is halted by the formation of a nearby star which heats the region, thus evaporating mantles from the warm grains and returning material to the gas phase. The model is described in section 2, the results are discussed in section 3, a note on the deuterium chemistry appears in section 4 and the general conclusions are made in the final section.

2. THE MODEL

The gas phase chemistry in the collapse phase will not be described here but is typical of cold, dense clouds (see Herbst, this volume and references therein). Accretion is allowed but, because of the uncertainty, no continually acting desorption mechanisms have been included. An atom, radical or molecule from the gas will collide with, and stick to, a dust grain at a rate given by:

$$\lambda_X = S_X n_X V_X \pi a^2 n_d \qquad \text{molecules cm}^{-3} \text{ s}^{-1}$$

where S_X is the sticking coefficient and is believed to be close to, if not, unity; n_X and n_d are the densities of the gas-phase species, X, and the dust grains respectively; πa^2 is the collisional cross section of a grain and V_X is the velocity of species X relative to the dust. This leads to an accretion timescale on the order of 3 x $10^9/n$ years, where n is the total H density. For a dense cloud this time is shorter than the time to reach chemical steady-state (Iglesias 1977).

Reactions on the surface have been limited to those with hydrogen atoms. This is because of:

(1) the high mobility of H atoms, which can quantum tunnel through surface potential barriers whereas heavier species may only move by the much slower process of thermal hopping;

and (2) the high abundance of H atoms. Even regions in which hydrogen is predominantly molecular it is expected that there will be sufficient atomic hydrogen to dominate the surface chemistry scheme. (Regions where this is not the case are under investigation).

This leads to the hydrogenation of O, C, N atoms and radicals to their saturated forms H_2O, CH_4 and NH_3, which along with accreted CO comprise the bulk of the mantle.

The gas and solid phase abundances are followed as a function of time as the cloud, of initial density n = 3000 cm^{-3}, undergoes an isothermal (T = 10 K), free-fall collapse.

The collapse is halted when the density reaches 10^7 cm^{-3}, after around 10^6 years. It is then assumed that some mechanism, associated with the formation of a nearby star, heats the grains, which then release the processed mantles back into the gas phase. The evaporation rate is determined by the vapour pressure of H_2O which comprises 60% of the mantle. The gas phase abundances in the hot core are then followed.

The parameters of the model are summarised in Table 1.

TABLE 1. Parameters of the model. Those parameters marked by an asterix (*) were constant throughout the collapse phase, the others varied as the density increased. All the parameters remained constant in the hot core phase.

	"collapse phase"	"hot core"
Kinetic Temperature (K)	10*	200
Cloud mass ($M_\odot$)	10*	10
Cloud radius (pc)	0.33	0.019
Cloud density (cm^{-3})	3×10^3	1×10^7
Visual extinction (mag.) to the centre.	1.64	~ 320

3. RESULTS AND DISCUSSION

Figure 1 shows the gas phase fractional abundance of certain species included in the model as a function of time. It can be seen from the diagram that there is a very steep fall in the abundances as accretion dominates. The effect is exaggerated over that determined for quiescent clouds because, in a free-fall collapse, the density rises rapidly in the latter stages, and the accretion timescale, proportional to 1/n, becomes very short. Magnetic fields and turbulence caused by rotation may slow the collapse. We have modelled cases in which the collapse is slowed by factors of 10 and 100 and found no qualitative difference from the results presented here. As already mentioned, the inclusion of accretion means that chemical steady-state is not reached, and the main gas phase species accreted onto the grains are O, C and N atoms, which then undergo reactions with hydrogen atoms. This leads to a mantle composition at the end of the collapse phase of:

H_2O	61.2%
CH_4	14.8%
NH_3	14.7%
CO	8.5%

This is similar to that found for cold, dense clouds (Brown and Charnley, 1988) and is in agreement with the fit of the 3 μm feature towards the BN object obtained by Knacke et al. (1982) using an ice mixture of water and ammonia in the ratio of 4:1. The CO is accreted from the gas phase; solid CO has been widely observed in the Taurus complex (Whittet et al., 1985).

At a grain temperature of 200 K the evaporation of the mantles is essentially instantaneous. There is no difference to the results for the case in which the grains are heated gradually. As can be seen from figure 1, hydrogenation in the grain mantle is an important process; the abundance of H_2O rises by two orders of magnitude over its abundance in the cold phase.

Other species which do not take part in the surface chemistry can also show enhancement with respect to their observed abundances in cold clouds. This is because their abundances, which are known from chemical models to show a peak at around 10^5 years, are effectively "frozen" by the molecules accreting onto the dust, and later being released. This may be the case for methanol, CH_3OH, which is overabundant in hot cores (Menten et al. 1986).

There is an absence of ions in the hot phase. The ions H_3^+ and He^+ react rapidly with the molecules returned to the gas phase and this is followed by dissociative recombination of the product ions. The density is sufficiently high that ionisation by cosmic rays is unimportant (Millar and Nejad, 1985) while X-rays and UV radiation can be neglected because of the very high extinction. Hence the abundance of ions in the hot core phase remains low.

The abundances of reactive atoms and radicals is also very small initially in the hot core phase since they have all been converted to stable hydrides on the surface. The chemistry in the hot core is thus dominated by neutral-neutral reactions - the timescale of which is very long, allowing the hot core phase abundances to survive for greater than 10^4 years.

The results are compared with the observations of the Orion Hot Core in Table II. From these results it can be concluded that the reactions

$$H + CN \dashrightarrow HCN$$

and $$H + C_2H \dashrightarrow C_2H_2$$

also occur on the surface, as one would expect since CN and C_2H have unsaturated bonds. There is also a good case for the surface

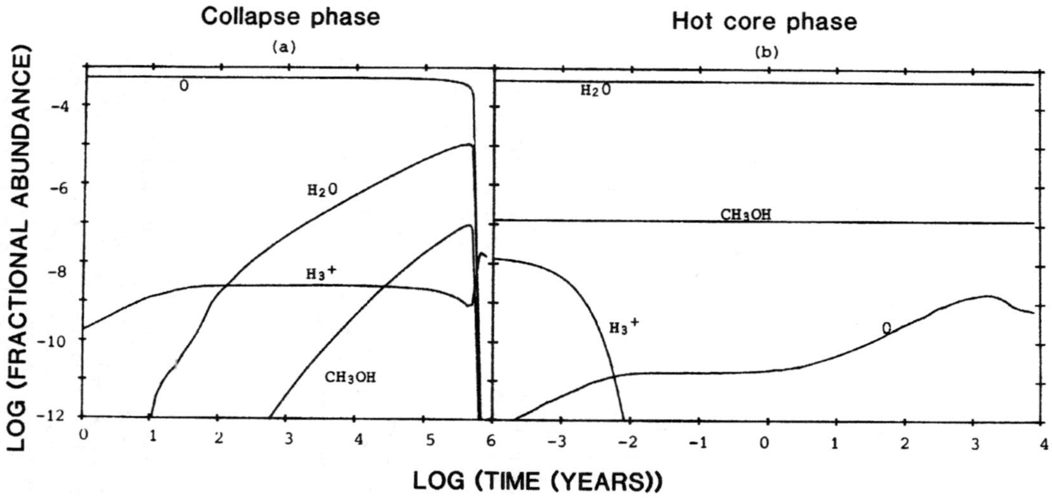

FIGURE 1. The fractional abundances of certain species included in the model are shown as a fraction of time. The sharp fall off in the collapse phase abundances is due to accretion onto the dust as the density rises rapidly towards the end of a free-fall collapse. The grains are heated following the formation of a nearby star and the grain mantles evaporate, returning material to the gas phase. The importance of surface reactions with hydrogen atoms is clearly seen in the enhancement of the H_2O abundance.

reaction:

$$H + HNC \dashrightarrow HCN + H$$

occurring. All other species show very good agreement with their observed abundances.

TABLE II. Calculated molecular abundances in the hot core phase and observed abundances in the Orion Hot Core.

Species	Calculated+	Observed
CO	6.7×10^{-5}	6.0×10^{-5}
C	3.9×10^{-11}	$<5.0 \times 10^{-7}$
CN	8.9×10^{-8}	$<5.0 \times 10^{-10}$
C_2H	4.0×10^{-7}	$<5.0 \times 10^{-10}$
H_2O	4.7×10^{-4}	-
CH_4	1.1×10^{-4}	-
HCO^+	1.6×10^{-14}	$<5.0 \times 10^{-10}$
H_2CO	1.7×10^{-7}	1.3×10^{-8}
HCN	8.6×10^{-8}	1.5×10^{-7}
HNC	2.4×10^{-7}	5.0×10^{-10}
NH_3	1.1×10^{-4}	$10^{-5} - 10^{-6}$
CH_3CN	3.9×10^{-8}	3.9×10^{-9}
HC_3N	1.3×10^{-9}	8.0×10^{-10}
HDO		1.0×10^{-7}
NH_2D		4.5×10^{-9}

+ Fractional abundances (relative to H nuclei) calculated after $t = 10^4$ years in the hot core phase. Table taken from Brown et al. 1988 - see references therein for sources of observational data.

4. DEUTERIUM

The observed enhancements of deuterated ammonia, NH_2D, and deuterated water, HDO, cannot be produced by gas phase chemistry at the high temperatures deduced for the hot core. In addition, the abundances cannot be reproduced by the "freezing" of cold phase peak abundances discussed in section 3. However, the enhanced abundances of deuterated species may be explained by including reactions with deuterium atoms on grain surfaces.

Deuterium atoms can quantum tunnel through surface potential barriers and so are mobile over the grain surface, albeit at a slower rate than hydrogen atoms, due to their larger mass and greater binding energy. Our model has therefore been extended to include surface reactions involving deuterium atoms.

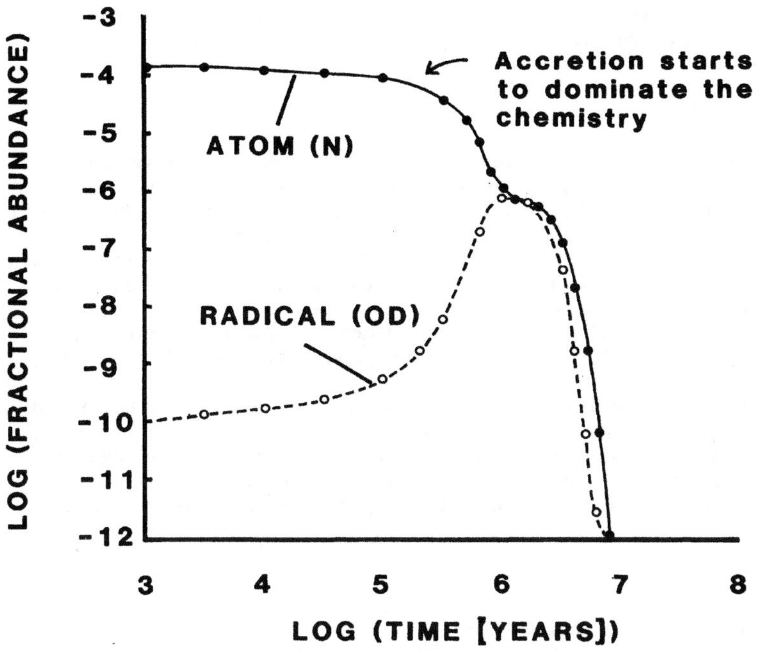

FIGURE 2. Radicals peak as accretion dominates. As atoms leave the gas phase by accreting onto dust grains, the main destruction route for radicals is removed. The radicals show a peak in their abundances before they too condense onto the grains.

However, even without such surface reactions, an enhancement can occur in the abundance of molecules containing deuterium as a result of the accretion process. The main destruction route for radicals in the gas phase is via reactions with atoms. As the atoms are removed from the gas by accretion, the main destruction route is reduced and many radicals, such as OD, CH, NHD and so on, show a peak in their abundance before they too condense onto the dust (see Figure 2). The deuterated radicals then undergo surface reactions with hydrogen leading to HDO, NH_2D and CH_3D (Brown, 1988). This enhancement, albeit significant, is less than that obtained when reactions with D atoms on the surface are included. However, the effect may be used to verify that accretion is dominating the chemistry in a particular region. If a large abundance of a radical, such as OD, can be observed, then the chemistry in the region may be dominated by accretion.

5. CONCLUSIONS

(1) By allowing accretion, surface reactions involving hydrogen, and the subsequent evaporation of grain mantles, the model successfully explains the high abundance of NH_3 and H_2O (implied from observations of HDO) seen in hot core regions.
(2) The model also accounts for the high abundance of species which take no part in the surface chemistry by "freezing" their peak cold phase abundances.
(3) The model can explain the observed enhancements of deuterated species - which is not possible by gas phase chemistry alone.

Finally we state that the chemistry seen in hot cores clearly shows that accretion occurs in dense clouds and support is given to the idea that mantles are removed from the grains sporadically rather than continuously. In view of the time dependent nature of accretion, steady-state chemistry is not applicable to dense clouds.

REFERENCES

Brown, P.D.: 1988 (in preparation).
Brown, P.D. and Charnley, S.B.: 1988. Mon. Not. R. astr. Soc., submitted.
Brown, P.D., Charnley, S.B. and Millar, T.J.: 1988. Mon. Not. R. astr. Soc., in press.
d'Hendecourt, L.B., Allamandola, L.J. and Greenberg, J.M.: 1985. Astron. Astrophys., 152, 130.
Iglesias, E.: 1977. Astrophys. J., 218, 697.
Menten, K.M., Walmsley, C.M., Henkel, C. and Wilson, T.L.: 1986. Astron. Astrophys., 157, 318.
Millar, T.J. and Nejad, L.A.M.: 1985. Mon. Not. R. astr. Soc., 217, 507.
Pauls, T.A., Wilson, T.L., Bieging, J.H. and Martin, R.N.:1983. Astron. Astrophys., 124, 123.
Whittet, D.C.B., Longmore, A.J. and McFadzean, A.D.: 1985. Mon. Not. R. astr. Soc., 216, 45P.

CHEMISTRY IN SHOCKED INTERSTELLAR GAS

D. R. Flower and T.S. Monteiro
Physics Department
The University
Durham DH1 3LE
U.K.

G. Pineau des Forêts and E. Roueff
DAMAP
Observatoire de Paris
F-92 195 Meudon Principal Cedex
France

ABSTRACT. Models of shocks in diffuse interstellar clouds are considered in which (a) the photodissociation of H_2 is neglected, and (b) this process is included. Remaining discrepancies between the results of independent computer model calculations (Pineau des Forêts et al. 1986, Draine 1986) in category (a) are shown to relate to the assumed products of the photodissociation of CH^+. We find that the rate of formation of CH^+ in shocked gas is not significantly enhanced by the rotational excitation energy of the H_2 molecule in the reaction $C^+(H_2(J),H)CH^+$ but is increased if the reaction $CH^+(H_2,H)CH_2{}^+$ proceeds at appreciably less than the Langevin rate at high effective temperatures.

1. INTRODUCTION

It has been recognised for many years that shocks propagate in the interstellar medium. Supernovae explosions, collisions between interstellar clouds, turbulence within clouds, and the expansion of compact H II regions can all be at the origin of supersonic motions which give rise to shocks. Observational support of the shock hypothesis is to be found in spectroscopic measurements of molecular line widths, which are often superthermal, of vibrationally excited H_2 (Beckwith et al. 1978, Nadeau and Geballe 1979, Simon et al. 1979, Ogden et al. 1979, Scoville et al. 1982, Davis et al. 1982), of highly rotationally excited CO (Watson et al. 1980, Storey et al. 1981, Stacey et al. 1982, Watson et al. 1985), and of vibrationally excited CO (Geballe and Garden 1987) in the Orion molecular cloud.

The propagation of a hydromagnetic shock in the interstellar gas was studied by Field et al. (1968), who considered the effects of a

T. J. Millar and D. A. Williams (eds.), Rate Coefficients in Astrochemistry, 271–280.

magnetic field orientated parallel to the shock front. Field at al. established that the magnetic field strongly influences the shock structure but made the simplifying approximation that the ionised and neutral fluids were fully coupled throughout the shock. In this case, the action of the magnetic field (on the ionised gas) is supposed to be transmitted simultaneously to the neutral fluid, an approximation which becomes increasingly inaccurate as the degree of ionisation of the gas decreases.

Effects arising from the partial decoupling of the flows of the ionised and neutral fluids were subsequently investigated by Mullan (1971) and by Draine (1980). The more general case of a magnetic field which is inclined relative to the shock front has been studied recently by Wardle and Draine (1987). The most important consequence of a magnetic field which is transverse to the flow is ion-neutral velocity drift, or ambipolar diffusion.

In shocked gas, endothermic ion-molecule reactions such as

$$C^+ + H_2 \rightarrow CH^+ + H - 4640\ K \qquad (1)$$

become important. Elitzur and Watson (1978, 1980) have suggested that this reaction is responsible for CH^+ formation in the diffuse interstellar gas. However, the model on which this suggestion was based did not include magnetic field effects. Elitzur and Watson studied the chemistry of the hot gas behind a hydrodynamic shock discontinuity, in which reaction (1) is driven by the thermal energy of the gas. More recent studies, using an updated and much extended chemistry (Mitchell and Watt 1985, Pineau des Forêts et al. 1986), have shown that the column densities of CH^+ generated in such shocks fall far short of the observed values. The inclusion of the magnetic field appears to be a prerequisite of any attempt to resolve the long-standing CH^+ problem on the basis of a shock model.

The formation of CH^+ has been the subject of two recent, independent studies of magnetohydrodynamic (MHD) shock propagation in diffuse interstellar clouds. On the one hand, Draine and Katz (1986b) and Draine (1986) have asserted that MHD models resolve the problem of CH^+ production, particularly along the line of sight to ζ Oph, whilst remaining in harmony with other observational constraints. On the other hand, Pineau des Forêts et al. (1986) were more cautious, concluding that models which generate sufficiently large column densities of CH^+ tend to overpopulate excited rotational states ($3 \leq J \leq 5$) of H_2, as compared with the observations.

In the present review, we discuss in Section 2 the differences in the models which lead to these conflicting conclusions. In Section 3, we present preliminary results of more recent calculations which, unlike those cited above, include the effects of the background ultraviolet radiation field on the H_2 molecules.

2. THE CH^+ PROBLEM REVISITED

In this Section, we shall consider in some detail the reasons for the discrepant conclusions of Pineau des Forêts et al. (1986) and Draine (1986) regarding the success of MHD models of the shock along the line of sight to ζ Oph. To initiate the discussion, we compare in Table 1 the parameters of the models which these authors chose to compare with the observed column densities, also listed in the Table.

Parameter	(a)	(b)	(c)	(d)	(e)
$n_o(H)$ cm^{-3}	8	2.86			
$n_o(H_2)$ cm^{-3}	6	8.57			
u_s kms^{-1}	16	9			
B_o μG	5	4.47			
χ	1	4			
log $N(CH^+)$	13.51	13.21		13.53	13.43
log N (CH)	13.15	12.66		13.53	$\lesssim$ 12.7 †
log N(OH)	13.36	11.90		13.71	$\lesssim$ 13.2 †
log $N[H_2(J=2)]$	17.35	17.39	18.03	18.56	18.56 ± 0.14
log $N[H_2(J=3)]$	17.40	17.09	17.16	17.07	17.07 ± 0.34
log $N[H_2(J=4)]$	16.29	15.57	15.72	15.68	15.68 ± 0.14
log $N[H_2(J=5)]$	15.66	14.68	14.97	14.77	14.63 ± 0.05
log $N[H_2(J=6)]$	13.93	12.65	13.76	13.58	13.69 ± 0.05

† This figure is Draine's estimate of the "shock" contribution to the total observed column density.

TABLE 1 Models and observations of the line of sight to ζ Oph. (a) Pineau des Forêts et al. (1986, Table 5c); (b) Draine (1986, Table 1, Model A); (c) Draine (1986, Table 1, Model A) augmented by the values of $N[H_2(J)]$ from the diffuse cloud model G of van Dishoeck and Black (1986, Table 5); (d) observed column densities, from sources cited by Pineau des Forêts et al. (1986); (e) observed column densities as cited by Draine (1986). The symbols are as follows: n_o is a preshock density, u_s is the shock speed, B_o is the preshock magnetic induction, χ is the assumed factor of amplification of the ultraviolet background radiation field, and $N(cm^{-2})$ is a column density.

First, it should be stated that the two models whose predictions are listed in Table 1 differ somewhat in their objectives. Whilst Draine (1986) attempted to tune his model parameters to optimise agreement with observations of ζ Oph, Pineau des Forêts et al. (1986) did not attempt such precise comparisons but limited themselves to more general statements based on a grid of models and observations of ζ Per and o Per, as well as ζ Oph. Nonetheless, the comparison of the models listed in Table 1 is meaningful and instructive.

The most significant differences in the model parameters which appear in Table 1 are: (i) the ratio of the preshock densities of molecular and atomic hydrogen, which is higher in the model of Draine

($n_o(H_2)/n_o(H)$ = 3) than in the model of Pineau des Forêts et al. ($n_o(H_2)/n_o(H)$ = 0.75). We note that the assumption of an appreciable fraction of atomic hydrogen in the preshock gas is open to criticism but defer consideration of this point to Section 3 below. (ii) Pineau des Forêts et al. adopt a larger value of the shock speed (16 km s^{-1}) than Draine (9 km s^{-1}). It should be borne in mind that the energy flux of the shock varies as u_s^3 for a given value of the total density ($n_H = n(H) + 2n(H_2)$) of the preshock gas. (iii) Draine adopts a substantially enhanced (χ=4) ultraviolet background radiation field.

Some of the factors giving rise to these differences in the chosen shock parameters have been discussed by Draine (1986). In particular, CH^+ formation is enhanced in his model by a rate coefficient for reaction (1) which is three times larger than assumed by Pineau des Forêts et al. - and twice the recent experimental values of Smith and Adams (1987) and Gerlich et al. (1987). The column density of CH^+ is further augmented by the enhanced radiation field (χ=4). In addition, the rates of molecular photodissociation reactions for χ=1 are typically twice as large in Draine's model as in the model of Pineau des Forêts et al., because of different standards for the intensity of the Galactic background radiation field in the relevant domain of ultraviolet wavelengths.

	CH^+	CH	OH	$H_2(J)$ J = 2	3	4	5	6
(a)	13.21	12.66	11.90	17.39	17.09	15.57	14.68	12.65
(b)	12.93	12.61	12.43	17.23	16.91	15.75	14.78	13.03
(c)	13.34	12.81	11.69	17.45	16.97	15.86	14.70	12.97

TABLE 2 Logarithmic column densities (cm^{-2}), as predicted by: (a) model A of Draine (1986); (b) a duplicate model, based upon the chemistry of Pineau des Forêts et al. (1986), in which $CH^+ + h\nu \rightarrow C + H^+$; (c) as (b), except that the products of the CH^+ photodissociation reaction have been changed to C^+ and H, to agree with (a).

In Table 2, row (b), are presented the results of a calculation in which the parameters of the model were chosen to match as closely as possible those of model A of Draine (1986). The use of a higher rate coefficient for the $C^+(H_2,H)CH^+$ reaction and of an enhanced background radiation field (χ = 4 x 2 = 8 was chosen to allow for the fact that the "standard" (χ=1) rates of photodissociation in Draine's model are typically twice as large as assumed by Pineau des Forêts et al.) certainly leads to an improved level of agreement with model A of Draine (row (a) of Table 2). However, discrepancies remain in $N(CH^+)$, which is larger by almost a factor 2 in Draine's model, in N(OH) and in $N[H_2(J)]$, the higher J levels being relatively overpopulated in our model, indicating a higher neutral temperature, T_n.

Such discrepancies are significant in the context of the comparison with the observations of ζ Oph. Whilst the value of $N(CH^+)$ predicted by model A of Draine (1986) might be considered to be in reasonable agreement with the observed column density, $N(CH^+)$ from the "duplicate" model certainly falls below the observational lower limit. In order to enhance $N(CH^+)$, a higher value of the shock speed is necessary, but this modification to the model has the undesirable consequences of increasing T_n and hence N(OH) and of exacerbating the problem of overpopulating the excited rotational states of H_2.

Only recently has the reason for these residual discrepancies been recognised. They stem from the assumed products of the photodissociation reaction

$$CH^+ + h\nu \rightarrow C + H^+ \; . \tag{2}$$

Quantum chemical calculations (Kirby 1980, Kirby et al. 1980) have shown that the products of this reaction are predominantly C and H^+ under interstellar conditions. However, in his work on ζ Oph, Draine (1986) assumed the products to be C^+ and H (cf. Table 6 of Draine and Katz 1986a). Whilst, at first sight, this may appear to be a minor error, in reality, it is significant and accounts for most of the residual discrepancies to which reference has been made. This may be seen from Table 2, by comparing model A of Draine (1986) with the results of our own calculations in which the products of reaction (2) are supposed to be C^+ and H (row (c) of Table 2). The agreement with Draine's model is much more satisfactory, from the viewpoint of $N(CH^+)$, N(OH), and of $N[H_2(J)]$, considering that the rate coefficients for collision-induced rotational transitions in H_2 are not identical in the two models.

The products of reaction (2) are critical to the model because the degree of ionisation of the gas within the shock is affected. When the correct photodissociation products (C and H^+) are adopted, $n(H^+) > n(C^+)$ in the vicinity of the maximum of T_n, whereas $n(H^+) \ll n(C^+)$ when the incorrect products (C^+ and H) are assumed. The increase in $n(H^+)$ leads to enhanced collisional coupling between the ionised and neutral fluids, a higher maximum of T_n and a reduced shock width, these changes being reflected in the computed column densities, as may be seen in Table 2. Unfortunately, the "correct" results (row (b) of Table 2) are not in good agreement with the observed column densities.

A criticism of all the models which have been discussed in this Section is that the photodissociation of H_2 has been neglected. At the same time, a significant fraction of atomic hydrogen is assumed to be present in the preshock gas. These assumptions might be considered to be contradictory, and so, in the following Section 3, we graduate to models which incorporate the process of H_2 photodissociation.

3. THE PHOTODISSOCIATION OF H_2 WITHIN MHD SHOCKS

It is well known that the photodissociation of H_2 in the diffuse interstellar gas proceeds through absorption of radiation in the Lyman and Werner bands, followed by decay into the continuum. Accordingly, a line transfer problem is involved in which self-shielding plays an important role.

Rotational transitions within the vibrational ground state of H_2 represent the major radiative cooling process in the shocks which we are considering. It follows that a change in the relative abundance of this coolant affects T_n and the relative populations of the rotational levels of H_2 within the shock. Furthermore, fluorescence following absorption of radiation in the Lyman and Werner bands modifies directly these level populations.

We have developed a numerical model of the dissociation and fluorescence of H_2 following excitation by the ultraviolet background radiation field (Monteiro et al. 1987). Previous work by Wagenblast and Hartquist (1987) was restricted to static clouds of uniform density and temperature, for which semi-analytic solutions to the transfer problem can be found (Federman et al. 1979). The existence of a substantial velocity gradient within the shock has the effect of Doppler shifting the self-shielded Lyman and Werner lines to adjacent wavelengths, where ultraviolet radiation is available for absorption, leading to both fluorescence and dissociation. These Doppler shifts may readily attain 5 or 10 km s^{-1} in the models under consideration, as may be seen from the work of Pineau des Forêts et al. (1986) and Draine (1986). In addition, the lines may be substantially broadened through the temperature rise in the shock, where the maximum of $T_n \approx 1000$ K.

				$H_2(J)$				
	CH^+	CH	OH	J = 2	3	4	5	6
(a)	13.23	14.11	13.15	17.80	17.36	16.17	14.86	12.91
(b)	13.34	13.90	13.70	17.72	17.44	16.12	15.14	13.44

TABLE 3 A comparison of logarithmic column densities (cm^{-2}) predicted by (a) the reference model of Pineau des Forêts et al. (1986), (b) the corresponding model in which the additional process of excitation of H_2 by the ultraviolet radiation field of Draine (1978) is included. Note that in (b) the rate coefficient of the $C^+(H_2,H)CH^+$ reaction is larger by a factor 1.5 than in (a), to accord with the more recent experimental values of Smith and Adams (1987) and Gerlich et al. (1987). The Doppler parameter in (b) is taken to be b = 1 kms^{-1}.

In Table 3, we compare the results of calculations which include the process of excitation of H_2 by the ultraviolet radiation field (row (b)) with the reference model (row (a)) of Pineau des Forêts et al. (1986, Table 6 and Figs. 1-9), in which this process was neglected. The main parameters of this model are $n_o(H_2) = 9.8$ cm^{-3}, $n_o(H) = 0.4$ cm^{-3}, $u_s = 12$ km s^{-1}, $B_o = 5$ μG, and $\chi = 1$, where the subscript 'zero' denotes the preshock gas. The rate coefficient of the $C^+(H_2,H)CH^+$ reaction is taken to be larger by a factor 1.5 in model (b) than in model (a), in accordance with the more recent experimental value (Smith and Adams 1987, Gerlich et al. 1987).

As might have been foreseen, the primary consequence of including the photodissociation of H_2 is a reduction of the H_2/H ratio within the shock, by about a factor of three in the vicinity of the maximum of T_n. Thus, gas which initially is largely molecular ($n_o(H_2)/n_o(H)$ = 24.5) acquires a substantial atomic component because of the photodissociation of H_2 but also because of chemical reactions, which tend to transform molecular to atomic hydrogen (cf. Fig. 1). Fig. 1 also shows that if the velocity shift of the neutral gas and the thermal broadening of the absorption lines within the shock are neglected, the H_2/H ratio is significantly overestimated, as would have been anticipated.

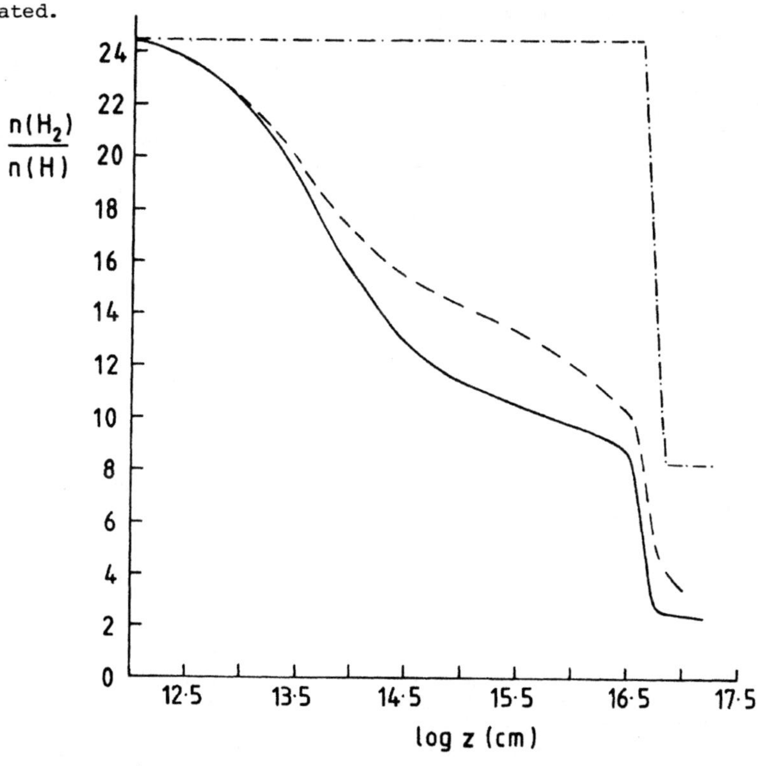

Figure 1. The variation of the H_2/H ratio in shock models in which the photodissociation of H_2 is neglected (dash-dot curve), this process is included but the velocity shift of the neutral gas and the thermal broadening of the absorption lines within the shock are neglected (dashed curve), and in which the velocity shift and thermal broadening are included (continuous curve).

The consequences of the reduction in the H_2/H ration in model (b) as compared with model (a) of Table 3 are: (i) the CH column density is lower in (b) than in (a), as N(CH) decreases with $n(H_2)/n(H)$ (cf. Pineau des Forêts et al. 1986, Table 5). (ii) The maximum of T_n in (b) exceeds by several hundred degrees that attained in (a), both because the density of the principal coolant (H_2) has fallen and because the photodissociation of H_2 heats the neutral gas through the injection of the kinetic energy of the product hydrogen atoms. (iii) The OH column density is substantially larger in (b) than in (a), because the barrier (2980 K) to the reaction $O(H_2,H)OH$ is more readily overcome at higher temperatures. (iv) The distribution of population over the rotational levels of H_2 has drifted to higher J-values in (b) than in (a), partly because of the higher T_n, and partly because of the process of ultra-violet pumping, which is particularly significant for J=6.

From the viewpoint of comparing such models with the observations of ζ Oph, the fall in N(CH) between (a) and (b) is to be welcomed, but the rise in T_n and its consequences are undesirable. However, as general conclusions should not be based on the results of a single model, we have attempted to tune the parameters of the shock, particularly the shock speed u_s and the radiation enhancement factor χ, to optimise the level of agreement with the observations of ζ Oph. The preliminary results of this optimisation procedure are compared with the observed column densities in Table 4.

				$H_2(J)$				
	CH^+	CH	OH	J = 2	3	4	5	6
(a)	13.43	≲12.7	≲13.2	18.56±0.14	17.07±0.34	15.68±0.14	14.63±0.05	13.69±0.05
(b)	13.08	13.0	13.1	17.34	17.11	15.90	15.18	13.82

TABLE 4 (a) Logarithmic column densities observed towards ζ Oph, as collated by Draine (1986); (b) a model with $n_o(H_2) = 9.8\ cm^{-3}$, $n_o(H) = 0.4\ cm^{-3}$, $u_s = 10\ km\ s^{-1}$, $B_o = 5\ \mu G$ and $\chi = 4$, in which H_2 photodissociation is included (the Doppler parameter $b = 1\ km\ s^{-1}$).

The agreement between the predictions of the model in Table 4 and the observations might be qualified as "marginal". The computed value of $N[H_2(J{=}2)]$ is smaller than the observed value by more than a order of magnitude, but additional contributions arise in the ambient gas along the line of sight. The calculated column density of CH falls above the "shock" contribution, as estimated by Draine (1986). Furthermore, $N(CH^+)$ is significantly less than observed.

Chemical factors which might lead to an increase of the CH^+ column density have been investigated. We have found that $N(CH^+)$ is not significantly enhanced by the effects of the rotational energy of the H_2 molecule on the rate of the reaction $C^+(H_2(J),H)CH^+$ (Gerlich et al. 1987). Neither is the rate of dissociative recombination, $CH^+(e,H)C$, a critical parameter, as CH^+ reacts less rapidly with e than with H_2 in these models. The rate coefficients for this latter reaction - $CH^+(H_2,H)CH_2^+$ - was reported to be 1.2×10^{-9} cm^3 s^{-1} by Smith and Adams (1977), the measurement having been made at room temperature. In MHD shocks, the reaction is driven essentially by the ambipolar diffusion at "effective" temperatures which attain several thousand degrees Kelvin. Were the rate coefficient at such high temperatures to be lower than the (essentially Langevin) room-temperature value, $N(CH^+)$ would increase. Further experimental work is need to clarify this issue.

We are currently computing a composite model of the line of sight to ζ Oph, which includes a "shock" component in addition to the ambient cloud. It remains to be seen whether this more complete model will satisfactorily account for the observations.

ACKNOWLEDGEMENTS

We are indebted to Dr. Bruce Draine for correspondence regarding his model of ζ Oph, in which he substantially confirms the results presented in Section 2. Our work is supported by a NATO research grant (0063/87) and by the UK Science and Engineering Research Council. The use of the computing facilities of NUMAC (Durham and Newcastle, UK) and CIRCE (Orsay, France) is gratefully acknowledged.

REFERENCES

Beckwith, S., Persson, S.E., Neugebauer, G. and Becklin, E.E., 1978, Ap. J., **223**, 464.
Davis, D.S., Larson, H.P. and Smith, H.A., 1982, Ap. J., **259**, 166.
Draine, B.T., 1978, Ap. J. Suppl., **36**, 595.
Draine, B.T., 1986, Ap. J., **310**, 408.
Draine, B.T. and Katz, N.S., 1986a, Ap. J., **306**, 655.
Draine, B.T. and Katz, N.S., 1986b, Ap. J., **310**, 392.
Elitzur, M. and Watson, W.D., 1978, Ap. J., **222**, L141 (erratum in 1978, Ap. J., **226**, L157).
Elitzur, M. and Watson, W.D., 1980, Ap. J., **236**, 172.
Federman, S.R., Glassgold, A.E. and Kwan, J., 1979, Ap. J., **227**, 466.
Field, G.B., Rather, J.D.G., Aannestad, P.A. and Orszag, S.A., 1968, Ap. J., **151**, 953.
Geballe, T.R. and Garden, R., 1987, Ap. J., **317**, L107.
Gerlich, D., Disch, R. and Scherbarth, S., 1987, J. Chem. Phys., **87**, 350.
Kirby, K., 1980, IAU Symp. No. 87, Interstellar Molecules, ed. B.H. Andrew, Reidel (Dordrecht), p. 283.
Kirby, K., Roberge, W.G., Saxon, R.P. and Liu, B., 1980, Ap. J., **239**, 855.
Mitchell, G.F. and Watt, G.D., 1985, Astr. Ap., **151**, 121.
Monteiro, T.S., Flower, D.R., Pineau des Forêts, G. and Roueff, E.,

1987, M.N.R.A.S., to be submitted.
Mullan, D.J., 1971, M.N.R.A.S., **153**, 145.
Nadeau, D. and Geballe, T.R., 1979, Ap. J., **230**, L169.
Ogden, P.M., Roesler, F.L., Larson, H.P., Smith, H.A., Reynolds, R.J. and Scherb, F., 1979, Ap. J., **233**, L21.
Pineau des Forêts, G., Flower, D.R., Hartquist, T.W. and Dalgarno, A., 1986, M.N.R.A.S., **220**, 801.
Scoville, N.Z., Hall, D.N.B., Kleinmann, S.G. and Ridgway, S.T., 1982, Ap. J., **253**, 136.
Simon, M., Righini-Cohen, G., Joyce, R.R. and Simon, T., 1979, Ap. J., **230**, L175.
Smith, D. and Adams, N.G., 1977, Int. J. Mass Spectrom. Ion Phys., **23**, 123.
Smith, D. and Adams, N.G., 1987, private communication.
Stacey, G.J., Kurtz, N.T., Snayers, S.D., Harwit, M., Russell, R.W. and Melnick, G., 1982, Ap. J., **257**, L37.
Storey, J.W.V., Watson, D.M., Townes, C.H., Haller, E.E. and Hansen, W.L., 1981, Ap. J., **247**, 136.
van Dishoeck, E.F. and Black, J.H., 1986, Ap. J. Suppl., **62**, 109.
Wagenblast, R. and Hartquist, T.W., 1987, M.N.R.A.S., in press.
Wardle, M. and Draine, B.T., 1987, Ap. J., **321**, 321.
Watson, D.M., Genzel, R., Townes, C.H. and Storey, J.W.V., 1985, Ap. J., **298**, 316.
Watson, D.M., Storey, J.W.V., Townes, C.H., Haller, E.E. and Hansen, W.L., 1980, Ap. J., 239, L129.

DYNAMICAL MODELS OF THE CHEMISTRY IN INTERSTELLAR CLOUDS

D. A. Williams
Mathematics Department
UMIST
Manchester M60 1QD
England

ABSTRACT. The physics of interstellar clouds affects the chemistry within them: one must consider the likely physical conditions before specifying the chemistry in detail. In many dark clouds, both observations and theory indicate that material is continually recycled between dense and tenuous states and that the dynamics of this recycling drastically affects the chemical evolution. In diffuse clouds, recent studies of interstellar grains indicate that there is a continual interaction between gas and dust which is limited by intermittent shocks. Thus, in both situations the timescales available for chemistry are relatively brief and so constrain the variety and nature of the chemical reactions to be included in chemical models.

INTRODUCTION

In their reviews in this volume, Herbst (1988) and van Dishoeck (1988) have described the procedures commonly adopted in the modelling of dark and diffuse interstellar molecular clouds. Flower (1988), in this volume, describes the chemical processes occurring in an interstellar shock. In this article, I argue that more realistic models are required because observations indicate that the situation is more complicated than the simple models imply. These more sophisticated models indicate that some chemical processes are likely to be important while others are insignificant.

Herbst (1988) adopts "pseudo time-dependence" in which the chemistry is followed in time within a static cloud in which density and temperature are fixed. Van Dishoeck (1988) has argued that chemistry in diffuse clouds is achieved so rapidly that the time dependence need not be followed. In each approach, the modeller chooses a set of reactions describing the chemistry of the species of interest, and uses an appropriate mathematical model to obtain either time dependent or steady state results to compare with observations. The comparison should then yield astrophysical information. The lack of uniqueness in such information has been

T. J. Millar and D. A. Williams (eds.), Rate Coefficients in Astrochemistry, 281–286.

discussed elsewhere (Mann and Williams 1986; Williams 1986). The failures of such models to agree in significant respects with the observations have been described in the reviews cited above. These failures are unlikely, however, to have their origin solely in poorly known rate coefficients. We claim that a fuller understanding of the physics is required before adjusting the fine detail of the chemistry.

DYNAMIC DARK MOLECULAR CLOUDS

Molecular line profiles in dark clouds are generally (though not always; cf. Myers and Benson 1983) observed to be broad ($\Delta\upsilon \sim 10$ km s^{-1}). Since the sound speed is ~ 0.3 km s^{-1} these observations indicate the presence of supersonic motions in the clouds. Such motions are soon damped unless maintained: thus, dark molecular clouds must generally contain substantial energy sources which drive these motions. Such sources are considered likely to be associated with the low mass stars which have been identified (as infrared point sources) by IRAS (Beichman et al. 1984) and in other observations (cf. Benson, Myers and Wright 1984) in many dark clouds. Molecular line profiles observed at or near these sources frequently show extended high velocity wings or asymmetrical profiles, suggesting directed or even bipolar flows originating with the source.

A characteristic object is the well-studied dark cloud Barnard 5 (B5). The IRAS survey (Beichman et al. 1984) detected four point sources in B5. Goldsmith, Langer, and Wilson (1986) mapped B5 in $^{12}C^{18}O$ showing that the cloud is composed of at least five fragments of density ~ 10^4 cm^{-3} embedded in a more tenuous medium. Goldsmith et al. also mapped $^{12}C^{16}O$ around the infrared sources and identified the presence of high velocity (~ 10 km s^{-1}) gas in the vicinity of the sources but extending ~ 1 pc. The interpretation of Goldsmith et al. is that the formation of a low mass star in a fragment may lead ultimately to the disruption of the fragment by the stellar wind. Thus, that fragment is prevented from giving rise to further star formation. The eroded molecular material swept up in the wind ultimately accumulates into new clumps from which new stars are formed. Therefore, Goldsmith et al. conclude that material is continually recycled between the dense fragments and the less dense material. They deduced that the cycle period is typically $\gtrsim 10^6$ yr, and they point out that in such a situation chemistry never achieves steady state. The mechanism by which gas in these clouds is accelerated to high velocities (probably involving shocks; Williams and Hartquist 1984) is likely also to have profound effects on the chemical balance of the gas.

Dynamical models of clouds can therefore account for large line widths and for long lifetimes compared to free fall collapse (cf. Norman and Silk 1980). The models make it necessary to study cloud chemistry by following a parcel of gas through a sequence of events. Since, in these models, each parcel of gas is probably shocked about once every million years, there is a natural mechanism limiting the growth of molecular mantles of H_2O and CO solid. Such mantles on grains in dark clouds are observed to be limited to appreciable, yet moderate, proportions (e.g. Whittet et al. 1983, 1985).

CHEMISTRY IN DYNAMICAL MODELS OF DARK CLOUDS

The chemistry in a dynamical model of a dark cloud has been explored in calculations by Charnley et al. (1988a). In this model, the wind of a T Tauri star blows a cavity in the molecular cloud, and impinges on clumps of dense molecular gas within the cavity. Close to the star, these impacts produce strong shocks in the wind and ionize it. The wind erodes the clumps and is mass loaded with atomic and molecular material. This mass loaded, partially ionized material is decelerated by a weak reverse shock at the cavity boundary, and accumulates there, becoming incorporated ultimately in a new clump from which a new star will form. In this model, the main chemical evolution takes place in this material at the cavity boundary: the chemical development is truncated when heavy atoms and molecules strike and stick to grain surfaces. This accretion time is on the order of $10^9/n$ yr, or $\lesssim 10^6$ yr in dark clouds. Thus, the chemistry has available to it, post shock, about one million years.

The chemistry in this gas is driven in various ways. Initially, the weak (~ 10 km s^{-1}) reverse shock at the cavity boundary induces a characteristic neutral atom-molecule chemistry in the hot post-shock phase. The ionization in this accumulated wind is relatively high, and reactions with ions modify the shock chemistry products. When the gas is cool, ion-molecule chemistry, familiar from earlier studies of dark clouds, plays its part. However, insufficient time is available for steady state to be achieved (this takes $\gtrsim 10^7$ yr; cf. Millar and Nejad 1985) because the heavy molecules such as CO, H_2O, NH_3 etc. are accreted on to the grain surface and lost from the gas phase.

In a typical calculation (see Charnley et al. 1988a) it is assumed that the wind is loaded with heavy molecules during its passage through the cavity; thus, the CO abundance is initially high. Other molecules, e.g. CH, form in the weak reverse shock at the bubble boundary (~ 10 yr) but are destroyed by the ions present. When the gas is cool (~ 100 yr), conventional ion-molecule chemistry begins to build a variety of molecules. Ultimately, all the molecules are lost from the gas (~ 10^6 yr) by accretion on to grain surfaces.

In this and similar calculations it is particularly interesting to note that at times of interest, (~ 10^6 yr), all of C^+, C, and CO may simultaneously take comparable values. This result is obtained for a variety of initial parameters and conditions, whereas in steady state calculations both C^+ and C are low when CO is abundant. Observational data (Phillips and Huggins 1981, Crawford et al. 1985, Keene et al. 1985) indicate that both C^+ and C may have abundances comparable to CO in some objects. These results thus have a natural explanation in dynamical models. On the other hand, the failure of conventional models to account for observations such as these is unlikely to be attributable to inadequacies of the chemistry adopted.

DIFFUSE CLOUD MODELS

A new model of interstellar grains (Duley 1987; Jones, Duley and Williams 1987), proposed to account for the variety of observed interstellar extinction curves, implies that grains are a chemically active component of the interstellar medium and that shocks occur in this medium about once every million years or so. The consequences for interstellar chemistry are that for a substantial fraction of the time, diffuse cloud material is far from a chemical steady state, and that elemental abundances in the gas are time dependent.

The model postulates that diffuse cloud grains are composed of cores of silicate (some large, ~ 2000 Å and some small ~ 100 Å) each with a mantle of amorphous carbon (typical thickness ~ 50 Å). Freshly deposited amorphous carbon contains much sp^3 bonded carbon, i.e. is diamond-like: this is the metastable form. Exposure to UV radiation promotes the conversion to the more stable graphite-like form (sp^2 bonding predominates). In both types of amorphous carbon, however, no long range order exists. The time scale for photodarkening (i.e. $sp^3 \longrightarrow sp^2$) conversion to occur in the interstellar medium (Duley and Williams 1988), inferred from laboratory data (Iida, Ohtaki, and Seki 1984) is on the order of a million years. The time scale for the deposition of the carbon on to grains is also on the same order in diffuse molecular clouds. Thus, we expect to find varying depletions of carbon (i.e. mantle thickness) and varying proportions of sp^2:sp^3 type. With depletions in the range (50 ± 25)% and sp^2:sp^3 varying from 0 to 1 but usually close to 0.5, then the entire family of interstellar extinction curves can be fitted (Jones, Duley and Williams 1987).

Since loss of gas phase carbon by accretion on to grains will continue indefinitely if unchecked, the range of depletion observed implies that carbon is returned to the gas phase at a rate comparable to the depletion rate. A straightforward interpretation of this observational result is that intermittent shocks return substantial amounts of mantle material to the gas phase (Duley and Williams 1984, Williams and Hartquist 1984).

We may, therefore, envisage a cycle of events in the diffuse interstellar medium: (i) typical diffuse material is shocked; (ii) the post shock cooling induces a substantial compression and densities ~ 10 - 100 times the pre-shock values may occur; (iii) a rarefaction wave penetrates the dense shell and tends to restore density and temperature to pre-shock values; (iv) the material maintains those values until a new shock arrives. It is assumed that the shock, (i), releases substantial amounts of carbon to the gas phase. Thereafter, the gas phase carbon content decreases, the accretion being most rapid in phase (ii). The enhanced accretion in phase (ii) may imply that the shock frequency be similarly increased. Although steady state in the chemistry is rapidly achieved (timescale $\lesssim 10^4$ yr), the dynamical situation varies continually throughout the cycle. On any line of sight, observations will include all phases described here. Although much detailed work has been performed on static clouds (cf. van Dishoeck 1988, this volume) and on hydrodynamic and magnetohydrodynamic shocks (Flower 1988, this volume) in diffuse clouds, a comprehensive model of the total line of sight, involving both compression and rarefaction in the gas, and both accretion of carbon and removal of mantles in shocks, has yet to be carried out.

CONCLUSION

Models and observations of both dark and diffuse molecular clouds indicate that cloud structure is modulated on a natural time scale on the order of a million years. In dark clouds, therefore, chemistry never achieves steady state, and chemistry must be studied in a time dependent fashion following the dynamical changes. If the conditions vary cyclically, then it is of interest to know whether the chemistry is repeated each cycle, or whether a chemical chaos ensues (Williams 1987). Exploratory calculations of this type have now been completed (Charnley et al. 1988b) and suggest that a limit cycle is reached. In diffuse clouds, the chemistry generally attains steady state (apart from during the hot, shocked phase). However, the dynamical situation continue to change. These considerations lead to two conclusions concerning the chemical network:

1. reactions with slow rates may not influence the chemistry significantly and may be excluded from the calculations;

2. the formation of very large species containing many atoms may be inhibited by the finite time scale.

One may, therefore, speculate that very large species - if they exist in the general interstellar medium - are the result of degradation of yet larger entities, i.e. grains.

ACKNOWLEDGEMENT

The author has benefitted from collaboration and discussion with many people including W W Duley, J E Dyson, A P Jones, T W Hartquist, and T J Millar.

REFERENCES

Beichman, C.A., Jennings, R.E., Emerson, J.P., Baud, B., Harris, S., Rowan-Robinson, M., Aumann, H.H., Gautier, T.N., Gillett, F.C., Habing, H.J., Marsden, P.L., Neugebauer, G. and Young, E., 1984. Astrophys. J. (Letters) 278, L45.

Benson, P.J., Myers, P.C. and Wright, E.L., 1984. Astrophys. J. 279, L27.

Charnley, S.B., Dyson, J.E., Hartquist, T.W. and Williams, D.A., 1988a. Mon. Not. R. astr. Soc. in press.

Charnley, S.B., Dyson, J.E., Hartquist, T.W. and Williams, D.A., 1988b. Mon. Not. R. astr. Soc. in press.

Crawford, M.K., Genzel, R., Townes, C.H. and Watson, D.M., 1987. Astrophys. J. 291, 755.

van Dishoeck, E.F., 1988. This volume.

Duley, W.W., 1987. Mon. Not. R. astr. Soc. 229, 203.

Duley, W.W. and Williams, D.A., 1984. Mon. Not. R. astr. Soc. 211, 97.

Duley, W.W. and Williams, D.A., 1988. Mon. Not. R. astr. Soc. in press.

Flower, D.R., 1988. This volume.

Goldsmith, P.F., Langer, W.D. and Wilson, R.W., 1986. Astrophys. J. (Letters) 303, L11.

Herbst, E., 1988. This volume.

Iida, S., Ohtaki, T. and Seki, T., 1984. American Inst. Conf. Proc., 120, 258.

Jones, A.P., Duley, W.W. and Williams, D.A., 1987. Mon. Not. R. astr. Soc. 229, 213.

Keene, J., Blake, G.A., Phillips, T.G., Huggins, P.J. and Beichman, C.A., 1985. Astrophys. J. 299, 967.

Mann, A.P.C. and Williams, D.A., 1986. Astrophys. and Space Science 121, 387.

Millar, T.J. and Nejad, L.A.M., 1985. Mon. Not. R. astr. Soc. 217, 507.

Myers, P.C. and Benson, P.J., 1983. Astrophys. J. 266, 309.

Norman, C. and Silk, J., 1980. Astrophys. J. 238, 138.

Phillips, T.G. and Huggins, P.J., 1981. Astrophys. J. 251, 533.

Whittet, D.C.B., Bode, M.F., Longmore, A.J., Baines, D.W.T. and Evans, A., 1983. Nature 303, 218.

Whittet, D.C.B., Longmore, A.J. and McFadzean, A.D., 1985. Mon. Not. R. astr. Soc. 216, 45P.

Williams, D.A., 1986. Quart. J. Roy. astr. Soc. 27, 64.

Williams, D.A., 1987. In 'Physical Processes in Interstellar Clouds', eds. G.E. Morfill and M. Scholer (D. Reidel Publishing Company) p.377.

Williams, D.A. and Hartquist, T.W., 1984. Mon. Not. R. astr. Soc. 210, 141.

CHEMISTRY IN EXPANDING CIRCUMSTELLAR ENVELOPES

T. J. Millar
Mathematics Department
UMIST
P.O. Box 88
Manchester M60 1QD
England

ABSTRACT. This review is concerned with the chemistry occurring in the expanding envelopes around cool, late-type stars. We show that, although many different chemical processes help determine the overall chemical composition of such envelopes, the detailed composition, as evidenced by the many molecules detected at radio wavelengths, is best explained by a complex ion-molecule chemistry driven by ionisation caused by the ambient external ultraviolet radiation field and by cosmic-rays.

1. INTRODUCTION

The expanding circumstellar envelopes (CSEs), or winds, of late-type stars have proven to be rich in molecular material. These stars which have evolved from main-sequence stars with masses in the range 1 - 8 solar masses (M_*) are extremely large and luminous with pulsating behaviour on a period typically on the order of 300 - 400 days. The cause of the winds, or mass-loss, is as yet unknown. It is reasonably certain that radiation pressure on dust grains which form in the outflow, followed by dust-gas collisions can drive the mass-loss to infinity but it is less clear as to the exact mechanism by which material is transported from the stellar surface to the the region in which the gas temperature is low enough, and yet the gas density large enough, for solid particles to condense. Although a number of suggestions have been made it is most likely that stellar pulsations is the major mechanism (Jura 1986).

In the past 10 years there has been a large number of observational studies of CSEs. These studies have been used to derive the physical properties of the outflows, in particular the expansion velocities, the mass-loss rates and molecular abundances. Infrared observations of species such as CO and NH_3 have shown that, close to the star,the velocity profile can be very complex with evidence for infall, ejection of shells and shocks, in addition to outflow (Wannier and Sahai 1985, Sahai 1987, Betz, McLaren and Spears 1979). In this region there is also evidence for asymmetries in the envelope structure (Ridgeway and Keady 1987, Bloemhof et al.

T. J. Millar and D. A. Williams (eds.), Rate Coefficients in Astrochemistry, 287–308.

1984). Information on the outer envelope is obtained mostly from millimetre-wave observations of CO. Over 130 CSEs have now been detected in CO (Knapp and Morris 1985, Zuckerman and Dyck 1986a,b, Knapp 1987) and accurate values for expansion velocities and mass-loss rates have been obtained. For carbon-rich CSEs, that is with a C/O ratio > 1, v ~ 15 km s^{-1} and $\dot{M} \sim 10^{-4} - 10^{-6}$ solar masses per year. The average values of these quantities in oxygen-rich stars, which have C/O < 1, are slightly less than those in C-rich CSEs. These observations show that the expansion velocity is constant throughout the outer envelope and that, in many cases the CSEs are remarkably symmetric: in IRC+10216 for example CO is almost spherically symmetric out to a radial distance of ~ 4000 R_*, or ~ 5 x 10^{17} cm (Wannier et al. 1979).

The radial temperature profile of the wind is more difficult to derive from the observations but theoretical calculations by Goldreich and Scoville (1976), Kwan and Linke (1982) and Slavsky and Scalo (1986) have shown that T(r) is proportional to r^{-a} with 0.6 < a < 0.8, typically. Finally, it should be noted that the dust-to-gas ratios in CSEs can be studied by combining radio observations of the gas, normally CO, with submillimetre observations which detect thermal emission from the grains. Although some uncertainty enters through the conversion of CO abundances to H_2 abundances, the ratios so-derived are generally comparable with, though often a factor of 2 - 4 less, than the standard interstellar ratio (Knapp 1985).

For simplicity we consider a spherically symmetric envelope expanding at constant velocity v. Then we can write some simple relations for the number density of H_2 molecules, n(r), at radial distance r, for the radial column density of H_2, N(r), from r to infinity, and for the radial extinction at 100 nm, $A_{UV}(r)$, of the external ultraviolet radiation field due to absorption and scattering by the circumstellar dust grains, as

$$n(r) = 10^5 \dot{M}_{-5}/r_{16}{}^2 v_{15} \text{ cm}^{-3}$$

$$N(r) = 10^{21} \dot{M}_{-5}/r_{16} v_{15} \text{ cm}^{-2}$$

$$A_{UV}(r) = 5.4 \dot{M}_{-5}/r_{16} v_{15} \text{ mag}$$

where $\dot{M}_{-5}$ is the mass-loss rate measured in units of 10^{-5} $M_\odot$ yr^{-1}, r_{16} is the radial distance measured in units of 10^{16} cm, and v_{15} is the expansion velocity measured in units of 15 km s^{-1}. These simple relations, together with that for the temperature profile, show that the physical parameters vary enormously throughout the envelope. As a result several different chemical processes can be important in different regions of the envelope; we shall discuss these in due course. Before doing so, it is worth noting that single-dish radio observations, which have provided most of our information on the chemical composition of CSEs, have spatial resolutions of 25 - 40 arcsec in general. For a star such as IRC+10216, for which we adopt a distance of 200 pc (6x10^{20} cm), this is equivalent to a radial distance of (7 - 12)x10^{16} cm. For this object $\dot{M}_{-5}$ = 5 and v_{15} = 1, n(r) = 5x10^3 cm^{-3} and $A_{UV}(r)$ = 2.7 mag at r = 10^{17} cm. Thus we have the ability to probe what is almost an unique situation in molecular astronomy - a region in which the H_2

abundance is large and the ultraviolet extinction is small.

In this review we shall consider only cool stars; in particular, we shall neglect any possible effects of chromospheric radiation to the chemistry. Glassgold and Huggins (1986) have discussed the chemistry in α Orionis, a star in which chromospheric as well as interstellar radiation must be considered.

2. CHEMICAL PROCESSES

In this section we shall present a brief review of the processes which contribute to the chemistry in CSEs before describing some detailed models developed recently for the external envelopes of carbon- and oxygen-rich stars. We shall begin by discussing the chemistry at the stellar surface followed by a description of the different chemical processes which occur as one moves outward from the star through the CSE.

2.1. LTE Chemistry

At the high densities, $n \sim 10^{12}$ cm^{-3}, and temperatures, $T \sim 1000$ - 2000 K which occur in the stellar photosphere, molecular processes are in local thermodynamic equilibrium (LTE) and molecular abundances are determined by elemental abundances and heats of formation. Although detailed calculations of LTE abundances are available for both carbon- and oxygen-rich stars (Tsuji 1973,1978, McCabe, Connon Smith and Clegg 1979, Lafont, Lucas and Omont 1982), it should be pointed out that many of the detailed results can be significantly in error because of errors in heats of formation, particularly for radicals and unstable species. Nevertheless, such calculations are expected to be much more accurate for stable species such as CO, H_2O, N_2 and HCN. Because of the exceedingly strong bond in CO this molecule takes up essentially all of the available oxygen and carbon in carbon-rich and oxygen-rich stars, respectively. In carbon-rich stars, therefore, the most abundant molecules are CO followed by C_2H_2, N_2, HCN and CH_4, while in oxygen-rich stars they are H_2O, CO and N_2. Indeed the detection of a molecule such as HCN has often been used to classify an object as carbon-rich although the recent detections of HCN in stars known to be oxygen-rich, discussed further in section 3.2, shows that such a procedure is not without pitfalls.

Normally, molecules produced in LTE chemistry are most easily observed through infrared studies which probe the inner parts of the CSE. Recently, however, Cernicharo and Guélin (1987) have detected 19 millimetre lines in IRC+10216 due to NaCl, AlCl and KCl together with 3 lines tentatively identified with AlF. The line widths and line profiles observed are consistent with a source size smaller than 15 arcsec and with an expansion velocity less than that observed for molecules in the outer envelope. For a 15 arcsec source size, the fractional abundance of NaCl with respect to molecular hydrogen, $X(NaCl) \sim 10^{-8}$, in reasonable agreement with

Tsuji's (1973) LTE calculations. The non-detection of these molecules, and also of SiS and SiO, at the terminal velocity of the wind has led to the suggestion that these refractory molecules have condensed out onto dust grains which form in the outflow (Cernicharo and Guelin 1987). Indeed it is possible that the formation of dust can itself perturb the gas phase abundances to allow unusual effects to occur, for example, the formation of HCN in an oxygen-rich CSE. This process is described elsewhere in this volume by Sharp.

2.2. Grain Surface Processes

We can get an estimate of the region in which grain surface processes may play a role in either the formation of molecules or in the removal of species from the gas phase through accretion by comparing the time-scale for the collision of a gas phase species with a dust particle to the expansion time of the outflow. For spherical dust particles of radius a and number density n_d moving through the gas at a relative velocity v_d (normally around 1 - 2 km s^{-1}), the collision time is given by

$$t_d = [\pi a^2 n_d v_d]^{-1} \sim 3\times10^{11} r_{16}^2 v_{15}/\dot{M}_{-5} \text{ sec}$$

while the expansion time is given by

$$t_e = [dn/ndt]^{-1} \sim 3\times10^9 r_{16}/v_{15} \text{ sec.}$$

For the case of IRC+10216, $t_d < t_e$ for $r_{16} < 0.05$, that is, grain accretion can only dominate inside $r = 5\times10^{14}$ cm.

It is of course extremely difficult to model the detailed surface chemistry which might occur on grains since the nature of the grain material is very poorly characterised. However there is some indication that surface processes might contribute to molecule formation in addition to the accretion effects discussed in 2.1. Infrared studies of CH_4 and SiH_4 absorption lines show that the molecules exist in a region in which the terminal velocity of the outflow has been reached but the molecules have no absorption components at velocities less than the terminal velocity as, for example, does CO. Indeed detailed models of the SiH_4 lines show that this species cannot exist in IRC+10216 closer to the star than 30 R_* $\sim 4\times10^{15}$ cm (Betz 1987).

2.3. Warm Gas Phase Chemistry

For an envelope with $\dot{M}_{-5} = 1$ and $v_{15} = 1$, and $r_{16} \sim 0.1 - 1.0$, then $n(r) \sim 10^5 - 10^7$ cm^{-3} and $T(r) \sim 200 - 300$ K. In this region the extinction due to dust is large enough to ensure that there is no ionisation from the external UV radiation field and the density large enough to ensure that the fractional ionisation caused by cosmic-rays is extremely small. In this case reactions between neutral species can be important but only if such reactions do not possess, or at most have small, activation energy barriers. Thus, reactions involving radicals can, in principle, occur although the abundance of such species is uncertain as discussed above. One important reactant is atomic hydrogen, as discussed by Slavsky and Scalo (1986), particularly in stars possessing an internal source of radiation or with high surface temperatures.

2.4. Cosmic-ray Ionisation

Cosmic-rays can penetrate the CSEs and provide a source of ionisation through collisions with molecuar hydrogen and helium, as well as providing a low level of ultraviolet photons through the excitation and susequent cascade of H_2 in collisions with secondary electrons (Prasad and Tarafdar 1983, Sternberg, Dalgarno and Lepp 1987, Gredel, Lepp and Dalgarno 1988). The details of cosmic-ray ionisation have been discussed by Glassgold, Lucas and Omont (1986) for carbon-rich envelopes and by Mamon et al. (1987) for oxygen-rich envelopes. Cosmic-ray ionisation of H_2 leads to H_3^+ which, because of the low proton affinity of H_2, quickly transfers its proton to other abundant molecules such as CO, C_2H_2 and NH_3 the last of which, because of its large proton affinity, takes up most of the protons - at least in carbon-rich stars; in oxygen-rich envelopes the major molecular ion deep in the envelope is H_3O^+. Glassgold et al. (1986) and Mamon, Glassgold and Omont (1987) have discussed the observability of various of these ions in CSEs.

The abundance of H_3^+ is roughly similar in both carbon- and oxygen-rich envelopes. Since it produced by cosmic-rays and destroyed in ion-molecule reactions, its abundance can be written (Mamon et al. 1987) as

$$n(H_3^+) = \zeta/\Sigma\ k_i x_i$$

where ζ is the cosmic-ray ionisation rate, and k_i is the rate coefficient of the reaction involving H_3^+ and a neutral species i which has fractional abundance x_i. Because the denominator in this expression is almost constant throughout the envelope the H_3^+ fractional abundance varies as r^2 while the column density becomes

$$N(H_3^+) = r_m/kx = 10^{13}(\zeta_{-17}r_{m18}/k_{-9}x_{-3})\ cm^{-2}$$

where r_m is the outer boundary of the envelope and all parameters are measured in cgs units with powers of ten indicated by subscripts (Mamon et al. 1987). For IRC+10216, we find $N(H_3^+) = 10^{13}\ cm^{-2}$, an order of magnitude below current infrared detection levels (Glassgold et al. 1986).

2.5. Photoprocesses

Jura and Morris (1981) and Huggins and Glassgold (1982a) realised independently that the external interstellar UV radiation field would modify the radial distribution of molecules in the outer envelope through photodissociation and photoionisation. Huggins and Glassgold (1982a), showed that molecular shells would result from the photochemical chains formed by the destruction of parent molecules flowing into the outer envelope. In particular, the observed spatial distributions of CCH and CN have been explained by the photodissociation of C_2H_2 and HCN (Huggins and Glassgold 1982a, Wootten et al. 1982, Huggins, Glassgold and Morris 1984, Trong Bach et al. 1987). The photons also cause ionisation in the outflow. For a carbon-rich star the most important molecule is C_2H_2 which has a large ionisation cross-section and which helps drive a complex ion-molecule chemistry. Nejad and Millar (1987) have shown that cosmic-rays only affect the ionisation balance in the region close to 10^{16} cm and that the fractional ionisation is relatively small. Photoprocesses limit the extent of the molecular envelopes around late-type stars and detailed models have been calculated for the dissociation of H_2 (Glassgold and Huggins 1983), and for CO taking into account the latest laboratory data on CO photodissociation (Mamon, Glassgold and Huggins 1988) as well as more general models describing both carbon- and oxygen-rich stars (Glassgold et al. 1987, Huggins and Glassgold 1982b).

The detailed results of the photochemical models are dependent on the adopted UV radiation field as well as on the photodissociation and photoionisation cross-sections, which in principle can be obtained by laboratory or theoretical studies, and on the nature of the dust particles for which the chemical composition, size and hence optical properties are poorly known. In the absence of definitive information, detailed models of the photochemistry have relied on some rather simplistic approaches to the radiative transfer of UV photons through the envelope. Nejad and Millar (1987,1988) have used the approach developed by Jura and Morris (1981) which assumes that the dust grains absorb but do not scatter and leads to a numerically simple equation for the intensity of UV radiation at any point in the envelope. On the other hand, Glassgold and co-workers have used a formalism devised by Gerola and Glassgold (1978) in which the scale-length, d_i, for photodissociation of any species is used to diminish the intensity incident on the CSE, that is, the intensity is reduced by a quantity $\exp(-d_i/r)$. These authors have also included the effects of both continuum and line self-shielding in their calculations.

2.6. Ion-molecule Chemistry

The ionisation provided by cosmic-rays and UV photons drives an ion-molecule chemistry in CSEs which leads to a rich variety of molecular species. In order to be important, chemical processes must occur on a time-scale faster than that of the expansion. Figure 1 from Nejad (1986) shows a number of such time-scales for the case of IRC+10216. One sees that a number of fast processes such as reaction with H_2 and dissociative recombination can occur faster than the expansion time but that grain surface processes will be unimportant in the outer envelope as discussed in 2.2. An important point to note in this figure is that fast reaction of a molecular ion with H_2 can dominate even dissociative recombination out to a radial distance of $\sim 10^{19}$ cm.

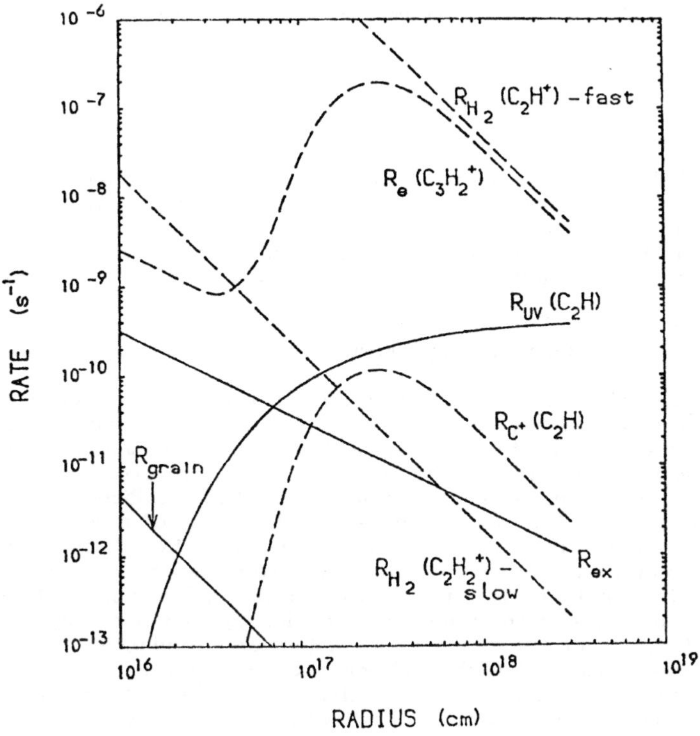

Figure 1. Rates of various processes occurring in IRC+10216 (Nejad 1986). The term $R_X(Y)$ is the rate at which Y is destroyed by X. The rate of expansion (R_{ex}) and the rate at which gas phase species collide with dust grains (R_{gr}) are also shown.

Infrared :	CH_4, SiH_4, C_2H_2, C_2H_4
Radio :	CO, CS, SiO, SiS, c-SiC_2, H_2S NaCl, KCl, AlCl, AlF(?) CN, HCN, HNC, NH_3, CH_3CN HC_3N, HC_5N, HC_7N, $HC_{11}N$ C_2H, l-C_3H, C_4H, C_5H, C_6H, c-C_3H_2 C_2S, C_3S, C_3N

Table I : Molecules detected in IRC+10216. Some molecules detected at radio wavelengths are observed also in the infrared.

In the following section we shall discuss some models which have been developed to describe the detailed chemistry in the outer CSEs.

3. CHEMICAL MODELS OF CIRCUMSTELLAR ENVELOPES

We shall divide our discussion into two sections, dealing firstly with carbon-rich envelopes, in particular that surrounding IRC+10216 for which most observational data has been obtained, before discussing the case of oxygen-rich envelopes for which there are less data available, though much more should be forthcoming in the near future.

3.1. The Carbon-rich Envelope around IRC+10216

The carbon-rich star IRC+10216 is an extremely faint visual object but at a wavelength of 5 microns is the brightest object in the sky outside the solar system due to emission from dust grains in its circumstellar envelope and its proximity to Earth, probably less than 200 pc. It has been well-studied at infrared and millimetre wavelengths and has proved to be a rich source of molecular material. Table I lists the molecules detected in IRC+10216. Neglecting isotopic variants a total of 32 different molecules have been observed, including 28 at radio wavelengths. These observations have been reviewed in two recent articles by Olofsson (1985) and Omont (1985) who also discuss the molecular abundances and the difficulties associated with deriving such abundances in sources in which the radial distribution of molecules can be very complex.

Detailed chemical models of IRC+10216 must address a number of observational points:

(i) Of the 28 molecules detected by radio telescopes, no fewer than 20 have been detected also in cold, dark interstellar clouds, specifically the dust cloud Taurus Molecular Cloud 1 (TMC-1). In particular, both IRC+10216 and TMC-1 are sources of long-chain, unsaturated hydrocarbon molecules.

(ii) Although accurate abundances are difficult to estimate in IRC+10216, the fractional abundances of many complex molecules are larger than in TMC-1.

(iii) The abundance ratio HNC/HCN is around 0.004 in the CSE (Johansson et al. 1984) but about 1.0 - 1.5 in cold clouds (Irvine and Schloerb 1984). The ratio in the CSE is much closer to that observed in warm interstellar clouds, such as Orion, in which Goldsmith et al. (1986) have found evidence for a correlation of the HNC/HCN ratio with temperature. The normal interpretation for the small ratio in IRC+10216 has been that both HCN and HNC have been formed in LTE and been frozen out (McCabe et al. 1979) but recently Glassgold et al. (1987) have shown that such a process cannot be resposible for the ratio in IRC+10216 for which the observed HNC/HCN ratio requires freeze-out at a temperature of 1400 K.

Species	Model(a)		Model(b)		V(obs)
	N(cm^{-2})	V	N(cm^{-2})	V	
CH	1.9(14)	4.7(49)	1.3(14)	1.7(49)	
CH_2	5.2(14)	1.5(50)	4.9(14)	1.3(50)	
C_2H	4.7(15)	3.1(50)	4.7(15)	3.2(50)	7.5(48)
C_2H_3	2.2(14)	8.5(48)	1.2(14)	5.4(48)	
C_3	1.0(14)	4.9(49)	6.1(13)	3.4(49)	
C_3H	5.5(13)	1.4(49)	9.8(12)	1.4(48)	1.0(48)
C_3H_2	4.8(13)	1.3(49)	3.2(13)	1.0(49)	2.0(47)
C_4H	3.5(14)	2.9(49)	4.6(14)	3.8(49)	1.0(49)
CN	1.5(15)	8.4(49)	1.4(15)	8.2(49)	
HNC	2.5(13)	2.1(48)	4.3(13)	2.3(48)	
HC_3N	1.4(13)	9.1(47)	9.8(13)	5.7(48)	1.2(49)
C_3N	1.8(13)	1.9(48)	2.1(14)	2.8(49)	3.3(48)
CH_3CN	2.3(13)	1.8(48)	2.5(13)	1.7(48)	3.6(46)
CCN	1.2(12)	2.2(47)	7.1(12)	1.2(48)	<4.5(47)
C_3O	7.8(12)	6.3(47)	1.4(12)	9.0(46)	
CH_2CO	2.3(11)	4.2(46)	1.3(11)	1.9(46)	
HCO^+	2.8(12)	7.8(46)	2.2(12)	9.4(46)	<1.8(46)
$C_2H_2^+$	5.3(13)	3.0(48)	6.7(13)	3.7(48)	
$H_2C_3N^+$	1.0(12)	1.7(46)	9.1(12)	1.6(47)	

Table II: Column density (N) and total number of molecules in the envelope (V) calculated for IRC+10216. Model (a) assumes a constant temperature, T = 10 K, throughout the outer envelope and uses non-polar rate coefficients; Model (b) has a radial temperature profile and uses polar rate coefficients.

However, the reactions

$$HNC + H \leftrightarrow CN + H_2 \leftrightarrow HCN + H$$

promote the conversion of HNC to HCN until the temperature has fallen to 500 K which would imply a much lower HNC/HCN ratio than observed (Glassgold et al. 1987).

(iv) Methyl cyanide, CH_3CN, is the only CH_3-containing species observed, as yet, in IRC+10216.

(v) Despite sensitive searches there has been no detection of any ion in IRC+10216. In particular, HCO^+ has an observed upper limit to its antenna temperature of only 20 mK (Lucas et al. 1986). Glassgold et al.(1987) have discussed the observability of the HCO^+ ion in some detail and shown that the results of their photochemical model are consistent with the observational limits to within the uncertainties involved.

Detailed chemical kinetic models of IRC+10216 have been presented recently by Glassgold et al. (1987) and Nejad and Millar (1987), following the original descriptions by Nejad, Millar and Freeman (1984) and Glassgold et al. (1986).

The model of Nejad and Millar (1987) assumes that parent molecules, namely H_2, CO, C_2H_2, N_2, HCN, CH_4 and NH_3 are injected into the outer envelope at r = 10^{16} cm with abundances which are taken from infrared observations, where available, or from LTE calculations. Through cosmic-ray ionisation and photoprocesses the parents give rise to a complex ion-molecule chemistry described by 85 species and around 270 reactions. The coupled system of first-order differential equations which describe the chemistry are integrated numerically to obtain abundances and column densities for each species. Table II presents radial column densities and total volume densities calculated for a number of species for the standard model of Nejad and Millar (1987). It can be seen that the abundances agree reasonably well with those derived from observations, with the exceptions of HC_3N and C_3N. However, Glassgold et al. (1987) have pointed out the importance of enhanced rate coefficients at low temperatures in increasing the HC_3N abundance, since in the region in which the ion-molecule chemistry is most active the gas temperature is low enough for the effects discussed by Adams, Smith and Clary (1985) to become important, although the temperature is still large enough to make the effects important only for species, such as HC_3N, which possess a large permanent electric dipole moment. The original results tabulated by Nejad and Millar (1987) assumed a constant gas temperature of 10 K throughout the envelope and no enhanced rate coefficients; columns 4 and 5 in Table II show the results when a radial temperature profile, taken from Glassgold et al. (1987), and rate coefficients calculated according to the prescription given by Herbst and Leung (1986) are used. As found by Glassgold et al., the HC_3N abundance increases, due to reactions involving $C_2H_2^+$ and CN and HCN, and is much more in agreement with observations. It is interesting to note that the abundances of protonated ions, such as $HCNH^+$ and $H_2C_3N^+$, also increase; a similar effect occurs also in dense interstellar clouds (Millar et al. 1985). Some species are actually less abundant. An example is C_3O whose column density decreases by a factor of 5.5 due to the adoption of a temperature profile in Model (b). C_3O is formed following the radiative association reaction between $C_2H_2^+$ and CO for which the rate coefficient decreases as the temperature increases. The region in which $C_2H_2^+$ is abundant has a higher temperature in Model (b) than in Model (a) leading to a less efficient formation of C_3O in Model (b).

The species C_2H_2 and $C_2H_2^+$ play an important role in determining the hydrocarbon chemistry which occurs in the envelope. Insertion reactions involving C^+, fixation reactions involving C and condensation reactions involving C_2H_2 and $C_2H_2^+$ form the C_3- and C_4-bearing hdrocarbons as well as those containing longer carbon chains. Although these latter species have not, as yet, been included in any detailed numerical model, Glassgold et al. (1986) have shown that they should be formed in an efficient manner.

The $C_2H_2^+$ ion also reacts with other species in the envelope. One important reaction is the radiative association

$$C_2H_2^+ + CO \longrightarrow H_2C_3O^+ + h\nu$$

which, followed by dissociative recombination with electrons, is thought to form C_3O in interstellar clouds (Herbst, Smith and Adams 1984). Such a process can lead to the formation of C_3O also in IRC+10216 and the detection of this molecule would be a strong indication that the type of chemistry outlined here is correct although, as noted above, the abundance of C_3O is sensitive to the adopted radial temperature profile.

The methyl ion, CH_3^+, forms in the envelope from the photodestruction of CH_4 and since it reacts slowly with H_2, it can react with more minor species in the gas. Its most important reaction is with HCN

$$CH_3^+ + HCN \longrightarrow CH_3CNH^+ + h\nu$$

which leads to the formation of methyl cyanide, CH_3CN. It also has a slow radiative association reaction with CO which can lead to ketene, CH_2CO, but this reaction is slow enough so that, although CO is more abundant than HCN, methyl cyanide is more abundant than ketene. In fact, the reaction of CH_3^+ with HCN is so rapid that other CH_3-containing species are much less abundant than CH_3CN.

The ion HCO^+ is formed and destroyed by proton transfer reactions in the inner region of the envelope and formed by other ion-molecule reactions at $r \sim 10^{17}$ cm and has a column density of $\sim 3\times10^{12}$ cm^{-2}. Glassgold et al. (1987) have presented an extensive discussion of the HCO^+ abundance and its sensitivity to parameters such as mass-loss rate and the cosmic-ray ionisation rate. In particular, their calculated antenna temperature for the $J = 1 - 0$ line is consistent with the upper limit obtained by Lucas et al. (1986) with the IRAM 30 m telescope and should be detectable with the Nobeyama 45 m dish, although the possibility exists that, if the electron abundance in the inner envelope is larger than that calculated, possibly from the ionisation of a small amount of atomic carbon left over from dust formation, then the HCO^+ abundance could be reduced significantly (Glassgold et al. 1987).

As mentioned above, the low HNC/HCN ratio appears not to be due to LTE chemistry. Rather the observed ratio arises naturally in the ion-molecule chemistry of the outflow (Glassgold et al. 1987, Nejad and Millar 1987). The $HCNH^+$ ion which forms by proton transfer close to 10^{16} cm can recombine to HNC as well as CN and HCN (Herbst 1978). In addition, the reaction

$$C^+ + NH_3 \longrightarrow H_2NC^+ + H$$

leads to HNC and CN, but not HCN, upon recombination and is the most important source of HNC at 10^{17} cm. Although the chemistry which forms HNC is similar to that occurring in dense interstellar clouds, the HNC/HCN ratio is much smaller in IRC+10216 because the reactions are efficient over a limited region in the outflow, or equivalently, they occur for a short time. Both Glassgold et al. (1987) and Nejad and Millar (1987) find good agreement with the observations without having to invoke LTE chemistry for HNC formation.

In general, the ion-molecule chemistry occurring in IRC+10216 is very similar to that which occurs in a dark cloud such as TMC-1. This explains the many species common to both sources, while the different time-scales and initial conditions explain the differences found for the fractional abundances. In the CSE species are formed as a result of the break-down of stable parents to atoms and atomic ions, while in interstellar clouds molecule formation leads to the build-up of stable species from atoms and atomic ions.

3.2. Chemistry in Oxygen-rich Envelopes

Until recently, there were relatively few observations of thermal molecular emission from the envelopes of oxygen-rich, late-type stars. The early chemical models were concerned generally with the production of maser molecules such as H_2O and OH (Goldreich and Scoville 1976, Huggins and Glassgold 1982b) and SiO (Clegg, Van IJzendoorn and Allamandola 1983). The models of Goldreich and Scoville and of Huggins and Glassgold included only a simple chemistry involving the photodissociation of H_2O but were able to account for several of the observational results including the observed relation between the size of the OH maser region and the mass-loss rate of the central star (Huggins and Glassgold 1982b). Clegg et al. (1983) included the chemistry of silicon in some detail, but within the context of a static atmosphere irradiated by chromospheric radiation.

Scalo and Slavsky (1980) were the first to present a detailed model of the chemical processes which occur throughout the CSE. Their model calculated molecular abundances from the LTE region out to the external envelope where photoprocesses dominate and included a detailed calculation of the thermal balence within the CSE. This model showed the importance of neutral chemistry in an oxygen-rich CSE, in contrast to the case in carbon-rich CSEs. Neutral chemistry is much more important because OH becomes the most abundant reactive species in the envelope. Since neutral radical chemistry is important, the detailed results of chemical abundance calculations are sensitive to the presence of activation energy barriers in several reactions and hence to the temperature profile assumed for the CSE. Since current experimental studies involving radical reactions are almost all carried out at room temperature, and in any case above 200 K, small activation barriers, for example less than 100 K, are not ruled out and can inhibit important reactions in the outer CSE where the temperature is lower than 100 K (see the article by I.W.M. Smith, this volume).

Slavsky and Scalo (1986) have discussed the chemistry of oxygen-rich CSEs in some detail and have, in particular, shown that the amount of atomic hydrogen in the CSE can have a profound effect on the chemical balance in the inner envelope. Reactions involving H atoms can prevent complete hydrogenation of species containing O, N and S atoms since H atoms abstract hydrogen from simple hydride molecules. The abundance of H atoms in CSEs is uncertain, although Glassgold and Huggins (1983) have discussed this theoretically. The showed that, in the absence of chromospheric radiation and for cool central stars, hydrogen will be in molecular form out to a radial distance of $\sim 10^{18}$ cm for $\dot{M} \sim 10^{-4}$ $M_{\odot}$ yr^{-1} due to self-shielding.

The chemistry in oxygen-rich envelopes depends primarily on two main species, OH produced in the photodissociation of H_2O and C^+ produced as the end-product of the CO photo-chain. We discuss these in turn.

3.2.1. The production of OH and its related chemistry. The photodissociation of H_2O has been studied by Hudson (1971) and summarised by Mamon et al. (1987). For a species such as H_2O, which has both a large column density and a large cross-section for photodissociation, self-shielding must be taken into account in addition to the usual dust shielding. Mamon et al. (1987) have assumed that H_2O is dissociated in three bands and have given detailed cross-sections, photo-rates and scaling lengths, d_i, for a mass-loss rate of 10^{-5} $M_\odot$ yr^{-1}, reproduced in Table III. The photodissociation and photoionisation processes in OH have been discussed by van Dishoeck and Dalgarno (1984) - see also the article by van Dishoeck in this volume - and these data are also given in Table III.

Species	$\sigma_d(cm^{-2})$	$\beta_d(s^{-1})$	$\sigma_i(cm^{-2})$	$\beta_i(s^{-1})$	$d_i(cm)$
H_2O	1.75(-17)	1.60(-10)	1.75(-17)	2.50(-11)	3.6(16)
	7.50(-18)	2.65(-10)			2.3(16)
	5.00(-18)	1.05(-10)			2.4(16)
OH		2.80(-10)		1.25(-11)	2.7(16)

Table III: Photodissociation and photoionisation rates for H_2O and OH with shielding lengths, d_i, (Mamon et al. 1987).

As is the case for carbon-rich CSEs, cosmic-ray ionisation produces H_3^+ and, via proton transfer reactions, HCO^+ and, most importantly, H_3O^+, in the inner envelope. Figure 2, taken from Mamon et al. (1987), shows the fractional abundance of H_3O^+ for a number of mass-loss rates. Deep in the envelope the almost linear dependence of $x(H_3O^+)$ with radial distance can be understood in terms of the formation and destruction reactions

$$H_3^+ + H_2O \longrightarrow H_3O^+ + H$$

and

$$H_3O^+ + e \longrightarrow \text{products.}$$

Using the expression for the H_3^+ abundance derived in section 2.4 and noting that, because of rapid proton transfer, $x(H_3O^+) = x(e)$, we find

$$x(H_3O^+) = [\zeta/n(r)k_e]^{1/2}$$

where k_e is the rate coefficient for the dissociative recombination of H_3O^+. This equation, which overestimates the abundance slightly, can be written as, with $\zeta = 10^{-17}$ s^{-1} and $k_e = 10^{-6}$ cm^3 s^{-1},

$$x(H_3O^+) = 10^{-8}r_{16}(v_{15}/\dot{M}_{-5})^{1/2}$$

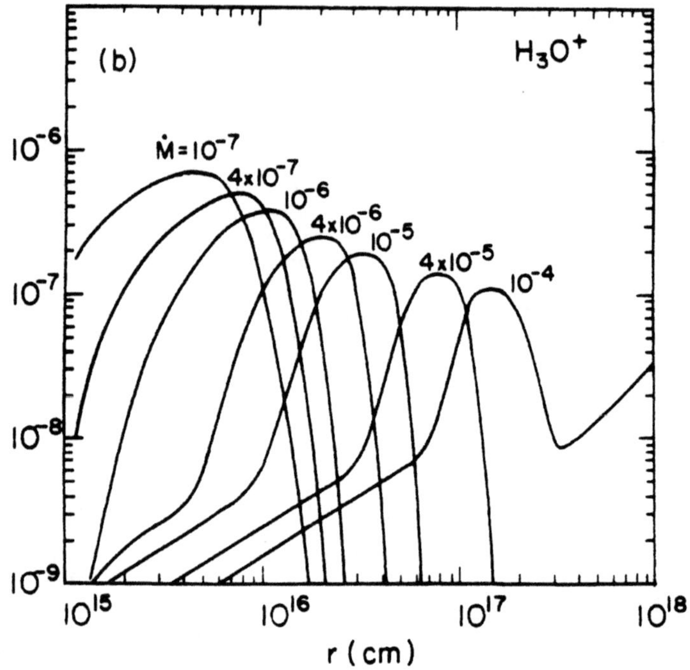

Figure 2. The fractional abundance of H_3O^+ is plotted as a function of radius for various mass-loss rates (Mamon et al. 1987).

which shows the dependence of $x(H_3O^+)$ on both radius and mass-loss rate, as discussed more fully by Mamon et al. (1987).

As the radial distance increases, the external radiation field dominates cosmic-ray ionisation and the radical OH becomes the dominant reactive species. In addition to loss through photoprocesses, OH can react with a large number of species and, in particular, with O and N atoms, the latter formed in the photodissociation of N_2, to produce O_2 and NO. The reaction

$$O + OH \rightarrow O_2 + H$$

is thought to occur without activation energy and enables a large amount of O_2 to form (Slavsky and Scalo 1986, Mamon et al. 1987, Nejad and Millar 1988). Carbon dioxide, CO_2, can be produced in the reaction

$$CO + OH \rightarrow CO_2 + H$$

which possesses an activation energy barrier of 80 K (Davis, Fisher and Schiff 1974) so that CO_2 formation is efficient only in gas which is both warm and contains a significant abundance of OH.

Several other reactions involving OH and O are important (Slavsky and Scalo 1986, Nejad and Millar 1988). The parent molecule H_2S is photodissociated to SH, S and S^+. Since these species are unreactive with H_2, reactions such as

$$S + OH \longrightarrow SO + H$$
$$O + SH \longrightarrow SO + H$$
and
$$SO + OH \longrightarrow SO_2 + H$$

form SO and SO_2, while if free silicon exists in the gas SiO forms via

$$Si + OH \longrightarrow SiO + H.$$

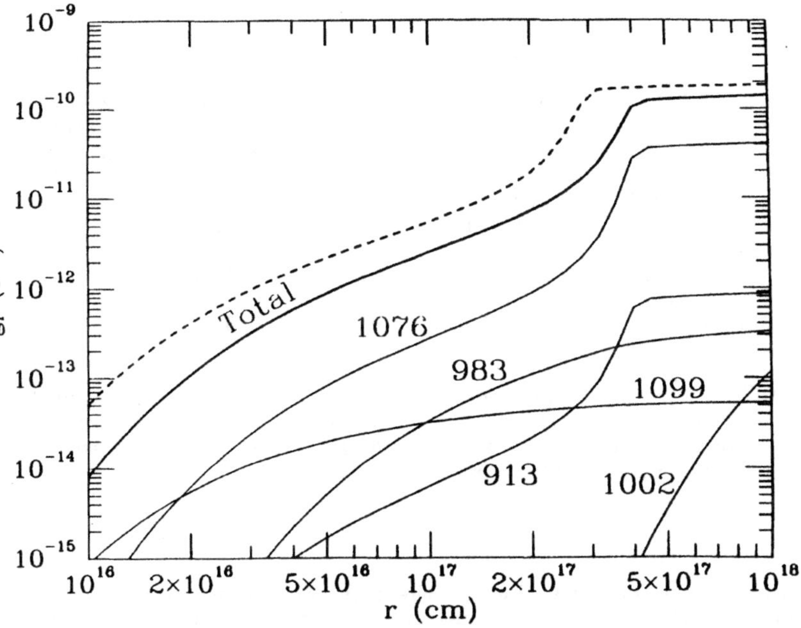

Figure 3. The photodissociation rate of individual CO bands, labelled by wavelength in Angstroms, as a function of radius for $\dot{M} = 10^{-5}$ $M_\odot$ yr^{-1} and v = 15 km s^{-1}. The total rate is shown both with and without (dashed line) H_2 blocking (Mamon et al. 1988).

3.2.2. The production of C^+ and its related chemistry. The photodissociation of CO is the major source of C^+ in an oxygen-rich envelope. It is now known that the process proceeds through line absorption (Letzelter et al. 1987) so that CO self-shields against photodissociation. Mamon et al. (1988) have discussed the photodissociation of CO in CSEs in some detail. Based on the laboratory results, they have calculated the rate by considering self-shielding in each of the relevant bands as well as mutual shielding by H_2 Lyman and Werner bands and also dust shielding. For cool stars with photospheric temperatures < 3000 K, hydrogen is mostly molecular (Glassgold and Huggins 1983); in warmer stars, atomic hydrogen can be abundant and gives rise to mutual shielding also.

Figure 3, taken from Mamon et al. (1988), shows the effects of self- and mutual shielding on several CO bands, labelled by their

Species	Model(a)	Model(b)	Species	Model(a)	Model(b)
OH	9.9(16)	9.8(16)	SH	2.0(15)	2.0(15)
O_2	4.3(15)	4.2(15)	SO	4.2(14)	4.4(14)
NO	2.1(15)	2.0(15)	SO_2	4.6(13)	4.7(14)
CO_2	3.6(14)	3.6(14)	CS	5.4(13)	8.9(13)
CH	2.2(14)	2.0(14)	NS	4.9(14)	4.3(14)
CH_2	8.1(14)	7.4(14)	S_2	8.7(13)	2.4(13)
CH_3	2.2(14)	2.2(14)	SO^+	4.0(13)	1.2(14)
CN	2.1(13)	2.0(13)	HCO^+	2.1(12)	1.9(12)
HCN	1.3(13)	1.2(13)	HCS^+	1.3(13)	5.5(13)
HNC	4.4(12)	5.8(12)	H_3O^+	2.6(13)	2.8(13)

Table IV: Calculated column densities, N(cm^{-2}) for an oxygen-rich CSE with $\dot{M} = 8 \times 10^{-5}\ M_{\odot}\ yr^{-1}$. Model (a) is the standard model of Nejad and Millar (1988); Model (b) incorporates polar rate coefficients.

wavelengths in Angstroms, and on the total photodissociation rate (dashed line) when H_2 blocking is ignored. The behaviour of these individual band dissociation rates with radius has been discussed by Mamon et al. (1988).

Since CO is fairly unreactive with most gas phase species, the radial abundance of CO can be determined by assuming that photodissociation is the only loss for CO. Figure 4, from Mamon et al.(1988), shows the resulting distribution for a variety of mass-loss rates. These authors have presented a number of approximate results which reproduce the full numerical results.

A detailed model of the chemistry in oxygen-rich envelopes has been constructed by Nejad and Millar (1988) and considers, in a fashion similar to that described above for IRC+10216, the chemistry of H,O,C, and N species when several parents are injected into the outer envelope. Table IV gives calculated column densities for a

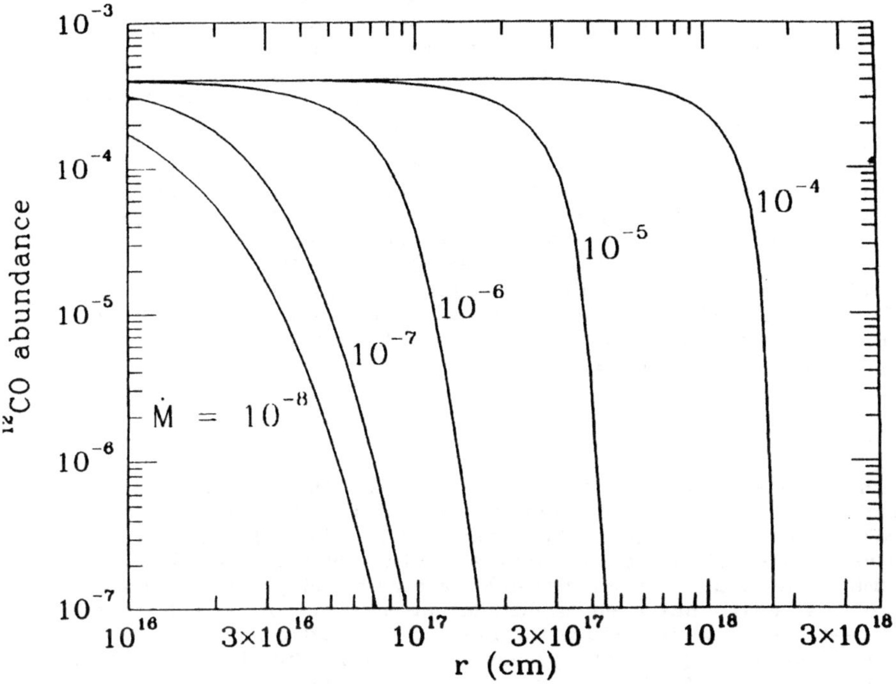

Figure 4. The fractional abundance of CO calculated as a function of radius for various mass-loss rates (Mamon et al. 1988).

number of species for the 'standard model' of Nejad and Millar (1988). In particular, we include the results of a calculation incorporating the effects of ion-dipolar molecule collisions using the approach described by Herbst and Leung (1986) which uses the trajectory scaling method to evaluate individual rate coefficients. As can be seen from Table IV, enhanced rate coefficients do not alter the calculated column densities to any great extent.

Nejad and Millar (1988) were primarily concerned with trying to explain the observations of HCN in oxygen-rich envelopes. C-N bonds can be formed in the outer envelope through ion-molecule reactions such as

$$C^+ + NH_3 \longrightarrow H_2NC^+ + H$$

and in neutral-neutral reactions such as

$$N + CH_3 \longrightarrow HCN + H_2.$$

As in carbon-rich envelopes, reactions occur which can 'shuffle' HCN <--> HNC. Nejad and Millar (1988) find that an appreciable column density of HCN can arise only if CH_4 is present as a parent species with an abundance much larger than that predicted by LTE calculations and if the photodissociation rate of N_2 is large, ~ 10^{-10} s^{-1} (van Dishoeck 1987). In addition, they find the abundance ratio HCN/HNC ~ 3, close to that observed in OH231.8+4.2 (Morris et al. 1987).

While the abundances of most neutral species are not much affected by enhanced rate coefficients, those of some protonated species such as HCS^+ do increase. The detection of H_3O^+ has been discussed in some detail by Mamon et al. (1988); here we simply note that in the case of enhanced rate coefficients the HCS^+/CS abundance ratio is ~ 0.6 while the observed upper limit in OH231.8+4.2 is 0.54 (Morris et al. 1988), although one must remember that, in this object, the observed line-widths may indicate that shock chemistry is important.

4. CONCLUSIONS

In this review we have attempted to show that the circumstellar envelopes of cool, late-type stars possess a rich chemistry which is similar in many respects to that occurring in interstellar clouds. In carbon-rich envelopes, cosmic-rays and ultraviolet photons drive a chemistry dominated by ion-molecule reactions and photo-reactions. Such a chemistry has been applied to the envelope of IRC+10216 and has been shown to reproduce the observations extremely well. In oxygen-rich envelopes these processes also occur but the presence of large amounts of OH make neutral chemistry more important. In both cases the effects of ion-dipolar collisions has little effect on abundances, with the exception of HC_3N and some protonated species (Glassgold et al. 1987, Millar 1987, unpublished).

Much work on circumstellar chemistry remains to be done. Mamon et al. (1988) have shown that self-shielding and mutual shielding play an important role in determining the spatial distribution of CO and hence of CI and C^+ in CSEs. The details of this process need to be included in a comprehensive chemical model. The recent detection of several sulphur species in IRC+10216 (Cernicharo et al. 1987) and in OH231.8+4.2 (Morris et al. 1987) should encourage the inclusion of sulphur chemistry in models. This may provide some insight into models of interstellar sulphur chemistry for which much uncertainty remains.

Models can also be developed which relax the assumption of spherical symmetry (Jura 1983) and indeed which can incorporate shock waves and internal radiation. Such models may give us some information on the evolution of planetary nebulae from the cool, late-type stars. In this regard it is of interest that Olofsson (private communication) has detected a thick CO shell which has detached itself from its central star, S Sct.

Photoreactions	Neutral-neutral
CH_3 + hυ	O + OH
HC_3N + hυ	CO + OH
C_2H + hυ	C_2H + C_2H
Ion-Molecule	**Dissociative recombination**
N + $C_3H_3^+$	$C_2H_2^+$ + e
C + $C_3H_3^+$	$H_2C_3N^+$ + e
S^+ + C_3H_2	$H_2C_3O^+$ + e

Table V: Some critical reactions for which laboratory information is needed.

Laboratory studies will continue to give insight on the importance of various chemical processes. In particular photoprocesses, especially involving radicals, need to be investigated and here, of course theory can make an important contribution (van Dishoeck 1987, this volume). Reactions involving acetylene and its associated ions and neutrals need to be studied at low temperature as do reactions of CI atoms which, as in interstellar clouds, can be important in the synthesis of complex molecules. Neutral reactions, particularly involving OH, need to be investigated at low temperatures in order to identify small activation energies which can prevent reaction at the low temperatures occurring in circumstellar envelopes. Table V gives a representative list of reactions for which laboratory information is crucial. All of these reactions occur also in interstellar clouds but they are much more important in CSEs.

The study of chemistry in CSEs has provided another 'laboratory' in which to explore the chemical processes thought to occur in interstellar clouds. The success of detailed models in reproducing the observed abundances, at least in IRC+10216 for which most information is available, gives one confidence that the basic chemical processes occurring in these disparate regions are essentially correct. As models of the chemistry and excitation of CSEs become more complex it may even be possible to constrain some unknown rate coefficients. For example, Nejad and Millar (1987) argued that the large photodissociation rate of CN advocated by Lavendy et al. (1984) was unlikely to be correct since the resulting CN distribution calculated for IRC+10216 was much smaller than that observed, a conclusion strengthened by recent theoretical calculations of CN photodissociation (van Dishoeck, this volume).

Finally, one should mention that molecules of increasing chemical complexity are being detected in CSEs. Carbon-rich envelopes, such as that surrounding IRC+10216, are presumably the birthplace of the polycyclic aromatic hydrocarbons (PAHs) and observational and theoretical studies of the chemistry of such envelopes may lead to a better understanding of the origin and significance of these exotic species. Indeed, since dust grains are known to form in circumstellar shells, studies of molecular line emission may provide information on the nucleation and subsequent growth of solid matter. In this context, the recent laboratory data on silicon reactions (Bohme, this volume) need to be included into models of oxygen-rich CSEs since such reactions may play a key role in the formation of silicate dust particles.

5. ACKNOWLEDGEMENTS

I am grateful to A. E. Glassgold and A. Omont for sending me preprints of their articles on CSEs and to L. A. M. Nejad for collaboration on much of the work described in this article.

REFERENCES

Adams, N.G., Smith, D. and Clary, D.C. 1985. Astrophys. J., **296**, L31.
Betz, A.L. 1987. In 'Astrochemistry' eds. M.S. Vardya and S.P. Tarafdar, D. Reidel (Dordrecht), p.327.
Betz, A.L., McLaren, R.A. and Spears, D.L. 1979. Astrophys. J., **229**, L97.
Bloemhof, E.E., Townes, C.H. and Vanderwyck, A.H.B. 1984. Astrophys. J., **276**, L21.
Cernicharo, J. and Guélin, M. 1987. Astron. Astrophys., **183**, L10.
Cernicharo, J., Guélin, M., Hein, H. and Kahane, C. 1987. Astron. Astrophys., **181**, L9.
Clegg, R.E.S., Van IJzendoorn, L.J. and Allamandola, L.J. 1983. MNRAS, **203**, 125.
Davis, D.D., Fisher, S. and Schiff, R. 1974. J. Chem. Phys., **61**, 2213.
Gerola, H. and Glassgold, A.E. 1978. Astrophys. J. Suppl., **37**, 1.
Glassgold, A.E. and Huggins, P.J. 1983. MNRAS, **203**, 517.
Glassgold, A.E. and Huggins, P.J. 1986. Astrophys. J., **306**, 605.
Glassgold, A.E., Lucas, R. and Omont, A. 1986. Astron. Astrophys., **157**, 35.
Glassgold, A.E., Mamon, G.A., Omont, A. and Lucas, R. 1987. Astron. Astrophys., **180**, 183.
Goldreich, P. and Scoville, N.Z. 1986. Astrophys. J., **205**, 144.
Goldsmith, P.F., Irvine, W.M., Hjalmarson, Å. and Elldér, J. 1986. Astrophys. J., **310**, 383.
Gredel, R., Lepp, S. and Dalgarno, A. 1988. Astrophys. J., in press.
Herbst, E. 1978. Astrophys. J., **222**, 508.
Herbst, E. and Leung, C.M. 1986. Astrophys. J., **310**, 378.
Herbst, E., Smith, D. and Adams, N.G. 1984. Astron. Astrophys., **138**, L13.
Hudson, R.D. 1971. Rev. Geophys. Sp. Phys., **9**, 305.
Huggins, P.J. and Glassgold, A.E. 1982a. Astrophys. J., **252**, 201.
Huggins, P.J. and Glassgold, A.E. 1982b. Astron. J., **87**, 1828.
Huggins, P.J., Glassgold, A.E. and Morris, M. 1984. Astrophys. J., **279**, 284.

Irvine, W.M. and Schloerb, F.P. 1984. Astrophys. J., **282**, 516.
Johansson, L.E.B., Andersson, C., Elldér, J., Friberg, P., Hjalmarson, Å., Höglund, B., Irvine, W.M., Olofsson, H. and Rydbeck, G., 1984. Astron. Astrophys., 130, 227.
Jura, M. 1983. Astrophys. J., 275, 683.
Jura, M. 1986. Astrophys. J., 303, 327.
Jura, M. and Morris, M. 1981. Astrophys. J., 251, 281.
Knapp, G.R. 1985. Astrophys. J., 293, 273.
Knapp, G.R. 1987. In 'Late Stages of Stellar Evolution' eds. S. Kwok and S.R. Pottasch, D. Reidel (Dordrecht).
Knapp, G.R. and Morris, M. 1985. Astrophys. J., 292, 640.
Kwan, J. and Linke, R.A. 1982. Astrophys. J., 254, 587.
Lafont, S., Lucas, R. and Omont, A. 1982. Astron. Astrophys., 106, 201.
Lavendy, H., Gandara, G. and Robbe, J.M. 1984. J. Mol. Spectros., 106, 395.
Letzelter, C., Eidelsberg, M., Rostas, F., Breton, J. and Thieblemont, B. 1987. Chem. Phys., 114, 273.
Lucas, R., Omont, A., Guilloteau, S. and Nguyen-Q-Rieu, 1986. Astron. Astrophys., 154, L12.
Mamon, G.A., Glassgold, A.E. and Huggins, P.J. 1988. Astrophys. J., in press.
Mamon, G.A., Glassgold, A.E. and Omont, A. 1987. Astrophys. J., 323, 306.
McCabe, E.M., Connon Smith, R.C. and Clegg, R.E.S. 1979. Nature, 281, 263.
Millar, T.J., Adams, N.G., Smith, D. and Clary, D.C. 1985. MNRAS, 216, 1025.
Morris, M., Guilloteau, S., Lucas, R. and Omont, A. 1987. Astrophys. J., 321, 888.
Nejad, L.A.M. 1986. Ph.D. Thesis, Univ. of Manchester.
Nejad, L.A.M. and Millar, T.J. 1987. Astron. Astrophys., 183, 279.
Nejad, L.A.M. and Millar, T.J. 1988. MNRAS, 230, 79.
Nejad, L.A.M., Millar, T.J. and Freeman, A. 1984. Astron. Astrophys., 134, 129.
Olofsson, H. 1985. In '(Sub)millimeter Astronomy' eds. P.A. Shaver and K. Kjär, ESO Publn. No.22, p.535.
Omont, A. 1985. In 'Mass Loss from Red Giants' eds. M. Morris and B. Zuckerman, D. Reidel (Dordrecht), p.269.
Prasad, S.S. and Tarafdar, S.P. 1983. Astrophys. J., 267, 603.
Ridgeway, S.T. and Keady, J.J. 1987. In 'Circumstellar Matter' eds. I. Appenzeller and C. Jordan, D. Reidel (Dordrecht), p.535.
Sahai, R. 1987. Astrophys. J., 318, 809.
Sahai, R. and Wannier, P.G. 1985. Astrophys. J., 299, 424.
Scalo, J.M. and Slavsky, D.B. 1980. Astrophys. J., 239, L73.
Slavsky, D.B. and Scalo, J.M. 1986. Preprint.
Sternberg, A., Dalgarno, A. and Lepp, S. 1987. Astrophys. J., 320, 676.
Trong-Bach, Nguyen-Q-Rieu, Omont, A., Olofsson, H. and Johansson, L.E.B. 1987. Astron. Astrophys., 176, 285.
Tsuji, T. 1973. Astron. Astrophys., 23, 411.

Tsuji, T. 1978. Astron. Astrophys., 62, 29.
Van Dishoeck, E.F. 1987. In 'Astrochemistry' eds. M.S. Vardya and S.P. Tarafdar, D. Reidel (Dordrecht), p.51.
Van Dishoeck, E.F. and Dalgarno, A. 1984. Astrophys. J., 227, 576.
Wannier, P.G., Leighton, R.B., Knapp, G.R., Redman, R.O., Phillips, T.G. and Huggins, P.J. 1979. Astrophys. J., 230, 149.
Wannier, P.G. and Sahai, R. 1986. Astrophys. J., 311, 335.
Wootten, A., Lichten, S.M., Sahai, R. and Wannier, P.G. 1982. Astrophys. J., 257, 151.
Zuckerman, B. and Dyck, H.M. 1986a. Astrophys. J., 304, 394.
Zuckerman, B. and Dyck, H.M. 1986b. Astrophys. J., 311, 345.

CONDENSATION CALCULATIONS IN CIRCUMSTELLAR SHELLS FOR DIFFERENT C/O RATIOS

C. M. Sharp
Max-Planck-Institut für Physik und Astrophysik
Institut für Astrophysik
Karl-Schwarzschild-Straße 1
8046 Garching bei München
Federal Republic of Germany

ABSTRACT. We have performed some equilibrium calculations using a modified form of the computer code SOLGASMIX to calculate the abundances of a number of important gas phase and condensed phase species at temperatures below 2000K which are relevant to very cool stellar atmospheres and circumstellar shells. Calculations have been done for a number of different C/O ratios from oxygen rich to carbon rich, and the condensation of grains can significantly alter the C/O ratio of the gas phase, which can have a large effect on a number of molecular species. It was found that for some moderately oxygen rich mixtures the condensation of silicates can substantially enhance the abundance of gas phase species like HCN which was looked at in some detail.

1. INTRODUCTION

At temperatures and densities of circumstellar shells, the chemistry is in general likely to be highly non-equilibrium. However, in the hotter and denser inner parts of stellar winds, circumstellar shells and in particular cool stellar atmospheres, equilibrium chemistry may be a reasonable approximation, and can at least give us a basis from which non-equilibrium chemistry could be considered.

As CO is the most tightly bound diatomic molecule, in a cool stellar atmosphere the chemistry is governed by the C/O ratio, such that whichever is the lesser of the two elements in abundance is almost completely locked up in CO with the other element free to form other compounds. In the carbon rich case, i.e. C/O>1, molecules such as HCN and C_2H_2 are very abundant species. At sufficiently low temperatures graphite and carbides will condense out removing carbon from the gas phase and decreasing the gas phase C/O ratio. However, of particular interest is the case when C/O<1 when silicates will condense out removing oxygen and increasing the gas phase C/O ratio. In this case species like HCN which would normally be negligible, could increase in abundance by a very large amount.

Because of the importance of grain condensation in the outer

T. J. Millar and D. A. Williams (eds.), Rate Coefficients in Astrochemistry, 309–313.

atmospheres and circumstellar shells of many stars ranging from oxygen rich to carbon rich, we have decided to investigate the effects of condensation on the gas phase chemistry for a large number of different C/O ratios. As the microwave emission of HCN has recently been detected in several circumstellar envelopes conventionally designated as "oxygen rich", Deguchi et al. (1986) and Jewell at al. (1986), we were particularly interested in looking at the effects of condensation on the abundance of that molecule for different C/O ratios.

2. CALCULATION

We have used a modified version of the computer code SOLGASMIX, Besmann (1977), where the equilibrium composition is computed by directly minimising the total free energy of the system. This code has the great advantage over others of solving simultaneously the equilibria of the gas phase and several condensed phases, when they exist.

The Gibbs free energies of formation of a number of gaseous and condensed species from their elements in their reference states were obtained from the JANAF tables, Chase (1982). These were then converted to the energies of formation from the elements in their neutral ideal monatomic gaseous state, and polynomials of the type

$$\Delta G_{pi}(T) = a/T + b + cT + dT^2 + eT^3, \qquad (1)$$

were fitted, where $\Delta G_{pi}(T)$ is the Gibbs energy of formation in calories per mole at temperature T of species i in phase p.

The total free energy of the system expressed as the dimensionless quantity G(T)/RT is given by

$$\frac{G(T)}{RT} = \left\{ \sum_{i=1}^{m_p} n_{pi} \left[\frac{\Delta G_{pi}(T)}{RT} + \ln P + \ln \left[\frac{n_{pi}}{N_p} \right] \right] \right\}_{p=1} + \sum_{p=2}^{s+1} \sum_{i=1}^{m_p} n_{pi} \frac{\Delta G_{pi}(T)}{RT}, \qquad (2)$$

where m_p is the number of species in phase p, n_{pi} is the number of moles of species i in that phase, N_p is the total number of moles in that phase, P is the total pressure in atmospheres, s is the number of condensed phases, and p=1 corresponds to the gas phase while p>1 corresponds to the condensed phases. In order to compute the composition, eq. (2) was minimised subject to the mass balance relationship

$$\sum_{p=1}^{s+1} \sum_{i=1}^{m_p} \nu_{pij} n_{pi} = b_j \qquad \text{for } j=1 \text{ to } N, \qquad (3)$$

where ν_{pij} is the stoichiometric coefficient of element j in species i in phase p, and b_j is the number of gramme-atoms of element j. In this work we considered only ideal gaseous and ideal distinct solid species, and no solution of one condensed species in another was considered, so additional terms in eq. (2) were omitted.

3. RESULTS

The calculations were performed with the 10 most abundant elements H, C, N, O, Mg, Al, Si, S, Ca and Fe, excluding the inert gases, using the solar Cameron (1973) abundances (excluding C and O which were varied). A total of 98 species were considered as follows: the 10 elements above as monatomic species, 33 diatomic and 36 polyatomic species in the gas phase, and 19 condensed phase species. The condensates considered with their mineral names in parentheses were as follows: C(graphite), Fe, AlN, SiC, CaO, CaS, FeO, FeS, MgO, Al_2O_3(corundum), Fe_2O_3(hematite), $MgSiO_3$(enstatite), $MgAl_2O_4$(spinel), Mg_2SiO_4(forsterite), Al_2SiO_5(kyanite), $CaMgSi_2O_6$(diopside), $Ca_2Al_2SiO_7$ (gehlenite), $Ca_2MgSi_2O_7$(akermanite) and $CaAl_2Si_2O_8$(anorthite).

Calculations were done at a total pressure of 10^{-4} atmospheres, i.e. about 100 dyne cm^{-2}. As mentioned above, it was assumed that the solid phases remained as distinct species and solutions were neglected. In fact gehlenite and akermanite which were treated separately form the solid solution melilite, but this approximation is not expected to have a major effect on the gas phase chemistry which we are looking at here.

Figure 1 shows the log relative abundance of HCN to H_2 plotted against 1000/T for the temperature range 2000 to 625K for 29 C/O ratios varying between 0.55 (solar) and 1.15 (slightly carbon rich). Also plotted are a number of curves labelled by letters that are the loci of points at which a condensed species appears or reacts to form other species as identified in table I that noticeably affect the HCN abundance, with other condensates like Fe being omitted. It is seen that a substantial enhancement in the relative HCN abundance can occur at about 1000K for a mixture that is still regarded as oxygen rich with a C/O ratio less than about 0.9 but greater than about 0.82, and is due to the depletion of oxygen in the gas phase by the condensation of a number of oxygen bearing minerals, in particular the two very important species Mg_2SiO_4 and $MgSiO_3$. It was also found that the abundance of C_2H_2 behaved very similarly to HCN as expected, and conversely H_2O and CO_2 decreased substantially at the corresponding temperatures.

Even for these nominally oxygen rich mixtures, the gas phase becomes sufficiently rich in carbon for graphite to condense at the points marked l. At lower temperatures graphite disappears at the points marked m as it reacts with H_2 to produce CH_4. The line on which the points l and m lie is in fact the vapour pressure curve of graphite when scaled in the appropriate units.

When the calculations were continued for larger C/O ratios, the chemistry became very much more complicated with the abundance being very sensitive to small changes in the C/O ratio and the condensation of species of relatively low abundance, such as Al_2O_3. At a C/O ratio greater than 0.97 rather than unity, the chemistry was seen to switch abruptly to the carbon rich case. Finally, for C/O ratios greater than 1.1 graphite is the first species to condense out, and its vapour pressure then determines the chemistry at lower temperatures. Calculations performed at a higher total pressure of 10^{-3} atmospheres produced qualitatively similar results, except that the relative HCN abundance increases in both the absence and presence of condensation.

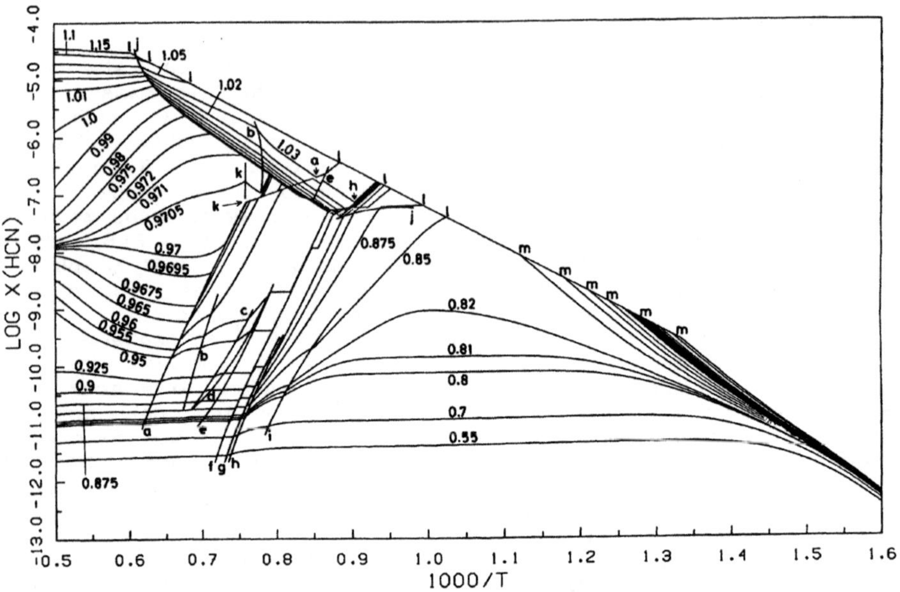

Figure 1. Relative abundance of HCN to H_2 for different C/O ratios at 10^{-4} atmospheres. The appearance or disappearance of condensates as identified in table I is marked a to m.

<u>TABLE I - LIST OF CONDENSATES WITH (A) = APPEARANCE AND (D) = DISAPPEARANCE OF A SPECIES WITH DECREASING TEMPERATURE</u>

a	Al_2O_3	(A)
b	CaS	(A)
c	$Ca_2Al_2SiO_7$	(A)
d	CaS	(D)
e	$MgAl_2O_4$	(A)
	Al_2O_3	(D)
f	$Ca_2MgSi_2O_7$	(A)
	$Ca_2Al_2SiO_7$	(D)
g	$CaMgSi_2O_6$	(A)
	$Ca_2MgSi_2O_7$	(D)
h	Mg_2SiO_4	(A)
i	$MgSiO_3$	(A)
j	SiC	(A)
k	AlN	(A)
l	C (graphite)	(A)
m	C (graphite)	(D)

When the C/O ratio of the gas phase was plotted against temperature for different total C/O ratios, it was indeed found that the effect of condensation made the gas phase C/O ratio approach unity from either the carbon or oxygen rich cases, but with graphite having by far the strongest effect.

As a check on our method and data we compared our results for a solar mixture at a total pressure of 10^{-3} atmospheres with those of table 3 in Lattimer et al. (1978). With the exception of $MgAl_2O_4$, $CaAl_2Si_2O_8$ and Al_2SiO_5 for which there were serious discrepancies, the agreement was good, and even though melilite was not considered, the first appearance and last disappearance of the constituent species which we treated separately were in reasonable agreement.

4. CONCLUSIONS

We have shown that with the assumption of equilibrium chemistry, the condensation of grains in cool stellar atmospheres and circumstellar shells can in certain cases substantially modify the chemistry of the gas phase. For the envelopes with positive detection of HCN, Deguchi et al. (1986) estimated the circumstellar abundance relative to H_2 at 10^{-8} to 10^{-7}. Reference to figure 1 indicates that for a C/O ratio close to solar only a minute increase of a very small HCN abundance can occur, but for C/O > 0.82, much larger enhancements in the abundance can take place, becoming comparable to the observed values.

Our calculations are meant to be applied to the photospheric or near photospheric environment where densities are high enough for us to calculate these equilibria on the basis of LTE. The equilibrium reactions of any species produced such as HCN carried out on a wind will eventually be "frozen out", due to the falling density , so could be observed at the much larger distances of microwave emission, neglecting the effects of other non-equilibrium processes.

Although Lattimer et al. (1978) have considered equilibria with a more extensive list of condensates, the addition of more condensed species to these calculations is not expected to qualitatively affect the gas phase C/O ratio or the behaviour of species like HCN for C/O < 0.9 and temperatures of about 1000K. However, we are making the assumption that equilibrium chemistry is valid, and in many cases it would be more realistic to take account of reaction rates. Short of that, the next level of approximation is to assume that below some temperature a condensed species no longer reacts with the gas phase or another condensed species.

5. REFERENCES

Besmann, T.M. (1977), *Oak Ridge National Laboratory*, Report No. TM-5775.

Cameron, A.G.W. (1973), *Space Science Reviews*, **15**, 121.

Chase, M.W., *JANAF Thermodynamic Tables*, Magnetic Tape Version (1982), The Dow Chemical Company.

Deguchi, S., Claussen, M.J., & Goldsmith, P.F. (1986), *Ap.J.*, **303**, 810.

Jewell, P.R., Snyder, L.E., & Schenewerk, M.S. (1986), *Nature*, **323**, 311.

Lattimer, J.M., Schramm, D.N.,& Grossman, L., (1978), *Ap.J.*, **219**, 230.

CHEMISTRY IN PRIMARY T TAURI WINDS

J.M.C. Rawlings
Mathematics Department
UMIST
P O Box 88
Manchester, M60 1QD

ABSTRACT. The chemistry of dustless T Tauri stellar winds is investigated. In the highly unusual conditions prevalent in the outflow it is found that the chemistry is, in some respects, similar to that of the early universe, although it is severely limited by certain effects common to most circumstellar chemistries. In particular, for the chemistry to be efficient, a significant fraction of hydrogen is required to be in the form of H_2 so that the chemistry is, to a large extent, controlled by the geometrical dilution of the ejecta and collisional dissociation. Indeed, if any molecular species (such as CO) are to be observable at all, then mass loading of the wind with H_2 is required. Thus, a positive detection of CO may provide useful information as to the physics and chemistry of T Tauri winds.

1. INTRODUCTION AND THE PHYSICAL MODEL

The study of the chemical and physical processes occurring in T Tauri outflows is of great importance as these stars have been proposed as the energy source of Herbig-Haro objects (Schwartz 1978; Cantó 1978) and turbulence in molecular clouds (Norman and Silk 1980, Franco 1983). In this work, the physical model of the outflow is taken from Hartmann, Edwards and Avrett (1982) in which the primary stellar wind (i.e. wind that has not interacted with its environment) is ionized and heated by Alfvén waves in the star's convection zone to reach a terminal velocity (of about 230 kms^{-1}) and a maximum temperature (of 20,000 K, cf. Hartmann et al. model no. 2) at z = 3 to 5, where z is the radial ordinate in units of stellar radii ($r_* \sim 2 \times 10^{11}$ cm). Thereafter the wind expands and cools radiatively and adiabatically. Other parameters for the model are the initial wind density at z = 1 ($n_0 \sim 10^{13}$ cm^{-3}), the density at z = 5 ($n \sim 5 \times 10^8$ to 10^9 cm^{-3}) and the stellar photospheric temperature ($T_* \sim 4000$ K). The cooling rate of the wind is obviously dependent on the physical conditions within the ejecta and in any case is by no means certain. Hartmann et al. suggest a

T. J. Millar and D. A. Williams (eds.), Rate Coefficients in Astrochemistry, 315–320.

radiative cooling function for which the timescale is roughly 10^4 s corresponding to a distance of 1 r_*, while adiabative cooling alone implies a temperature drop from T = 15000 K at z = 10 to T = 5000 K at z = 23. Molecular cooling (especially H_2) is not expected to be significant but in order to test the sensitivity of the chemistry an additional power law term was included in the cooling function together with the adiabative and radiative terms discussed above.

The ionization structure of the wind is also very important in determining the chemistry. Since the photospheric temperature is so low, the initial ionization is simply taken to be that of material in LTE at z = 3 to 5 where the ejecta temperature is at a maximum, so that the ionization is simply given by the balance of radiative recombination with the collisional ionization rates. Once the ejecta cools below about 15000 K, recombination occurs. For T Tauri ejecta temperatures and densities (at z = 10 where T ~ 15000 K) this recombination will occur over a characteristic recombination length of $\Delta z \sim 6$. Thus the conditions within the ejecta are changing from being hot, dense and ionized to being relatively cool, diffuse and neutral.

2. CHEMICAL MODELLING

In order to account for the temporal and spatial variations of the ejecta conditions discussed above, it has been necessary to develop a time-dependent chemical model (Rawlings, Williams and Cantó, 1988).

The T Tauri chemistry is highly unusual in that the radiation field originating from the T Tauri star is very strong in the IR but very weak in the UV (typically at z = 10 8 orders larger than that applicable to the ISM in the IR and of comparable size to that of the ISM in the UV). This essentially means that, in the absence of dust and the large densities required for three body H_2 formation, the only viable H_2 formation route is via H_2^+ i.e.

$$H + H^+ \longrightarrow H_2^+ + h\nu$$
$$H_2^+ + H \longrightarrow H_2 + H^+$$

limited by:

$$H_2^+ + h\nu \longrightarrow H + H^+$$
$$H_2^+ + e^- \longrightarrow H_2 + h\nu.$$

The main destruction route for the H_2 thus formed is via collisional dissociation which totally controls the H_2 fractional abundance when $T \gtrsim 2500$ K (cf. Roberge and Dalgarno, 1979). The more efficient H_2 formation route via H^- is highly suppressed by the strong IR radiation field which results in the photodetachment of the intermediate, H^-:

$$H + e^- \longrightarrow H^- + h\nu$$
$$H^- + h\nu \xrightarrow{IR} H + e^-$$
$$[H^- + H \longrightarrow H_2 + e^-]$$

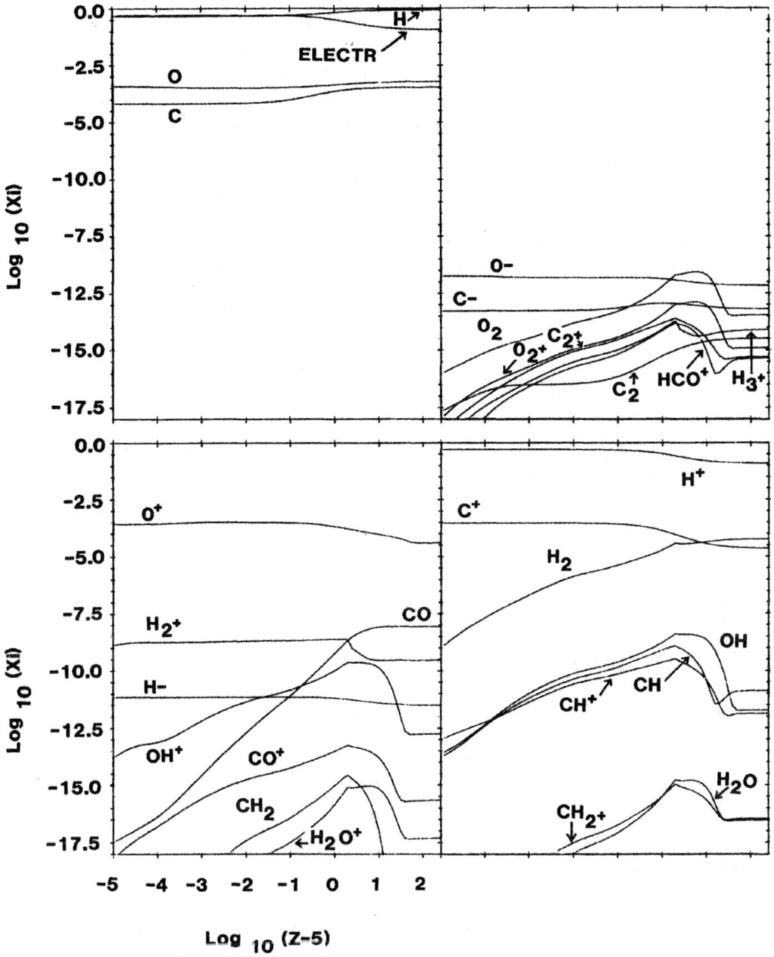

Figure 1. Results from the basic time-dependent chemical modelling. The results are presented as $\log_{10}[X_i]$ vs. $\log_{10}[Z - 5]$ where X_i is the fractional abundance of species i relative to hydrogen and Z is the radial ordinate in units of stellar radii.

For typical ejecta densities it is unlikely that H^- opacity or any other flux shielding mechanisms are likely to be significant. The time-dependent chemical model included 546 reactions involving 56 species and was limited to molecules containing the elements H, C and O which were assumed to be present in the ejecta with cosmic abundances. An example of the results is presented in Fig. 1 from which it can be seen that the ejecta is primarily atomic with a few species, such as H_2 and CO, rising to appreciable fractional abundances. It has also been possible to examine the chemical formation and destruction routes for the various species involved in the chemistry which are, generally speaking, high temperature and neutral-neutral reactions. Part of this network is illustrated in Fig. 2 for the species H_2 and CO.

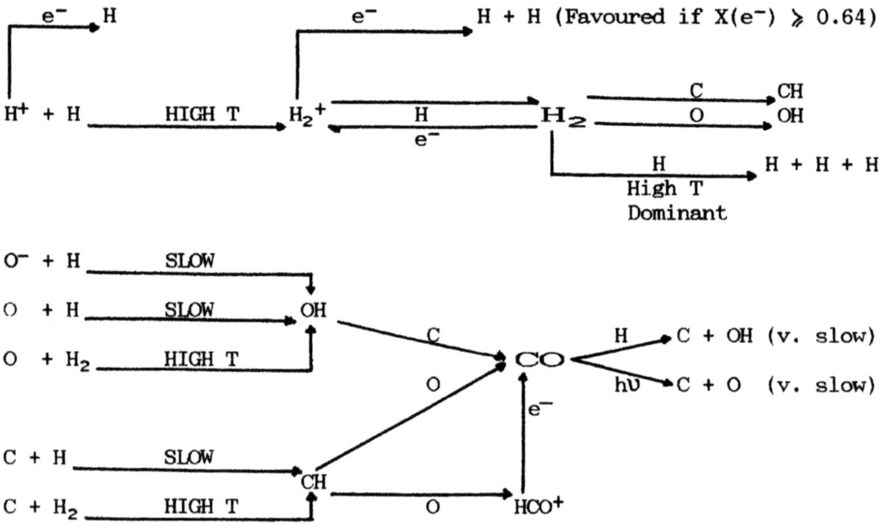

Figure 2. H_2 and CO formation/destruction network in the T-Tauri wind. Only the principal formation and destruction routes for H_2 and CO are shown. H_2 is formed only by the H^+ route shown and is primarily destroyed by the temperature sensitive collisional dissociation reaction. There are two CO formation routes: high temperature fast reactions involving H_2 and temperature-independent slow reactions involving H. Once formed, the CO is chemically highly stable.

The most important point to note is that the chemical species are essentially stable once formed so that in the case of H_2, say, for which the formation route via H_2^+ is very slow, the abundance is limited by the ejecta expansion and cooling so that it is essentially "frozen out" at a certain value. Similarly for CO; at high temperatures the intermediate CH is rapidly formed:

$$C + H_2 \;-->\; CH + H$$

but as the temperature falls the only CH formation route is the relatively inefficient:

$$C + H \;-->\; CH + h\nu$$

so that the CO formation rate is severely reduced.

These features are clearly visible in Fig. 1 so that at $\log_{10}(z-5) \sim 0.5$ the CH and OH abundances drop dramatically and the CO abundance is frozen. If the temperature of the ejecta was larger then the CH formation reaction via H_2 would be faster but conversely the fractional abundance of H_2 would be lower (limited by collisional dissociation). To investigate this and the effect of varying other parameters, a large grid of runs was performed. The maximum CO abundance that could be predicted this way was $X(CO) \sim 10^{-8}$.

One possible way in which the chemistry may be enhanced is if some dusty, molecular material (say from a pre-existent circumstellar disc, cf. Chevalier 1983) is mixed into the wind. The model was adapted to include a 1% mixing (so that initially $x(H_2) \sim 10^{-2}$). Results from this model show that the early presence of a significant amount of H_2 boosts the CH and hence the CO formation rates. However, the H_2 is unstable against collisional dissociation and is essentially destroyed on some characteristic timescale t_d after which time the CO abundance is frozen as in the basic model. Reducing the initial ejecta temperature results in a slower CO formation rate as before but it also has the effect of increasing the "lifetime" of the pre-existent H_2 (t_d) since the H_2 collisional dissociation reaction:

$$H_2 + H \;-->\; H + H + H$$

is highly sensitive to the temperature. The net result is for the molecular formation to be somewhat enhanced by having a lower initial ejecta temperature. If the initial temperature is taken to be 4000 K (at z = 5) then the model predicts a final (i.e. "frozen out") CO fractional abundance of $X(CO) \sim 10^{-6}$ implying that about 1% of all carbon is converted into CO.

3. CONCLUSION AND DISCUSSION

It has been shown that the primary T Tauri wind is basically atomic with little molecular formation. The only route of H_2 formation is via H_2^+ which is a mechanism that has been proposed in models of the chemistry of the early universe (e.g. Lepp and Shull 1984).

The limited chemistry that does occur is dominated by high temperature and neutral-neutral reactions (and collisional dissociation in the case of H_2) with most molecular species being stable to their environment once formed. However, the formation reactions are slow so that molecular abundances are limited by the geometrical dilution of the ejecta. Loss via photoreactions is not an important limiting factor to the chemistry, although data for photoreactions (particularly in the case of a non-standard interstellar radiation field) is extremely sparse and is a major uncertainty of the model. In the case where a small amount of molecular material (say 1%) is mixed into the wind, the model predicts a fractional abundance of CO : $X(CO) \sim 10^{-6}$ (under favourable conditions) which, if the primary wind can be regarded as an unbounded uniform outflow, corresponds to a column density of $N(CO) \sim 2 \times 10^{14}$ cm^{-2} which may well be observable. Thus observations of high velocity CO may tell us something about the physics as well as the chemistry of T Tauri outflows.

REFERENCES

Cantó, J., 1978. Astr. Ap., 70, 111.
Chevalier, R.A., 1983. Astrophys. J., 268, 753.
Franco, J., 1983. Astrophys. J., 264, 508.
Hartmann, L., Edwards, S. and Avrett, E., 1982. Astrophys. J., 261, 279.
Lepp, S. and Shull, J.M., 1984. Astrophys. J., 280, 465.
Norman, C. and Silk, J., 1980. Astrophys. J., 238, 158.
Rawlings, J.M.C., Williams, D.A. and Cantó, J., 1988. Mon. Not. R. astr. Soc., 230, 695.
Roberge, W.G. and Dalgarno, A., 1979. Astrophys. J., 233, L25.
Schwartz, R.D., 1978. Astrophys. J., 223, 884.

THE PHOTOCHEMISTRY OF PLANETARY ATMOSPHERES

A. Dalgarno
Harvard Smithsonian Center for Astrophysics
60 Garden Street
Cambridge, Massachusetts 02138
U.S.A.

ABSTRACT. The photochemistry of the atmospheres of the planets and their satellites is summarized.

1. INTRODUCTION

The compositions of the atmospheres of the planets and their evolution since the formation of the solar system are the result of a complex array of interacting chemical and dynamical processes driven by the absorption of photons and energetic particles from the sun and of cosmic rays from the galaxy. Photochemistry and plasma chemistry exercise a major influence on the composition of planetary clouds and on the nature of the liquid and solid surfaces.

Astrophysics does not include the planet Earth in its domain of inquiry and I exclude it from consideration other than to note that the study of planetary and satellite atmospheres whose histories have not been modified by biological processes provides important insights into their significance in the evolution of the terrestrial atmosphere and into the potentially catastrophic impact of human activities.

The atmospheres of the planets can be divided into two main groups. The terrestrial planets Mars and Venus lie close to the Sun, have masses comparable to the Earth and atmospheres dominated by the heavy elements, oxygen, nitrogen and carbon. The Jovian planets Jupiter, Saturn, Uranus and Neptune, are large, massive bodies, distant from the Sun, with atmospheres composed mostly of hydrogen and helium. The remaining planets Mercury and Pluto are small bodies. Mercury orbits close to the sun and has a tenuous atmosphere in which atomic sodium and potassium have been seen. Pluto lies at a great distance from the Sun and its surface is frozen. It may have an atmosphere containing methane, similar perhaps to those of the satellites Titan of Saturn and Triton of Neptune, which have atmospheres of nitrogen with an admixture of methane.

In addition to Titan and Triton, an atmosphere of sulphur dioxide has been discovered on the Jupiter satellite Io. Associated with Io is an ionized torus containing oxygen and sulphur in several ionization stages.

T. J. Millar and D. A. Williams (eds.), Rate Coefficients in Astrochemistry, 321–338.

2. MARS AND VENUS

The atmospheres of Mars and Venus are mainly carbon dioxide with a small admixture of molecular nitrogen and trace amounts of other species. They differ considerably in their surface temperatures and pressures. On Mars, the temperature increases from 214K at the surface to values up to 350K in the daytime mid-latitude exosphere, and on Venus the temperature diminishes from 735K at the surface to a daytime exospheric temperature averaging about 300K. The surface pressure on Mars is 6.4 bars and on Venus 92 bars. In the past few years, the photochemistries of Mars and Venus have been reviewed by Wayne (1985), Barth (1985), Prinn (1985) and Krasnopolsky (1986).

The principal constitutent CO_2 is photodissociated,

$$CO_2 + h\nu \rightarrow CO + O \ , \tag{1}$$

by solar radiation. The products recombine in a three-body process

$$CO + O + M \rightarrow CO_2 + M \tag{2}$$

where M is either CO_2 or O. Reaction (2) involves a change in spin multiplicity and is very slow with a rate coefficient of 2.10^{-37} $cm^6 s^{-1}$. The three-body reaction

$$O + O + M \rightarrow O_2 + M \tag{3}$$

has a much larger rate coefficient of $2.44 \cdot 10^{-34} \exp(900/T)$ $cm^6 s^{-1}$, so that the reactions (1), (2) and (3) tend to produce an atmosphere of CO and O_2. The actual abundances of CO and O_2 are small, particularly on Venus for which an upper limit of O_2 to CO_2 of $5\ 10^{-7}$ above the cloud tops has been measured (Traugher and Lunine 1983). Other mechanisms must exist that lead to the effective recombination

$$CO + \tfrac{1}{2}O_2 \rightarrow CO_2 \quad . \tag{4}$$

On Mars, CO_2 is formed by

$$CO + OH \rightarrow CO_2 + H \quad . \tag{5}$$

In a wet atmosphere, OH is created by the photodissociation of H_2O,

$$H_2O + h\nu \rightarrow O + OH \tag{6}$$

and by the reaction of metastable $O(^1D)$ atoms with H_2O,

$$O(^1D) + H_2O \rightarrow OH + OH \quad . \tag{7}$$

The $O(^1D)$ atoms are products of the photodissociation of O_3,

$$O_3 + h\nu \rightarrow O(^1D) + O_2, \tag{8}$$

ozone having been formed by

$$O + O_2 + M \rightarrow O_3 + M \quad . \tag{9}$$

Ozone is destroyed by photodissociation and by

$$O_3 + H \rightarrow O_2 + OH \quad . \tag{10}$$

Metastable $O(^1D)$ atoms are also a product of the photodissociation of H_2O,

$$H_2O + h\nu \rightarrow H_2 + O(^1D) \quad . \tag{11}$$

The metastable atoms are quenched in collisions with CO_2,

$$O(^1D) + CO_2 \rightarrow O(^3P) + CO_2 \tag{12}$$

but nevertheless play a significant role in producing odd hydrogen atoms.

Hydroperoxyl is formed by

$$H + O_2 + M \rightarrow HO_2 + M \tag{13}$$

and then reacts with O,

$$HO_2 + O \rightarrow OH + O_2 \tag{14}$$

and OH,

$$HO_2 + OH \rightarrow H_2O + O_2 \quad . \tag{15}$$

Further hydrogen interchange reactions occur such as

$$HO_2 + HO_2 \rightarrow H_2O_2 + O_2 \tag{16}$$

$$H_2O_2 + h\nu \rightarrow OH + OH \quad . \tag{17}$$

Molecular hydrogen is formed from

$$H + HO_2 \rightarrow H_2 + O_2 \tag{18}$$

as well as reaction (11).

In a dry atmosphere, as occurs over the polar ice caps, on Mars H_2 is the main source of odd hydrogen through photodissociation of H_2, and the reaction

$$O(^1D) + H_2 \rightarrow OH + H \quad . \tag{19}$$

Water is produced through reaction (15).

The photochemical cycle is sensitive to the abundance of H_2O and the CO and O_2 content may be substantially larger during periods of low obliquity and the composition of the upper atmosphere may be drastically modified (Lindner and Jakosy 1985).

The chemical sequence reforming CO_2 is less effective on Venus. On Venus, O_2 is scarce and the sources of $O(^1D)$ are small. But Venus has a small component of a few tenths of ppm of hydrogen chloride and HCl is readily photodissociated

$$HCl + h\nu \rightarrow H + Cl \quad . \tag{20}$$

Then

$$Cl + H_2 \rightarrow HCl + H \tag{21}$$

and its reverse

$$HCl + H \rightarrow H_2 + Cl \tag{22}$$

occur. Water can be created through

$$HCl + O \rightarrow OH + Cl \tag{23}$$

$$HCl + OH \rightarrow H_2O + Cl \tag{24}$$

and in the absence of $O(^1D)$ survives a long time.

On Venus, CO_2 can also form through the sequence

$$Cl + CO + M \rightarrow ClCO + M \tag{25}$$

followed by

$$ClCO + O_2 \rightarrow ClO + CO_2 \quad . \tag{26}$$

The reaction

$$HO_2 + Cl \rightarrow HCl + O_2 \tag{27}$$

reforms HCl and may branch to form ClO,

$$HO_2 + Cl \rightarrow ClO + OH \quad . \tag{28}$$

The rate coefficient of the possibly significant reaction

$$ClO + CO \rightarrow Cl + CO_2 \tag{29}$$

appears to be unknown.

The photochemistry of neutral carbon has been investigated by Fox (1982a) and Paxton (1985). Photodissociation of CO is a major source of C atoms (Fox 1982a) and through photoionization of C a major source of C^+ (Fox 1982b). A reconsideration of the carbon photochemistry may be needed in light of the recent spectroscopic data on the CO molecule (Letzelter et al. 1987, Yoshino et al. 1988). The consequences to the circumstellar (Mamon, Glassgold and Huggins 1988) and interstellar diffuse (van Dishoeck and Black 1987; Viala et al. 1988) and dense (Gredel, Lepp and Dalgarno 1987) cloud chemistry are considerable.

Sulphur compounds have been detected in the atmosphere of Venus. Their chemistry has been discussed in detail by Yung and De More (1982), Krasnopolsky and Parshev (1983), Prinn (1985) and De More et al. (1985). Photolysis of SO_2 is a major source of atomic oxygen in the stratosphere of Venus (Winick and Stewart 1980) and possibly in the mesophere (Leu and Yung 1987). A complex series of reactions occurs and the chemical sequence leads to the formation of sulphuric acid, which exerts a powerful influence on the nature of the Venus clouds (cf. Krasnopolsky 1985). The sulphur and chlorine cycles are intimately related and sulphuryl chloride SO_2Cl_2 may be a major reservoir of chlorine (De More et al. 1985).

The thermal balance on Venus has been studied (Hollenbach, Prasad and Whitten 1985, Dickinson and Bougher 1986, Fox 1988). For Venus, there are difficulties in reconciling the observed thermal structure with the calculated heating efficiencies (Fox 1988). Important uncertainties attend the energetics of the quenching reaction.

$$O(^1D) + CO_2 \rightarrow O(^3P) + CO_2) \qquad (29)$$

and the vibrational excitation and de-excitation processes

$$O + CO_2(v) \rightarrow O + CO_2(v') \quad . \qquad (30)$$

The major ions on Mars and Venus are O_2^+ and O^+. The essential steps in the chemistry of the ionospheres are

$$CO_2 + h\nu \rightarrow CO_2^+ + e \qquad (31)$$

$$CO_2^+ + O \rightarrow O_2^+ + CO \qquad (32)$$

$$\rightarrow O^+ + CO_2 \qquad (33)$$

$$O + h\nu \rightarrow O^+ + e \qquad (34)$$

$$O^+ + CO_2 \rightarrow O_2^+ + CO \qquad (35)$$

$$O_2^+ + e \rightarrow O + O \quad . \qquad (36)$$

The presence of N_2 leads to the terrestrial reactions

$$N_2 + h\nu \rightarrow N_2^+ + e \qquad (37)$$

$$N_2^+ + O \rightarrow NO^+ + N \qquad (38)$$

$$O^+ + N_2 \rightarrow NO^+ + N \qquad (39)$$

$$O_2^+ + N \rightarrow NO^+ + O \qquad (40)$$

and the additional reaction

$$N_2^+ + CO_2 \rightarrow N_2 + CO_2^+ \quad . \tag{41}$$

Agreement with the limited data for Mars is readily achieved (Fox and Dalgarno 1979). For Venus the agreement between the model ion densities and the measurements is broadly satisfactory when reactions of metastable ions are included (Fox 1982b) though there is a tendency for the models to underestimate the C^+ densities (Paxton 1985). The charge transfer reaction

$$O^+ + C \rightarrow O + C^+ \tag{42}$$

may be a large source of C^+ ions, supplementing those from photoionization of C,

$$C + h\nu \rightarrow C^+ + e \tag{43}$$

from

$$He^+ + CO \rightarrow C^+ + O + He \tag{44}$$

and from reactions of O^{2+} with CO and CO_2 (Fox, 1982a,b). Many of the rate coefficients are uncertain and the agreement between models and observations is tentative.

The energetics of the ionospheric reactions are critical to the escape of nitrogen from Mars and of oxygen and hydrogen from Venus. The contribution to the escape flux of nitrogen from Mars through dissociative recombination

$$N_2^+ + e \rightarrow N + N \ , \tag{45}$$

depends upon the distribution of vibrationally excited N_2^+ ions in the Martian atmosphere (Yung et al. 1977, Fox and Dalgarno 1983), and the excitation of the end products (Queffelec et al. 1985).

On Venus, hydrogen atoms with sufficient energy to escape the gravitational attraction of the planet are produced in the reactions

$$O^+ + H_2 \rightarrow OH^+ + H \tag{46}$$

$$OH^+ + e \rightarrow O + H \tag{47}$$

but the fraction of the available energy that is converted into kinetic energy of the products is unknown. Similar uncertainties attend the reactions

$$CO_2^+ + H_2 \rightarrow CO_2H^+ + H \tag{48}$$

$$CO_2H^+ + e \rightarrow CO_2 + H \quad . \tag{49}$$

However, the largest escape flux may arise from charge transfer of H with hot thermal protons (Hodges and Tinsley 1986).

3. THE JOVIAN PLANETS

The atmospheres of the Jovian planets have some of the characteristics of dense interstellar clouds in consisting almost entirely of molecular hydrogen with an envelope of atomic hydrogen produced by photodissociation. Helium is present probably at near the solar abundance ratio, but because it is more massive than H and H_2 its concentration falls rapidly with increasing altitude and it plays but a minor role in the chemistry. The Jovian planets are distant from the Sun and ultraviolet photons are less important and cosmic rays are more important in driving the chemistry than for the terrestrial planets.

The compositions of Jupiter and Saturn are summarized in Table 1 which lists the volume mixing ratios of these constitutents that have been detected (Strobel 1983, Noll et al. 1986, Bjorcker et al. 1986). Some of the abundances are uncertain and some of the identifications are tentative. Much less is known about Uranus and Neptune, though there is evidence for the presence of CH_4 in addition to hydrogen. Ammonia is expected to condense out in the tropopause cold traps.

TABLE 1

Compositions of the atmospheres of Jupiter and Saturn

Constituent	Volume Mixing Ratio	
	Jupiter	Saturn
H_2	0.90	0.94
He	0.10	0.06
CH_4	3 10^{-3}	2 10^{-3}
NH_3	3 10^{-4}	variable
H_2O	1-30 10^{-6}	--
C_2H_6	5 10^{-6}	5 10^{-6}
PH_3	7 10^{-7}	2 10^{-7}
CH_3D	2 10^{-7}	2.3 10^{-7}
C_2H_2	1 10^{-7}	5 10^{-6}
C_2H_6	3 10^{-7}	7 10^{-6}
CO	1 10^{-9}	--
HCN	2 10^{-9}	--
GeH_4	7 10^{-10}	--
C_2H_4	7 10^{-9}	--
CH_3C_2H	2 10^{-9}	tentative
C_3H_8	--	tentative
C_6H_6	2 10^{-9}	--

The upper atmospheres of the Jovian planets are composed of

atomic and molecular hydrogen. Helium becomes a major component at lower altitudes and slightly modifies the ionic distribution in the ionospheric regions. At still lower altitudes, methane is a major constituent and its photochemistry initiates sequences leading to hydrocarbons. The complexity of the hydrocarbons is limited by reactions with H_2 and with H. On Jupiter, photochemical reactions involving phosphine and ammonia are significant. On Saturn, because of the lower temperatures at a given pressure, the abundance of ammonia in gaseous form is much less than on Jupiter.

The ionospheres result from the photoionization and electron impact ionization of H_2

$$H_2 + h\nu \rightarrow H_2^+ + e \qquad (50)$$

$$e + H_2 \rightarrow e + H_2^+ + e \quad . \qquad (51)$$

The H_2^+ ions react with H to form H^+,

$$H_2^+ + H \rightarrow H_2 + H^+ \ , \qquad (52)$$

and with H_2 to form H_3^+ ,

$$H_2^+ + H_2 \rightarrow H_3^+ + H \quad . \qquad (53)$$

In most models, it has been supposed that the H_3^+ ions undergo rapid removal through dissociative recombination

$$H_3^+ + e \rightarrow H_2 + H \qquad (54)$$

$$\rightarrow H + H + H \qquad (55)$$

so that in equilibrium H^+ is the major ion, removed only slowly by radiative recombination.

$$H^+ + e \rightarrow H + h\nu \qquad (56)$$

or by downward diffusion followed by three-body recombination

$$H^+ + H_2 + H_2 \rightarrow H_3^+ + H_2 \quad . \qquad (57)$$

The determination by Smith and Adams (1984) that dissociative recombination of H_3^+ is slow for ions in the ground vibrational state modifies the chemistry. The H_3^+ ion may be the major constituent in the ionosphere (McConnell and Majeed 1987), depending upon the value of the recombination coefficient. The recombination coefficient depends also on the vibrational distribution of the ionospheric H_3^+ ions, so that vibrational excitation and de-excitation processes are a significant part of the chemistry of the Jovian ionospheres. Also significant is the vibrational distribution of molecular hydrogen. The process

$$H^+ + H_2(v \geq 4) \rightarrow H + H_2^+ \qquad (58)$$

may be an important sink for H^+ ions (McElroy 1973). Cravens (1987) has shown that in auroral regions the inclusion of (58) reduces the electron density by an order of magnitude.

Carbon monoxide has been detected in the atmospheres of Jupiter and Saturn, suggesting that there exist extraplanetary sources of oxygen, the oxygen ultimately being converted into CO. On Saturn, H_2O and OH may be sputtered from the rings (Connerney and Waite 1984). Then H^+ is removed by

$$H^+ + OH \rightarrow H + OH^+ \tag{59}$$

$$H^+ + H_2O \rightarrow H_2O^+ + H \tag{60}$$

(Connerney and Waite 1984) and H_3^+ by

$$H_3^+ + OH \rightarrow H_2O^+ + H_2 \tag{61}$$

$$H_3^+ + H_2O \rightarrow H_3O^+ + H_2 \quad . \tag{62}$$

The theoretical models are fairly successful in reproducing Voyager data for Jupiter (McConnell and Majeed 1987, Cravens 1987), but less so for Saturn, in part because there appear to occur large variations in the electron density profiles (Atreya et al. 1984).

The photochemistry of Jupiter has been reviewed by West, Strobel and Tomasko (1986) and Atreya (1986), and of Saturn by Atreya et al. (1984), Prinn et al. (1984) and Atreya (1986). Methane is photodissociated

$$CH_4 + h\nu \rightarrow {}^1CH_2 + H_2 \ , \tag{63}$$

mostly by absorption of Lyman alpha radiation. Then

$${}^1CH_2 + H_2 \rightarrow CH_3 + H \tag{64}$$

produces CH_3 which by three-body association

$$CH_3 + CH_3 + H_2 \rightarrow C_2H_6 + H_2 \tag{65}$$

leads to ethane. Ethane is shielded from photodissociation by CH_4 and it is transported downwards into the stratosphere where it accumulates.

There are several pathways to form acetlyene. Photodissociation of CH_4 also yields ${}^3CH_2$,

$$CH_4 + h\nu \rightarrow {}^3CH_2 + H_2 \tag{66}$$

and ${}^3CH_2$ reacts with CH_3

$${}^3CH_2 + CH_3 \rightarrow C_2H_4 + H \tag{67}$$

to form C_2H_4, which is photodissociated.

$$C_2H_4 + h\nu \rightarrow C_2H_2 + H_2 \quad . \tag{68}$$

Alternatively (Nava, Mitchell and Stief 1987), the reaction

$$CH_2 + CH_2 \rightarrow C_2H_2 + H_2 \qquad (69)$$

is a source of acetylene. Acetylene is also photodissociated

$$C_2H_2 \rightarrow C_2H + H \qquad (70)$$

$$\rightarrow C_2 + H_2 \qquad (71)$$

and C_4H_2 may be made through

$$C_2H + C_2H_2 \rightarrow C_4H_2 + H \quad . \qquad (72)$$

Reactions similar to (72) can build long chain molecules but their formation is inhibited by

$$C_2H + H_2 \rightarrow C_2H_2 + H \qquad (73)$$

$$C_2H + CH_4 \rightarrow C_2H_2 + CH_3 \quad . \qquad (74)$$

A small supply of complex hydrocarbons can be produced by three-body processes. Propyne (methylacetylene) has been tentatively identified. It can be formed by

$$^3CH_2 + C_2H_2 + H_2 \rightarrow CH_3C_2H + H_2 \quad . \qquad (74)$$

Acetylene is also attacked by O,

$$O + C_2H_2 \rightarrow CH_2 + CO \qquad (75)$$

in one of many paths for the conversion of oxygen into CO.

Ammonia is shielded by H_2, CH_4, and C_2H_6 from photons shorter than 165 nm. Photolysis of NH_3 by longer wavelength photons leads to NH_2 and H,

$$NH_3 + h\nu \rightarrow NH_2 + H \quad . \qquad (76)$$

Then NH_2 can form hydrazine,

$$NH_2 + NH_2 + H_2 \rightarrow N_2H_4 + H_2 \qquad (77)$$

which probably condenses as a haze.

Phosphine behaves similarly. A sequence initiated by

$$PH_3 + h\nu \rightarrow PH_2 + H \qquad (78)$$

leads to the solid condensation product P_2H_4. Jupiter and Saturn may behave differently. On Jupiter, hydrogen atoms are supplied by the photodissociation of ammonia and

$$PH_3 + H \rightarrow PH_2 + H \quad (79)$$

and

$$H + P_2H_4 \rightarrow P_2H_3 + H_2 \quad (80)$$

occur which may lead on Jupiter to the formation of red phosphorus P_4 (Kaye and Strobel 1984).

The presence of HCN is unexpected. It has been attributed (Kaye and Strobel 1984) to the combination of the photochemistries of NH_3 and C_2H_2. They suggest the HCN is a photodissociation product of C_2H_5N formed by

$$H + C_2H_2 + H_2 \rightarrow C_2H_3 + H_2 \quad (81)$$

$$NH_2 + C_2H_3 + H_2 \rightarrow C_2H_5N + H_2 \ . \quad (82)$$

Then

$$C_2H_5N + h\nu \rightarrow HCN + CH_3 + H \quad (83)$$

or

$$C_2H_5N + h\nu \rightarrow H_2CN + CH_3 \quad (84)$$

and

$$H + H_2CN \rightarrow HCN + H_2 \ . \quad (85)$$

Ferris and Itikawa (1987) suggest a modified scheme of reactions of C_2H_5N with H leading to

$$CH_3CN + H \rightarrow HCN + CH_3 \quad (86)$$

whereas Stribling and Miller (1987) advocate chemistry initiated by an electrical lightning discharge.

4. SATELLITE ATMOSPHERES

Io has an atmosphere of SO_2. Photodissociation

$$SO_2 + h\nu \rightarrow SO + O \quad (87)$$

$$SO + h\nu \rightarrow S + O \quad (88)$$

and photoionization

$$SO_2 + h\nu \rightarrow SO_2^+ + e \quad (89)$$

$$S + h\nu \rightarrow S^+ + e \quad (90)$$

$$O + h\nu \rightarrow O^+ + e \quad (91)$$

occur. The ion S^+ does not react with SO_2 but O^+ does to produce O_2^+,

$$O^+ + SO_2 \rightarrow O_2^+ + SO \quad . \tag{92}$$

The ion O^+ also may charge transfer in collision with S

$$O^+ + S \rightarrow O + S^+ \quad . \tag{93}$$

The neutral and ionic constituents of the atmosphere are the source of the material in the ionized torus that is associated with Io. The torus contains oxygen and sulphur in neutral and ionized forms and their distribution appears to be a result of electron impact ionization, charge transfer and electron recombination (Moreno and Barosa 1986, Smith and Strobel 1986, Sittler and Strobel 1987, Shemansky 1987). The charge transfer reactions that participate are

$$S^{2+} + O \rightarrow S^+ + O^+ \tag{94}$$

$$O^{2+} + S \rightarrow O^+ + S^+ \tag{95}$$

$$S^{3+} + O \rightarrow S^{2+} + O^+ \tag{96}$$

$$\rightarrow S^+ + O^{2+} \tag{97}$$

$$S^{3+} + S \rightarrow S^+ + S^{2+} \quad . \tag{98}$$

There are conflicting views (Smith and Strobel 1986, Shemansky 1987) on the importance of

$$O^{2+} + S^+ \rightarrow O^+ + S^{2+} \tag{99}$$

$$O^{2+} + S^{2+} \rightarrow O^+ + S^{3+} \tag{100}$$

in controlling the loss of O^{2+} ions. Estimates of the rate coefficients are given by Johnson and Strobel (1982).

Titan and Triton have massive atmospheres. Their gravitational attraction is weak and H and H_2 escape easily. The principle constituent is N_2. At the troposphere, the volume mixing ratio of N_2 exceeds 0.97. In the absence of large amounts of H and H_2, the photodissociation of CH_4 leads to the formation of complex hydrocarbons. Table 2 is a list of the volume mixing ratios in the stratosphere, taken from Yung, Allen and Pinto (1984), who have presented a comprehensive discussion of the photochemistry of Titan. Briefer reviews are given by Hunten et al. (1984), Strobel (1985) and Wayne (1985).

TABLE 2

Stratospheric volume mixing ratios on Titan

CH_4	1-3	10^{-2}	H_2	2	10^{-3}
C_2H_6	2	10^{-5}	C_2H_2	2	10^{-6}
C_2H_4	4	10^{-7}	C_3H_8	3	10^{-6}
CH_3C_2H	3	10^{-8}	C_4H_2		10^{-8}-10^{-7}
HCN	2	10^{-7}	HC_3N		10^{-8}-10^{-7}
C_2N_2	10^{-8}-10^{-7}		CO	6	10^{-5}
CO_2	2	10^{-9}			

Jones and Lewis (1987) have drawn attention to the possible influence of shock induced chemistry on Titan, particularly in the early phase of its evolution, the shock energy coming from infall of meteoritic and cometary debris. The N_2 may have been produced by the shock-driven conversion of NH_3. The chemistry is driven now by photons and by magnetospheric electrons. The fast electrons ionize and dissociate N_2 to produce N, N_2^+ and N^+. Many of the nitrogen atoms produced by electron-impact dissociation and by dissociative recombination of N_2^+ ions are in the long-lived metastable $N(^2D)$ state which can react with CH_4

$$N(^2D) + CH_4 \rightarrow NH + CH_3 \quad . \tag{101}$$

The NH is mostly converted back to N_2 by

$$NH + N \rightarrow N_2 + H \tag{102}$$

but some release atomic nitrogen

$$NH + H \rightarrow N + H_2 \quad . \tag{103}$$

Then HCN can be formed by

$$N + CH_3 \rightarrow HCN + H_2 \tag{104}$$

and CN by

$$HCN + h\nu \rightarrow H + CN \quad . \tag{105}$$

The reaction of CN with HCN,

$$CN + HCN \rightarrow H + C_2N_2 \ , \tag{106}$$

is slow, but cyanogen can be produced alternatively by

$$N(^2D) + C_2H_2 \rightarrow CHCN + H \tag{107}$$

and

$$CHCN + N \rightarrow H + C_2N_2 \quad (108)$$

(Yung 1987). Cyanoacetlyene can be made by

$$C_2H + HCN \rightarrow HC_3N + H \quad (109)$$

and dicyanoacetlyene by

$$CHCN + CHCN \rightarrow H_2 + C_4N_2 \quad . \quad (110)$$

The acetlyene in reaction (107) and the ethynyl in reaction (109) are products of a chemistry initiated by (101) and by photodissociation of CH_4 as in reactions (63) and (66). On Titan, the excited $^1CH_2$ is quenched by N_2 to the ground state $^3CH_2$,

$$^1CH_2 + N_2 \rightarrow {}^3CH_2 + N_2 \quad . \quad (111)$$

Then

$$CH_2 + CH_2 \rightarrow C_2H_2 + H_2 \quad (112)$$

$$CH_2 + CH_3 \rightarrow C_2H_4 + H \quad (113)$$

$$CH + CH_4 \rightarrow C_2H_4 + H \quad . \quad (114)$$

More complex hydrocarbons can be built by three-body associations such as

$$CH_2 + C_2H_2 + M \rightarrow CH_3C_2H + M \quad . \quad (115)$$

Thus propane is formed by

$$CH_3 + C_2H_5 + M \rightarrow C_3H_8 + M \quad , \quad (116)$$

following

$$C_2H + C_2H_6 \rightarrow C_2H_2 + C_2H_5 \quad , \quad (117)$$

the ethane arising from reaction (65). Still more complex hydrocarbons can be created and high abundances of C_4 hydrocarbons are predicted (Yung Allen and Pinto 1984). The predicted mixing ratio for butane, C_4H_{10}, is $3 \cdot 10^{-7}$. The formation of amino acids is discussed by Khare et al. (1986).

The complex hydrocarbons may cause the extensive haze layers. They are finally deposited on the surface. The ethane is liquid and forms an ocean in which is dissolved methane, nitrogen and other compounds. Acetylene is deposited as a solid.

Water is probably brought steadily into the atmosphere through meteoric bombardment. It will be broken down into O and OH. Reactions such as

$$O + CH_2 \rightarrow CO + H + H \quad (118)$$

$$O + CH_3 \rightarrow H_2CO + H \quad (119)$$

$$OH + CH_2 \rightarrow CO + H_2 + H \quad (120)$$

$$OH + CH_3 \rightarrow CO + H_2 + H_2 \quad (121)$$

$$CO + OH \rightarrow CO_2 + H \quad (122)$$

follow. The chain is terminated by the sublimation of CO_2 to produce a layer of dry ice on the surface.

Electron impact dissociation of CO

$$e + CO \rightarrow e + C + O(^1D) \quad (123)$$

produces metastable $O(^1D)$ atoms which react with CH_4

$$O(^1D) + CH_4 \rightarrow OH + CH_3 \quad (124)$$

Dissociation of H_2O is also a source of $O(^1D)$ atoms.

A torus of hydrogen atoms has been detected near Saturn. The source of the hydrogen atoms is probably Titan, as argued most recently by Richardson and Eviatar (1987). Titan may also be the source of a torus of nitrogen atoms (Barbosa 1987). Thus the chemistry of the Titan atmosphere profoundly influences the distribution of material throughout the entire Saturnian system.

The atmosphere of Triton is similar to that of Titan. It is predicted also to have a deep ocean of hydrocarbons with dissolved N_2. The chemistry is driven by cosmic rays and magnetospheric electrons and by photon absorption. Because the CH_4 to H_2 ratio is high, hydrocarbons of considerable complexity can be produced as on Titan.

Much of the discussion of Yung, Allen and Pinto (1984) applies with minor modifications to Triton. The chemistry on Triton has been discussed recently by Delitsky and Thompson (1987).

The chemistry of planetary atmospheres is now a well-established discipline in which studies of any one planet enhance our understanding of them all. Chemical processes dominate the evolution of the atmospheres of the planets, determine the nature of the clouds and the composition of the oceans and are the major source of the escaping atoms and ions that occupy the space between the planets and satellites.

Enormous uncertainties remain in the theoretical models that have been proposed and major changes in them may yet be demanded by new observations. A major obstacle to further progress is the inadequacy of our knowledge of the rate coefficients of reactions involving the large molecules that are significant in the photochemistry of the planets.

ACKNOWLEDGMENT. This work has been supported by the National Aeronautics and Space Administration under Grant NSG-7421.

REFERENCES

Atreya, S.K. 1986 Atmospheres and ionospheres of the outer planets and their satellites (Springer).

Atreya, S.K., Waite, J.H., Donahue, T.M., Nagy, A.F. and McConnell, J.C. 1984 in Saturn, Ed. T. Gehrels (University of Arizona Press).

Barosa, D.D. 1987 Icarus **72**, 53.

Barth, C.A. 1985 in The photochemistry of atmospheres, Ed. J.S. Levine (Academic Press).

Bjoraker, G.L., Larson, H.P. and Kunde, V.G. 1986 Icarus **66**, 579.

Connerney, J.E.P. and Waite, J.H. 1984 Nature **312**, 136.

Cravens, T.E. 1987 J. Geophys. Res. **92**, 11083.

Delitsky, M.L. and Thompson, W.R. 1987 Icarus **70**, 354.

Demore, W.B., Leu, M-T., Smith R.H. and Yung, Y.L. 1985 Icarus **63**, 347.

Dickinson, R.E. and Bougher, S.W. 1986 J. Geophys. Res. **91**, 70.

Ferris, J.P. and Ishikawa, Y 1987 Nature **326**, 777.

Fox, J.L. 1988 Planet. Spa. Sci. in press.

Fox, J.L. 1982a J. Geophys. Res. **87**, 9211.

Fox, J.L. 1982b Icarus **51**, 248.

Fox, J.L. and Dalgarno, A. 1979 J. Geophys. Res. **84**, 7315.

Fox, J.L. and Dalgarno, A. 1981 J. Geophys. Res. **86**, 629.

Fox, J.L. and Dalgarno, A. 1983 J. Geophys. Res. **88**, 9027.

Gredel, R., Lepp S. and Dalgarno, A. 1987 Ap. J. Lett. **323**, L137.

Hodges, R.R. and Tinsley, B.A. 1986 J. Geophys. Res. **91**, 13649.

Hollenbach, D.J., Prasad, S.S. and Whitten, R.C. 1985 Icarus **64**, 205.

Hunter, D.M. Tomasko, M.G., Flasar, F.M., Samuelson, R.E., Strobel, D.F. and Stevenson, D.J. 1984 in Saturn, Ed. T. Gehrels (University of Arizona Press).

Johnson, R.E. and Strobel, D.F. 1982 J. Geophys. Res. **87**, 10385.

Jones, T.D. and Lewis, J.S. 1987 Icarus **72**, 381.

Kaye, J.A. and Strobel, D.F. 1984 Icarus **55**, 314.

Kaye, J.A. and Strobel, D.F. 1983 Icarus **54**, 417.

Khare, B.N., Sagan, C. Ogino, H. Nagy, B., Cevat Cer, Scharn, K.H., and Arakawa, F.T. 1986 Icarus **68**, 176.

Krasnopolsky, V.A. 1986 Photochemistry of the atmospheres of Mars and Venus (Springer).

Krasnopolsky, V.A. 1985 Planet. Spa. Sci. **33**, 109.

Krasnopolsky, V.A. and Parshev, V.A. 1983 in Venus Ed. D.M. Hunten, Colin, L., Donahue, T.M. and Moroz, V., (University of Arizona Press).

Letzelter, C., Eidelsberg, M., Rostas, F., Breton, J. and Thieblemont, B., 1987 Chem. Phys. **114**, 273.

Leu, M-T. and Yung, Y.L. 1987 Geophys. Res. Lett. **14**, 949.

Lindner, B.L. and Jakosy, B.M. 1985 J. Geophys. Res. **90**, 3435.

Maimon, G.A., Glassgold, A. E. and Huggins, P.J., 1987, Ap. J. in press.

McConnell, J.C. and Majeed, T. 1987 J. Geophys. Res. 92, 8570.

McElroy, M.B., 1973, Spa. Sci. Rev. **14**, 460.

Moreno, M.A. and Barosa, D.D. 1986 J. Geophys. Res. **91**, 8591.

Nava, D.F., Mitchell, M.B. and Stief, L.J. 1987 J. Geophys. Res. **91**, 4585.

Noll, K.S., Knacke, R.F., Tokanuga, A.T., Lacy, J.H., Beck, S. and Serabyn, E. 1986 Icarus **65**, 257.

Paxton, L.J. 1985 J. Geophys. Res. **90**, 5089.

Prinn, R.G. 1985 in The Photochemistry of atmospheres Ed. J.S. Levine (Academic Press).

Prinn, R.G., Larson, H.P., Caldwell, J.J. and Gautier, D. 1984 in Saturn Ed. T. Behrels (University of Arizona Press).

Queffelec, J.L. Rowe, B.R., Morlais, M., Gomet, J.C. and Vallee, F. 1985 Planet. Spa. Sci. **33**, 263.

Richardson, J.D. and Eviatar, A. 1987 Geophys. Res. Lett. **14**, 999.

Shemansky, D.E. 1987 J. Geophys. Res. **92**, 6141.

Sittler, G.C. and Strobel, D.F. 1987 J. Geophys. Res. **92**, 5741.

Smith, D. and Adams, N.G. 1984 Ap. J. Lett. **269**, 329.

Smith, R.A. and Strobel, D.F. 1986 J. Geophys. Res. **91**, 9469.

Stribling, R. and Miller, S.L. 1987 Icarus **72**, 48.

Strobel, D.F. 1985 in The photochemistry of atmospheres Ed. J.S. Levine (Academic Press).

Strobel, D.F. 1983 Int. Rev. Phys. Chem. **3**, 145.

Traugher, J.T. and Lunine, J.I. 1983 Icarus **55**, 272.

Van Dishoeck, E.F. and Black, J.H. 1987 private communication.

Viala, Y.P., Letzelter, C., Eidelsberg, M. and Rostas, F. 1988 Astron. Ap. in press

Wayne, R.P. 1985 Chemistry of atmospheres (Clarendon Press: Oxford).

West, R.A., Strobel, D.F. and Tomasko, M.G. 1986 Icarus **65**, 161.

Yoshino, K., Stark, G., Smith, P.L., Parkinson, W.H. and Ito, K. 1988 J. Physique in press.

Yung, Y.L. 1987 Icarus **72**, 468.

Yung, Y.L., Allen, M. and Pinto, J.P. 1984 Ap. J. Suppl. **55**, 465.

Yung, Y.L. and DeMore, W.B. 1982 Icarus **51**, 199.

Yung, Y.L. Strobel, D.F., King, T.Y. and McElroy, M.B. 1977 Icarus **30**, 26.

THE UMIST RATE FILE FOR ASTROCHEMISTRY

Andrew Bennett,
Mathematics Department,
UMIST, P.O. Box 88,
MANCHESTER, M60 1QD,
United Kingdom.

ABSTRACT. The need for large numbers of accurate reaction rate coefficients, for a great diversity of reaction types, when constructing useful chemical models is well known. In this paper I will be discussing the ways in which the measurements and theoretical calculations of rate files are currently used - more particularly how models can be made which acknowledge the need for the ease of changing reaction schemes.

1. INTRODUCTION

Many useful chemical results can be obtained as the result of "back of the envelope" calculations, in particular, cases where the most important chemical formation and destruction routes of a species in a particular circumstance are clearly known. These calculations can lead to clear chemical insights without the burden of heavy numerical labour.

However, as chemical models have necessarily increased in complexity, the need to use numerical methods to study astrochemistry has become stronger. We are also becoming increasingly interested in the time dependent development of abundances, rather than just steady state abundances. This means the use of computers. Indeed, in this paper I have mainly restricted the discussion to the numerical calculation of time dependent models.

To be useful, a numerical chemical model must include a set of species and a reaction scheme which adequately describes the complexity of the astrophysical circumstance. The physical conditions must also be included - preferably in as sophisticated a way as possible, since with our observations we cannot sample points in space, but rather lines of sight, where the physical conditions are changing (for example, even in dark clouds, the major column densities of some species are a product of the chemistry of the edge regions).

It is also obvious that we need an accurate and complete library of chemical reactions between many different species, from

T. J. Millar and D. A. Williams (eds.), Rate Coefficients in Astrochemistry, 339–346.

which we can abstract a more limited reaction scheme, that is suited to the requirements of our model. It is equally obvious that such a complete library is a forlorn hope at the moment.

However, we can start by searching through the chemical and astronomical literature, and strive compile an up-to-date list of measured and calculated rate constants, to which we add guesses for unknown rate coefficients which we believe to be significant.

It is important to note that such a list will contain many reaction rate coefficients which have been measured between species that are relevant in interstellar chemistry, but which are essentially pointless, since we believe that they form neither a major formation or destruction route for either the reactants or the products, for example

$$CN^+ + CO_2 \; --> \; CO_2^+ + CN \qquad k = 4.5x10^{-10} \; cm^3 \; s^{-1}.$$

This is not an argument against measuring or including such rate constants, but rather, a recognition of the fact that many reaction schemes used in research are sometimes rather magpie like in approach.

For instance, at UMIST, we currently have a rate file of some 2273 reaction between 226 species, excluding isotopes. The reactions attempt to describe the chemistry of the elements H, He, C, N, O, S, P, Cl, Si, Mg, Na, and Fe. No attempt is made to exclude reactions between species which are not particularly relevant.

How can we extract a relevant scheme (species and reactions) from the library, so as to minimise computer resources, and shorten the development time of the model? The large number of reactions involved also makes checking the rate file for inconsistencies manually time-consuming and prone to error. Below I describe an approach that has been examined at UMIST with a view to make the development and calculation of models faster and more accurate. Before that however, I summarise what information could usefully be included in the "master" rate coefficient file.

2. THE CHEMICAL LIBRARY

Reactants, products and *rate coefficients* are of course the essential ingredients to any rate file. Several compilations of rate coefficients likely to be useful in interstellar chemistry are available, most notably Prasad and Huntress (1980). Other work deal with more specific classes of reaction, such as Anicich and Huntress (1986) for positive ion-molecule reactions. The chemical literature can also be of help, such as the Manchester Elementary Reaction Bibliography, a computer database of references to theoretical and experimental studies of neutral-neutral reactions, maintained by J.C. Whitehead at the University of Manchester. References to the rate coefficients used are very important in tracking down misprints and errors. Temperature dependence is of particular interest, since the power law and activation energy representations often used do not adeauately represent the temperature dependance of some, important reactions.

Accuracy - A scale of accuracy that is sufficiently broad to cope with the high measured accuracy of some ion-molecule reactions, and the complete guesses still considered acceptable for some other types of reaction, is bound to be somewhat arbitrary, but it can help in assessing the reaction scheme. We currently use the approximate scheme:

Category	Approximate Accuracy (Claimed)
1	0-25%
2	25-50%
3	50%-x2
4	x2-x10 ish
5	Guess

N.B. that this categorisation is used for both theoretically estimated and measured rate constants; a flag indicating measurement is also included as a separate item.

Relevancy - The abundance and chemistry of many species depend on a limited subset of all the species in the model with which it reacts. If we want to use a reduced chemical network, say for ease of computation, we want to retain such important reactions at the expense of the unimportant ones. Which ones are important will depend, obviously, on the physical circumstances of the chemistry. For instance, in a shock calculation, reactions with high temperature barriers will obviously be more important than the same reactions in a cold cloud. A routine to impose an upper activation energy on the reactions used in the model will obviously take care of this, the major way in which the importance of a reaction changes relative to the others. We therefore assign, somewhat arbitrarily, a "relevancy", based on our assumptions about the chemistry at a temperature which is high enough for the reaction barrier not to eliminate the reaction.

Reaction Type - For some purposes, it is very useful to have the type of reaction available. Although this can be calculated by the computer from a careful examination of the reactants and products, it is very time-consuming, and so this is done on the master file only. This information can help us in calculating the statistical distribution of atoms in the products of reactions involving deuterium, say, or to give statistical information about the database.

3. SETTING UP THE EQUATIONS : DELOAD

In this section we discuss a method developed by L.A.M. Nejad at UMIST which substantially reduces the execution time of the chemical models written in FORTRAN over the methods previously used. It also provides some facilities for the editing of reaction schemes, through a FORTRAN program called DELOAD (Differential

Equation LOADer)

In general, to find the time-dependent behaviour of the chemistry we must solve a set of non-linear differential equations, one for every species. Suppose we have N_s species, including elements. For each we can write an equation describing the terms which contribute to the formation and loss for that species. For species X we have:

$$\frac{dn_X}{dt} = \sum_A \sum_B k_{AB}\, n_A n_B + \sum_A k_A n_A - n_X \sum_A k_{AX} n_A - n_X \sum_x k_X$$

where k_{AB} is the rate of two body reactions leading to the formation of X, A + B – – > X + products,

k_{AX} is the rate of two body reactions leading to the destruction of X, A + X – – > products,

k_A is the rate of one body processes (e.g. photoprocesses and cosmic rays) which lead to the formation of X, A – – > X + products,

and k_X is the rate for one body destruction processes for X, X – – > products.

The n_A, n_B, and n_X are the number densities of species of A, B, and X in cm^{-3}. The rate constants are known and fixed (at least for the purposes of the computation !). The positive terms in the equation above represent formation routes to species X; the negative ones represent the destruction pathways. We have assumed that the density is sufficiently low that three body reactions can be ignored. This is not always the case in astrophysical circumstances (e.g. Rawlings, 1986), but we can, of course, modify the equations accordingly. For each element included in the model we also have a conservation equation, each of which we may used to replace one equation of the form given above.

We must solve this stiff, initial value problem numerically, and we use the GEAR software package (Hindmarsh, 1974). This is coded in FORTRAN, which I shall assume to the computational language of most astrophysicists.

The GEAR package expects to be supplied with the differential equations in a user-written subroutine, called DIFFUN. Now we could code the formation terms of the equation above as:

```
DO I=1,NOSPEC
  DO J=1,NOSPEC
    DY=DY+K(I,J)*Y(I)*Y(J)
  NEXT J
NEXT I
```

for each species, and code the destruction terms similarly. However, this is a very inefficient use of computer time, because at every pass through the inner (J) loop, at most one term will be added to DY, though usually none. In practice, this means the computer spends a lot of its time in the DIFFUN subroutine.

This bottleneck can be relieved by writing the equations of formation and destruction for each species in full. Doing this by hand would be far too time consuming, and so this subroutine is written by computer. DELOAD is an equation "hard-wirer", which automatically produces such FORTRAN code for the subroutine DIFFUN, required by GEAR (Nejad, 1986). This produces up to 3000 lines of FORTRAN in 5 minutes on a current scientific workstation computer. DELOAD takes account of the various terms listed above, and also restrictions imposed by the FORTRAN language, such as length of line and number of continuation lines allowed in one statement.

There are however, some problems with DELOAD, all of which basically relate to shortcomings in the implementation of FORTRAN compilers. Although the code produced by DELOAD is valid FORTRAN 77, some compilers, even on mainframes, have memory limitations. For instance, due to memory limitations during compilation, the DIFFUN subroutine may have to be split into 2 or more pieces for even comparatively simple models (this is true even on some "Virtual Memory" machines, where the operating system effectively creates an infinite user memory for running programs).

The program relies on the character handling capabilities added to FORTRAN in the 1977 standard, and many mainframes are not optimised for character operations, and so the time DELOAD takes to run is not always insignificant. Hence we have developed a "workstation approach" to the problem of preparing and editing the reaction schemes and codes, whereby a personal interactive computer is used in this stage, and the models are later transferred to mainframes for processing (currently CDC7600s and a Cyber 205). Small models, with around 50 species, have been developed and successfully run on personal computers, however, taking around 10 minutes to run, which is quite acceptable given the shorter overall development time.

The use of DELOAD results in a speed increase of about 3 times, at least. It removes the bottleneck of the computation from the DIFFUN subroutine to elsewhere in the GEAR package. Hence, when using DELOAD, for a given number of species, extra reactions do not necessary slow the model down, and so one does not have to be so critical in weeding out reactions which one believes to be unimportant.

One class of mainframe now available is the vector processor, which is optimised for operations on consecutive elements of arrays, which, in FORTRAN, often means DO loops. Might not these processors provide a speeding up of the DO loops? The answer is no, because each DO loop only involves perhaps one addition to the abundances. But nevertheless the speed of these processors is often such that they are to be preferred to normal, scalar machines.

As a final aside, it may also be noted, that although DELOAD was written for use with GEAR, and hence time-dependent models, both have also proved useful in setting up and executing steady state models, using the Newton-Raphson technique (Rawlings, 1986).

4. SETTING UP THE CHEMISTRY : SMERSH

SMERSH stands for Software for Manipulating and Editing Reaction Schemes, and is a suite of mostly short programs that have been written at UMIST to maintain the master rate file, RATE87, and to create subset chemistries.

As rate files are typed in by many people over the course of several years, they consequently become rather haphazard in coverage and accuracy. Maintaining the rate coefficient file has become more time consuming as it has grown in size, to the point where some form of data base management is needed. Currently, SMERSH can do several things to help the chemical model constructor, and hopefully these will increase in number and usefulness in time.

Firstly, SMERSH can check that all the reactions in the library are balanced, that is they neither destroy nor create elements. With complex species, it can be remarkably difficult to spot typing errors such as this in the editing of reaction files. It can also order reactions, by reaction type, and check for redundant and duplicated equations.

It can also create subset chemistries from species lists, and number them, which DELOAD requires. SMERSH checks to make sure that the schemes are closed, in that every species formed is also destroyed, and vice versa. Lists of species, and their formation and destruction pathways can be created.

Given the uncertainties in reaction rates, one sometimes wishes to replace the current "best guess" with new estimates, as an experiment, for instance, changing all the ion-high dipole moment reaction rate coefficients using an analytic formula. To facilitate this, we have a species file, which contains information about each species, such as ionisation potential, dipole moment and polarisability.

Automatic creation of large hydrocarbon chemistries, and of isomer chemistries, are also areas where the computer can save large amounts of time. SMERSH is designed to make it easier to try new chemistries (the record, from choice of species, to a hard copy plot of the results for a 130 species time dependent chemistry, is currently two and a half hours).

5. APPLICATIONS AND DISCUSSION

As an example of the use of the techniques outlined above, I will give an overview of a deuterium chemistry developed by E. Herbst, T.J. Millar, and the author. To be treated fully, a model of deuterium chemistry requires that all reactions involving H are duplicated with ones using D, if we are answer such questions as when, in chemical synthesis, is D inserted into a complex species. For instance, when does D become substituted in DC_3N?

This naturally means much typing of an extended reaction scheme, especially for a complex set of species. In this study, therefore, a program automatically creates deuterated analogues of all reactions. Knowing the reaction type, we assume that the total rate coefficient for the D reaction is the same as that for the H. (We could be more sophisticated as DELOAD calculates the molecular weights of the species in the scheme). Branching ratios are found statistically, taking reaction type into account, so that when we believe a reaction takes place by a certain pathway that forbids the transfer of H from radical to radical, then the transfer of D is also forbidden.

RATE87, when deuterated, produces a scheme with around 350 species and 5750 reactions, many of which will be pointless. The scheme is too large to be practical.

In practice, SMERSH can output of "relevant" reactions for around 125 species of H, He, C, N, O and Si. When fully deuterated, this scheme yields 225 species, including 8 conserved species, and 1700 reactions. This size of chemistry can be readily handled on mainframes. The chemistry used treats the isotopomers of methanol (CH_3OD and CH_2DOH), and their associated chemistries, as separate species.

Of course, this has only taken the typing out of preparing the chemistry; the real science lies in choosing appropriate deuterium fractionation reactions, which are added to the automatically produced reaction list. We now have a reaction scheme, and may use DELOAD, and construct our model.

During the study, it was considered desirable to see what effect using "Batesian" and non-Batesian dissociative recombination products has on the chemistry. However, these schemes have a different number of reactions, but preparing two models automatically requires a minimum of further work.

On a more general note, this work could be readily extended to other isotopic variants such as ^{13}C or ^{17}O, extending this chemistry throughout the entire model. Because of the lower energy differences between these isotopes and their more common counterparts, the effect of isotope chemistry will, however, be more limited.

The point that I want to make here is the ease of updating and changing the reaction scheme makes it much more likely that updated rate coefficients, and additional species will be added to the scheme, and many more possibilities tried during the study.

There are two main points that I would like to make in conclusion. They both, in some degree, relate to the fact that reaction schemes are becoming increasingly complex as our knowledge of interstellar chemistry improves. The first is that an up-to-date and accurate library of chemical reactions is an essential prerequisite to building chemical models.

The second is that software to manipulate and produce reaction schemes and aid chemical model production can significantly cut the effort required to produce new chemistries, and so make studies broader and more flexible than might otherwise be the case.

ACKNOWLEDGEMENTS

I thank Tom Millar for his advice during this work, and also Eric Herbst for his enthusiasm. I also thank all the astrochemists at UMIST for their welcome and much needed feedback concerning this project.

REFERENCES

Anicich, V.G. and Huntress, W.T. Jr.: 1986. Astrophys. J. Suppl. Ser., **62**, 553.

Hindmarsh, A.C.: 1974. 'GEAR; Ordinary Differential Equation System Solver', Lawrence Livermore Laboratory, Report UCRL-30001, Rev. 3.

Nejad, L.A.M.: 1986. Ph.D. Thesis, University of Manchester.

Prasad, S.S. and Huntress, W.T. Jr.: 1980. Astrophys. J. Suppl. Ser., **43**, 1.

Rawlings, J.M.C.: 1986. Ph.D. Thesis, University of Manchester.

NAME INDEX

A

B

I

J

K

Q

R

S

SUBJECT INDEX

GPSR Compliance

The European Union's (EU) General Product Safety Regulation (GPSR) is a set of rules that requires consumer products to be safe and our obligations to ensure this.

If you have any concerns about our products, you can contact us on ProductSafety@springernature.com

In case Publisher is established outside the EU, the EU authorized representative is:

Springer Nature Customer Service Center GmbH
Europaplatz 3
69115 Heidelberg, Germany

Batch number: 10815974

Printed by Printforce, the Netherlands